Bradford Hill's Principles of Medical Statistics

Twelfth Edition

Austin Bradford Hill and I. D. Hill

Edward Arnold
A division of Hodder & Stoughton
LONDON MELBOURNE AUCKLAND

First published in Great Britain 1937 as *Principles of Medical Statistics*
Ninth edition 1971
Published as *A Short Textbook of Medical Statistics* (Tenth edition) 1977
Eleventh edition 1984
Published as *Bradford Hill's Principles of Medical Statistics* (Twelfth edition) 1991

British Library Cataloguing in Publication Data

Hill, Sir Austin Bradford
Bradford Hill's medical statistics. - 12th ed.
I. Title II. Hill, I.D.
610.28

ISBN 0-340-53739-6

Typeset in 10 on 11pt Times by Wearside Tradespools, Fulwell, Sunderland.
Printed and bound in Great Britain for Edward Arnold, a division of Hodder and Stoughton Limited, Mill Road, Dunton Green, Sevenoaks, Kent TN13 2YA by Biddles Ltd, Guildford and King's Lynn.

Preface

If any man, who shall desire a more particular account of the several Alterations . . . shall take the pains to compare the present Book with the former; we doubt not but the reason of the change may easily appear.

Preface to Book of Common Prayer of the Church of England. 1666.

When the first edition of this book was published in 1937 it dealt with medical statistical matters of the early twentieth century. Now that the century nears its close there is, regrettably, no need to change its main message: that statistics are widely misused. We are indeed bombarded with figures nowadays—some invaluable, some poor, some bad, and an almost endless number that are misleading, sometimes deliberately but more often through lack of adequate thought. Fallacies are as widespread as ever.

This is not a textbook of statistical methods, and never has been—it is a text book of *medical* statistics, that is the use of statistics and statistical methods applicable to medicine. Medicine is everywhere concerned with sick persons and with those who may become sick in the absence of preventive methods, but medical problems vary widely from country to country. In the United Kingdom we now have virtually no problems with diphtheria, poliomyelitis, typhoid fever or tuberculosis; cases that we see here are usually imported, but the fact that they can be imported shows that they still exist elsewhere. Smallpox seems to be the only disease in which all countries are alike, the efforts of the World Health Organisation having apparently eradicated it.

Knowing that the book is read and used in many other countries, either in English or in translations, we have therefore not attempted to make all the examples deal with current medical problems here in Britain. The arithmetical procedures remain the same in any case, but the practitioner must always be aware of the different medical background that needs to be taken into account in interpretation.

In this edition, we have removed some concepts that are now little used, and thus found space to introduce others that are more important in present-day practice. These include more detail on the normal and binomial distributions, and introduction of the Poisson distribution,

Fisher's test for 2×2 tables, Kendall's rank correlation and the logrank test for comparing survival rates in different groups.

We include introductory sketches of elementary analysis of variance and of Bayesian methods, in each case not attempting to give details of the best ways of performing them—computers will usually look after the mechanics nowadays—but seeking to give an insight into their underlying purposes, so that the doctor can read with profit papers that use them.

We have also followed the modern fashion in giving more attention to confidence intervals and less to tests of significance. This is not merely a matter of fashion, however; it is genuinely a better approach and the gains are considerable.

We thank Professor D.V. Lindley for reading the first draft of Chapter 19. His comments have considerably improved it, but he is not responsible for any faults.

It seems to be necessary nowadays to apologise for the standard device of the English language of using the masculine forms of personal pronouns, it being understood that women as well as men are indicated. We wish that gender-free personal pronouns existed, but in their absence none of the suggested alternatives is satisfactory, and many are barbaric. It is perfectly possible to sympathise with the aim of those who dislike the usage, and yet to prefer to write decent English so far as we are able.

Austin Bradford Hill
I.D. Hill

Postscript

This is the final edition of *Bradford Hill's Principles of Medical Statistics* to be supervised by himself. Up to the eleventh edition (1984), which appeared when he was aged 87, he had done all the work of revision himself (with a small amount of help in a few places). For this edition I did the revision, but sent copies of everything to him for approval. He looked forward to its appearance and, the last time I saw him, his first words were 'What news of the book?' I hope that he would have liked the result, but he died on 18 April 1991 at the age of 93.

In the Preface to the seventh edition (1961), he jokingly wrote:

> I am greatly indebted to Dr P. Armitage and Mr I.D. Hill for reading my expanded text and for their advice and criticisms. Contrary to custom, for the faults that remain I trust sincerely that the reader may hold them largely responsible.

Now comes the time to return the compliment and hope that the reader will hold him responsible for any faults in this twelfth edition

I.D. Hill

Contents

Chapter 1

The aim of the statistical method

'Is the application of the numerical method of the subject-matter of medicine a trivial and time-wasting ingenuity as some hold, or is it an important stage in the development of our art, as others proclaim?' In 1921 that was a very reasonable question to be asked by the writer of an article on statistics in a medical journal. It could hardly be asked in the same setting today. Medical historians of the future may well note that in the second half of the twentieth century there was a very noticeable increase in the number of papers published in medical journals of which the essence was largely statistical, i.e. in the planning, the analysis and the presentation of observations, clinical trials or experiments. However, with the medical curriculum already overflowing with subjects essential to the would-be doctor, it was inevitable that many of the writers and readers of these papers would have had little, or no, training in statistical methods.

Experience shows that many such medical writers and readers find mathematical methods of approach obscure and even repellent. On the other hand, an introduction to simpler ways of statistical thinking and working might be welcome, and this, indeed, is now the practice in most medical schools. In future, therefore, there should be few doctors without some knowledge of the statistical approach and ways of thinking and of the simpler statistical methods. How effective, in actual practice, the critic might ask, is this teaching likely to prove? Can we in solving the problems of medicine reach satisfactory results by means of such relatively simple numerical methods? In other words, can we satisfactorily test hypotheses and draw deductions from data that have been analysed by means of such simple methods? The answer is undoubtedly *yes*, that many of the figures included in medical papers can by relatively simple statistical methods be made to yield information of value. Sometimes the yield may be rather less than that which might be obtained by more erudite methods which are not at the worker's command but the best should never be made the enemy of the good, and even the simplest statistical analysis carried out logically and carefully is an aid to clear thinking with regard to the meaning and limitations of the original records. If these conclusions are accepted, the question immediately at issue becomes this: are simple methods of the interpretation of figures only a synonym for common sense or do they involve an art or knowledge which can be imparted? Familiarity with medical statistics leads inevitably to the conclusion that common sense is *not* enough. It seems that many people are not capable of using common sense in the handling and interpretation of numerical data until they have been instructed in quite elementary ideas and techniques. Mistakes which when pointed out look extremely foolish are quite frequently made by intelligent people, and the same mistakes, or types of mistakes, continue to crop up again and again. There is often lacking what has been called a 'statistical tact, which is rather more than simple good sense'. This tact the majority of people must acquire (with a minority it is undoubtedly innate) by a study of the basic principles of statistical thought and method.

One object of this book is to discuss these basic principles in an elementary way and to show, by representative examples taken from medical literature, how these principles are frequently forgotten or ignored. There is no doubt that the discussion will often appear too simple and that some of the mistakes to which space is given will be thought too futile to need attention. That such is not the case is revealed by the recurrence of these mistakes and the neglect of these elementary principles, a feature with which every professional statistician is familiar in published papers and in those submitted to him by their authors for 'counsel's opinion'.

Another object is to describe, in simple terms, some methods of which the details would be beyond the book's scope, in the hope that those who would not wish themselves to employ the methods may nevertheless gain some idea of the principles involved so that they may more easily read published papers that use them.

Definition of statistics

There have been many definitions of statistics over the last 200 years, many of which are not recognisable as the subject that we know today. Probably the best definition nowadays is 'numerical data involving variability, and the treatment of such data'. It is because variability is so widespread in all living matter that few biological studies can be free of statistical thinking.

The laboratory worker, and the worker in clinical or preventive medicine, must constantly deal with such variability, but the former is in a very much stronger position than the latter, because he can frequently exclude variables in which he is not interested and confine his attention to one or more controlled factors at a time. In other words he can *experiment*.

The great advantage of experimenting is as follows: suppose we observe two features X and Y that tend to occur together. We should like to be able to say that X causes Y, but there are, in fact, four possible explanations: (1) X causes Y; (2) Y causes X; (3) some other, unobserved, feature (Z say) causes both X and Y; (4) pure coincidence, which would be unlikely to happen again if we were to repeat the study. With a well-planned experiment, explanations (2) and (3) can at once be ruled out. We can say 'I know that X was not caused by Y, or by some unknown Z, because I know what did cause X. I caused X, through my experimental plan.' We still need to weigh the evidence carefully in deciding whether we can rule out explanation (4) or not. In an observational study, however, where we have not been able to experiment, all four possible explanations always have to be taken very seriously. It is because he knows that a complex chain of causation is so often involved that the statistician may appear to be an unduly cautious and sceptical individual.

For example, suppose we have a number of persons all of whom have been in contact with a case of infectious hepatitis and to a proportion of them is given an injection of gamma globulin. The others are observed as 'controls'. We wish to know whether the injection prevents the development of a clinical attack. It is possible that the risk of developing an attack after exposure is influenced by such factors as age and sex, social class and all that that denotes, duration and intimacy of contact, and so on. So far as is possible a statistical analysis necessitates attention to *all* such influences. We must endeavour to equalise the groups we compare in every possible influential respect except in the one factor at issue—namely, the prophylactic treatment. If we have been unable to equalise the groups *ab initio* we must equalise them to the utmost extent by the mode of analysis. As far as possible it is clear, however, that we should endeavour to eliminate, or allow for, these extraneous or disturbing causes *when the observations are planned*; with such planning maybe we can determine not only whether the treatment is of value but whether it is more efficacious in one situation than another, at one age than another, etc. It is a serious mistake to rely upon the statistical method to eliminate disturbing factors at the completion of the work. *No* statistical method can compensate for badly planned observations or for a badly planned experiment. But a knowledge of it can contribute considerably to the design of an experiment.

Planning and interpretation of experiments

It follows that the statistician may be able to advise upon the statistical lines an experiment such as that referred to above should follow. Elaborate experiments can be planned in which quite a number of factors can be taken into account statistically at the same time (see, for example, *Experimental Designs* by W.G. Cochran and G.M. Cox and *Planning of Experiments* by D.R. Cox). It is not the intention of this book to discuss these more complex methods of planning and analysis; attention is mainly confined to the simpler types of experimental arrangement which are so frequently required in medicine. Limitation of the discussion to that type must not be taken to mean that it is invariably the best form of experiment in a particular case.

The essence of the problem in a simple experiment is, as emphasised above, to ensure beforehand that, as far as possible, the groups to be compared (e.g. a control and a treated group) are the same in all *relevant* respects. The word 'relevant' needs emphasis for two reasons. First, it is obvious that no statistician, when appealed to for help, can be aware of all the factors that are, or may be, relevant in particular medical problems. From general experience he may well be able to suggest certain broad disturbing causes which should be considered in planning the experiment (such as age and sex in the example above), but with factors which are narrowly specific to a particular problem he cannot necessarily be expected to be familiar. The onus of knowing what is likely to be relevant in such a situation must rest more upon the experimenter, who is, presumably, familiar with that narrow field. Thus, when the statistician's advice is required it may be his task to suggest means of allowing for the disturbing causes, either in planning the experiment or in analysing the results, but not, invariably, to determine what *are* the relevant disturbing causes. At the same time no statistician who is wise will advise at all upon a medical problem with which he is totally unfamiliar. Successful collaboration demands that the statistician should learn all he can of the problem at issue and the experimenter (clinician, community physician, etc.) all he can of the statistical approach. Without substantial knowledge on both sides the blind may well lead the blind.

The second point that must be observed as regards the equality of groups in all relevant respects is the caution that must attend the interpretation of statistical results. If we find that Group *A* differs from Group *B* in some characteristic, say an attack-rate, can we be certain that that difference is due to the fact that Group *A* was inoculated (for example) and Group *B* was uninoculated? Are we certain that Group *A* does not differ from Group *B* in some other character relevant to the issues as well as in the presence or absence of inoculation? For instance, in a particular case, inoculated persons might, on the average, belong to a higher social class than the uninoculated and therefore live in surroundings in which the risk of infection was less.

If we are to experiment, by treating two groups of people differently, we must always be aware of the ethical problems of doing so. Later in the book, we shall return to these problems in detail. For the present, let us assume that the circumstances are such that we are able, ethically, to perform such an experiment—then we are duty-bound to make it a good one. It is certainly unethical to perform an invalid experiment.

The need for randomisation

In such a study, we should equalise so far as we can the disturbing factors of which we are aware. We might therefore divide the subjects into pairs, so that those within each pair are as similar as we can make them in terms of age, sex, social class, etc. However, we need to try to equalise for all other, unknown, disturbing factors too. Without knowing what they are it is obviously impossible to do so precisely, but there is a method available that will do so on the average, whatever these unknown factors may be.

The method is that, within each pair, we must choose one subject to have the experimental treatment and the other to be the control *at random*. The human mind is not capable of doing this without help—the decision must be handed over to a random process, such as tossing a coin, or its equivalent in the use of a table of random numbers.

To appreciate the difficulty of doing it unaided, think of the analogy of a sporting event in which an initial decision is made before the match can start by tossing a coin. Imagine that you were given the job of deciding at random which team has won the toss, without actually using a coin. Would you be able to do so, giving each team an equal chance, ignoring all the other information you might have? Even if you could do so once (which is most unlikely) could you do so on further occasions, when you must now not allow your past decisions to have any effect? Even if you could, would you then be able to convince other people that you had always acted fairly?

The use of a randomising device removes all these difficulties. It is basic to all experimental design that, having first equalised what you can, randomisation is essential for the final assignment of treatments to subjects. Informal approximate 'randomness' is not good enough.

Statistics in clinical medicine

The essence of an experiment in the treatment of a disease lies in comparison. To the dictum of Helmholtz that 'all science is measurement' we should add, as that great experimenter Sir Henry Dale pointed out, a further clause, that 'all true measurement is essentially comparative'. On the other hand there is a common catch-phrase that human beings are too variable to allow the contrasts inherent in a controlled trial of a remedy. Yet if each patient is 'unique' it is difficult to see how any basis for treatment can be sought in the previous observations of other patients—upon which clinical medicine is founded. In fact, of course, physicians must, and do, base their 'treatment of choice' upon what they have seen happen before—whether it be in only two or three cases or in a hundred.

Although, broadly speaking, human beings are not unique in their responses to some given treatment, there is no doubt that they are likely to be variable, and sometimes extremely variable. Two or three observations may therefore give, merely through the customary play of chance, a favourable picture in the hands of one doctor, an unfavourable picture in the hands of another. As a result, the medical journals become an arena for conflicting claims—each in itself, maybe, perfectly true of what the doctor saw but insufficient to bear the weight of the generalisation placed upon it.

Far, therefore, from arguing that the statistical approach is impossible in the face of human variability, we must realise that it is *because* of variability that it is essential. It does not follow, to meet another common criticism, that the statistical approach invariably demands large numbers. It may do so; it depends upon the problem. But the responses to treatment of a single patient are clearly a statement of fact—so far as the observations were truly made and accurately recorded. Indeed that single case may give, in certain circumstances, evidence of vital importance.

If, for example, we were to use a new drug in a proved case of tuberculous meningitis and the patient made a complete, immediate and indisputable recovery, we should have a result of profound importance. The reason underlying our acceptance of merely one patient as illustrating a remarkable event—not necessarily of cause and effect—is that long and wide experience has shown that in their response to tuberculous meningitis human beings are *not* variable. In days past, in the absence of modern drugs (e.g. streptomycin) they one and all failed to make complete, immediate and indisputable recoveries. Therefore, although it would clearly be most unwise upon one case to pass from the particular to the general, it would be sheer madness not to accept the evidence presented by it.

If, on the other hand, the drug were given to a patient suffering from acute rheumatic fever and the patient made a complete, immediate and indisputable recovery, we would have little basis for remark. That recovery may clearly have followed the administration of the drug without the slightest probability of related cause and effect. With this disease human beings *are* variable in their reactions—some may die, some may have prolonged illnesses but recover eventually with or without permanent damage, some may make immediate and indisputable recoveries—whatever treatment we give them. We must, therefore, have more cases before we can reasonably draw inferences about cause and effect. We need a statistical approach and a designed experiment (the details are discussed in Chapter 23).

While, therefore, in many instances we do need larger numbers for a sound assessment of a situation, it certainly does not follow—as is sometimes asserted—that the statistician would have rejected some of the original and fundamental observations in medicine on the grounds of their small number. To take a specific example, *fragilitas ossium* was originally described on two cases and this, a later writer said, statisticians would regard as useless evidence. But why should they? If exact descriptions and illustrations were given of these two cases, then, of course, they form part of the body of scientific knowledge. They are undeniable evidence of an occurrence. What *can* happen, what *does* exist, quite regardless of the *frequency* of occurrence and irrespective of *causation* or association, may be observed, as already stated, even on a sample of one. It can only be in relation to an appeal from the particular to the general that a statistician—and, equally, anyone else—could object. If on the basis of the two cases the clinician, in practice, let us say, near the London meat market, should argue that the condition was specific to butchers, then one might suggest that the experience was too limited in size and area to justify any such generalisation.

In short, there is, and can be, no magic number for either clinician or statistician. Whether we need one, a hundred, or a thousand observations turns upon the setting of our problem and the inferences that we wish to draw.

It must be clear, too, that almost without statistics, and certainly without accurate measurement, the mental, or quite rough, contrasting of one treatment (or some other course of action) with another will give a truthful, if not precise, answer, *if* the treatment has a very real and considerable effect. Without a strictly controlled trial the merits of penicillin could not fail to come to light. Its effects would have been incompatible with past experience. With such 'winners' it is easy for the critics of the often relatively slow statistical approach to be wise after the event, and to say that the general evidence available at the start of a long, and perhaps tedious, trial made it unnecessary or pedantic. They forget the many occasions when the trial has shown that a vaunted treatment has little, if any, value—in spite of all the general 'evidence' that was available. Without a trial it might well have lingered on, to the detriment of patients. Further, it is difficult to determine through general impressions whether some drug is quite useless or of some slight but undoubted value—and to reduce, say, a relapse rate from 6 to 3 per cent would not be unimportant. It is even more difficult to determine with uncontrolled and uncoordinated observations whether one powerful drug is more valuable than another in particular situations. Only a carefully designed clinical trial is likely to serve this purpose. But that is not by any means to say that the statistically guided experiment is the *only* profitable means of clinical investigation or invariably the best way of advancing knowledge. It is merely one way.

One difficulty of the variability of patients and their illnesses is in classifying the patients into, at least, broad groups, so that we may be sure that like is put with like before treatment. Even in the absence of such groups, valid answers can be reached by the strict use of randomisation, but grouping before randomisation is obviously preferable where feasible.

Even if the treatment is not of general value but apparently of benefit in relatively isolated cases, satisfactory evidence of that must lie in statistics—namely, that such recoveries, however rare, do not occur with equal frequency amongst equivalent persons not given that treatment. Sooner or later the case is invariably based upon that kind of evidence, but in the absence of a planned approach it is often later rather than sooner. As Lord Platt (when President of the Royal College of Physicians) emphasised, in a more general setting, records in clinical research are likely to be disappointing 'unless they have been kept with an end in view, as part of a planned experiment . . . Clinical experiments need not mean the subjection of patients to uncomfortable procedures of doubtful value or benefit. It means the planning of a line of action and the recording of observations designed to withstand critical analysis and give the answer to a clinical problem. It is an attitude of mind.'

Returning to the problems of classification, by the statistical process of condensation of the individual items of information into a few groups, and, further, into average and other values briefly descriptive of the data, we are clearly sacrificing some of the original detailed information. We must be particularly careful, therefore, that we sacrifice nothing relevant to the issues or more than is essential to clarity and ease in handling, interpreting and presenting the data. It is rarely feasible in practice, however, to publish the full case histories of a large number of patients specially treated and similar details of an equally large control group; and even were it feasible, that material alone cannot supply the reader (or writer) with the information he needs until it has been appropriately condensed. The question is, was the special treatment of value, i.e. the elucidation of cause and effect? The elucidation must normally be achieved by the construction, from the original mass of recorded data, of relatively short tables and statistical values based upon them, to show the relevant position before and after treatment of the specially treated and orthodoxly treated groups.

Other examples of the application of statistics in clinical medicine are:

(1) observations on the natural history of a disease—what are its presenting signs and symptoms, what is its course, how variable is it in its manifestations from patient to patient, with what characteristics, e.g. age or sex, is it associated?;
(2) the follow-up and assessment of patients treated in some particular way—particularly, perhaps, in surgery;
(3) the definition of 'normal'—at what point does the measure of some bodily characteristic become pathological?;
(4) the accuracy of laboratory procedures constantly used in clinical medicine, e.g. blood-counts.

In short, in statistics must lie the answers to many of the fundamental questions posed by clinical medicine—of diagnosis, of treatment and of prognosis. We need the careful collection of statistical information, its analysis by appropriate methods and its presentation in the literature. Initially it calls for precise and accurate observation of the patient. By such means the general body of medical knowledge is built up. Furthermore, within the clinical field, contributions to the epidemiology and aetiology of disease can be made. The clinician is primarily concerned with the individual patient and the problem of restoring him to health. But if he is aware of some of the social and other features of his patients (e.g. occupation) he may, by simple arithmetic, note that a certain disease appears to be associated with a certain feature more often than would appear to be due to chance; and thus pave the way to more extensive observations. Thus, for example, did Gregg in Australia associate the feature 'rubella in pregnancy' with the diseases 'eye and other defects' in the babies brought to him.

Statistics in the field of public health

In public health work we may sometimes be concerned with a similar planning of experiments and the analysis of their results, e.g. in a test of the efficacy of a vaccine as a means of preventing an attack from, say, whooping cough; or in measuring the effects on mother and infant of supplementing the diet of pregnant women in some particular way. Frequently, however, we have to deal in this field with statistics that come from no deliberate experiment but that arise, and are collected, from a population living and dying in an everyday course of events. Thus we have the general death rate of the population in a given period of time; its death rates at particular ages—in infancy, in childhood, in the prime of life, and in old age; its death rates from particular causes—respiratory tuberculosis, cancer, violence, etc. For some diseases—the infectious notifiable diseases—we may have figures relating to the number of attacks occurring from time to time.

The object of these statistics and the statistical methods applied to them may be regarded as twofold. On the one hand we shall use them as simple numerical assessments of the state of the public health, to show by contrasts between one place and another, or between one period of time and another, whether the death rates of the population, for example, are relatively high or low. It is only on the basis of such evidence that we can effectively consider the problems with which preventive medicine is faced and where and when remedial measures are most needed. For instance, we find that the frequency of death in the first year of life, i.e. the infant mortality rate, becomes greater as we pass down the social scale from the professional classes to the general labourer. Regardless of the factors that lead to this result, we know at least that such a problem exists and needs attention. Again, we may observe that cases of typhoid fever are more frequently notified from one type of area than from another. The cause is unknown but a problem is defined. Or, finally, the records may show that the death rates of young children from common infectious diseases are higher in the more crowded urban communities of a country than in the less crowded. Can we counteract that unfavourable experience of the overcrowded areas?

The initial use of such statistics, as accurately and completely collected and compiled as is possible, is therefore to *direct attention* to the problems of health and ill-health presented by the population under study. Without such figures we can have but little knowledge of the most important fields for action, and the collection and tabulation of statistical data are, therefore, fundamental to public health work. In other words, the certificates of death and sickness that the doctor in his daily work is required to complete are not merely ephemeral bits of paper to satisfy legal demands. They may well make serious contributions to the problems of preventive medicine. Needless to say, there will be difficulties. Diagnosis and accurate certification are not easy tasks. The aids to them and their resulting accuracy must change from time to time and vary from place to place. Every medical statistician must be aware of that. But if we await perfection we shall wait for ever. So long as we are not ignorant of the imperfections much can be learnt from these imperfect records of mortality and sickness. For instance, in spite of some undoubtedly wrong diagnoses we are perfectly well aware of the marked seasonal distribution of influenza in Great Britain and the problem that the distribution raises in the epidemiology of this disease. In spite of errors in determining the cause of death we are aware that relatively more men than women die of cancer of the lung (in the United Kingdom at this time, but not to the extent that used to be the case).

The second, and of course closely associated, object in the collection of such figures is the determination of the basic *reasons* for the contrasts observed. For unless we can determine those reasons, the development of effective preventive measures must obviously be hampered and may be misdirected. Why, to take the examples given above, do infants of the more impoverished classes die at a higher rate than those of the wealthier? To what extent (if any), for instance, is it due to the malnourishment of the mother and child, to what extent to overcrowding in the home and a more frequent risk of specific infections, to what extent to ignorance of how to care for the infant, or even, sometimes, to frank neglect of it? Is a higher incidence, or an epidemic, of typhoid fever in a particular area or type of area due to a defective water supply, to milk-borne infections, or to some other form of transmission? Does the pre-school child die more readily under conditions of overcrowding because such conditions expose it to a greater risk of infection early in life, or does it succumb more easily to the infection it has acquired through factors associated with overcrowding, such as malnutrition or, possibly, lack of skilled attention in the early stages of illness?

Here the statistical method comes into play, endeavouring to disentangle the chain of causation and allowing us, sometimes, to determine the most important factors in need of correction. Since we are dealing with uncontrolled observations, often liable to errors, the task may be very difficult; the effects of the multiplicity of causes often cannot be completely distinguished, e.g. the effects on health of overcrowding *per se* as apart from the features of poverty which invariably accompany it. But the original vital statistics having indicated the problem and their analysis having, at least, suggested a cause and effect, we may be able to progress further by a more deliberate collection of additional data or sometimes by a specifically designed experiment.

However that may be, good vital statistics must be the essential forerunner of the development of preventive measures designed to promote the health and well-being of any population, or of some particular fraction of it, and must serve as one of the fundamental 'yardsticks' for determining the success or failure of such measures. They are fundamental to the study of epidemiology in the modern sense of that word, i.e. the community characteristics of every disease. If such statistics are lacking, or too inexact or too incomplete to be useful, then it may be essential to seek them in special field studies on a sampling basis.

The use of computers

Since the mid-1950s the automatic electronic computer has taken over more and more of the calculations involved in the statistical analysis of data. It can thus relieve the worker (with or without a calculating machine) of a heavy task. This development has both advantages and disadvantages.

Among the advantages are the accuracy and reliability of computer calculation. In spite of the impression often given (gleefully) by the popular press, the computer is a very reliable machine. But just as a car, no matter how reliable, cannot be expected to behave sensibly when in the hands of an incompetent driver, so the usefulness of the results from a computer depend critically upon the competence of the programmer.

When programs have been written for specific uses and stored as a package, it is possible for a would-be user to request a particular analysis and, then, by merely supplying his numerical data, to be presented with the results of using it—even though he may not have known how to calculate those results. This is a reasonable thing to do *provided that the user has sufficient knowledge of what problem he is himself trying to solve, and, equally, of exactly what problem the computer program was devised to solve*. In short, he must have sufficient knowledge to be sure that the one is precisely relevant to the other. The use of an unsuitable program for a particular task is, alas, all too common.

The computer has also made possible the use of statistical methods that were virtually impossible in the days of hand computation—simply because of the sheer quantity of calculation involved. Such methods are, in general, beyond the scope of this book.

Perhaps the main disadvantage of computer methods is that the computer can do no more than that which it has been instructed to do. The human mind, on the other hand, when working through a collection of facts and figures, has an astonishing ability to detect some totally unexpected combination of events. By looking at, and brooding over, basic data the worker may, with a flash of inspiration, emerge with some idea or some discovery which bears no relation whatever to the original object of the work. At present nobody has any idea of how to program a computer to do that, to achieve such inspiration. Maybe this will always be so—but it is dangerous to prophesy. But the importance of studying one's basic data cannot be over-emphasised.

Summary

The statistical method is required in the analysis and interpretation of figures which are at the mercy of numerous influences. Its object is to determine whether these individual influences can be isolated and their effects measured. The essence of the method lies in the determination that we are really comparing like with like, and that we have not overlooked a relevant factor which is present in Group *A* and absent from Group *B*. The variability of human beings in their illnesses and in their reactions to them and to their treatment is a fundamental reason *for* the planned clinical trial and not *against* it. Large numbers are not invariably required and it is clear that in particular circumstances even one or two cases well observed may give information of vital importance.

Vital statistics and their analysis are essential features of public health work, to define its problems, to determine, as far as possible, cause and effect, and to measure the success or failure of the steps taken to deal with such problems. They are fundamental to the study of epidemiology.

Chapter 2

Collection of statistics: sampling

Present-day readers of the early volumes of the *Journal of the Royal Statistical Society* would be struck by one marked characteristic. In their surveys of the state of the housing, education, or health of the population in the 1830s, it was the aim of the pioneers of that time to study and enumerate *every* member of the community with which they were concerned—the town in Lancashire, the borough of East London, the country village, whatever it may have been. That aim was frequently brought to nought by the very weight of the task. Sometimes the collection of the data was beyond their capacity in time, staff and money; sometimes, having done their best to collect them, they were weighed down by the statistical analysis that the results demanded. In contrast, the worker today would (or should) instinctively reflect on the possibility of solving such a problem by means of sampling.

By the method of sampling he may make these, and many other, tasks practicable in terms of cost, personnel, speed of result, etc., but he will also, quite often, render the results more, rather than less, accurate. He will, of course, be introducing an additional error, the sampling error due to the fact that he has studied only a proportion of the total. However well the sampling may be carried out, that is inevitable. But since the work of observation and recording is thus made so much lighter, it may well be that it can be carried out with more precision and more uniformly by a smaller number of workers and, perhaps, by more highly skilled workers. Further, with a sample of, say, 1000 it may be possible to pursue and complete the records for all, or very nearly all, the persons included. The attempt to enumerate the whole population may lead, through the practical difficulties, to a loss of an appreciable number of the observations required. With such an *incomplete* 'whole' population we are then, in fact, left not only with a sample but with one that raises doubts that we cannot resolve as to whether it is representative. With the completed random sample of 1000 we can, on the other hand, justifiably infer the values that exist in the whole population—or, more strictly, the limits between which they are likely to lie. Such estimates from a properly chosen sample are adequate in nearly all circumstances. In particular, sampling methods to provide vital statistics may be specially appropriate in developing countries where total information on health aspects of the population through birth and death registration may be unobtainable.

It follows that in preparing to make a survey or setting in train the collection of statistical data to illuminate some problem the first questions that the worker must ask himself are: Precisely what data do I need? Can I investigate the problem by means of a sample? If so, how shall I set about obtaining a sufficiently large and representative sample?

Absolute, or proportional, sample size

Before discussing how to choose the size of a sample, or methods of drawing a sample, one point must be emphasised, because it is probably one of the most misunderstood points in all statistics, and lack of knowledge of it constantly leads to false statements. It is that the amount of information that can be gained from a sample depends upon its *absolute* size, not upon its size as a proportion of the population size. Thus to refer to a sample of size 50, of size 500 or of size 5000 tells one how much that sample is worth (provided it is a worthy sample in other respects too), but to say that a sample is 1 in 10, or 1 in 100, or 1 in 1000 of the population tells one virtually nothing of its worth (no matter how worthy it is in other respects).

This fact, which seems so obvious when one has lived with statistics for a few years, seems to be counter-intuitive when first met. Thus there are frequent references in the press to a sample of 1000, say, being useless because the population of the country is over 50 million and 1000 is such a tiny percentage of that; but, as has been well-said elsewhere, to decide whether the wine is good you need only a sip, whether the bottle is a litre or half a litre, and you do not need to know which size it is. It may be argued that that is because wine is so homogeneous, but the reply to that is that, to get the equivalent of that homogeneity is exactly why we need to sample at *random*.

It may further be argued that to sample 99 out of 100 must, surely, tell one more about the 100, than 99 out of 10 000 will tell one about the 10 000. Yes, indeed, but for practical purposes we are not normally concerned with samples that almost exhaust the population, but only with samples that are a small proportion (say less than 20%) and for these it is indeed true, to a very close approximation, that 99 out of 1 000 000 tells you as much about the 1 000 000, as 99 out of 1000 tells you about the 1000.

Drawing a sample

Let us suppose then that there is a population which can be readily sampled—whether, for example, it be of institutions in a country, houses in a town, clinical records in a hospital, or medically qualified men and women on a register. Experience has shown that an *apparently* quite haphazard method that leaves the choice to the worker is very unlikely to be truly haphazard. He will unconsciously pick too many (or too few) houses at the corner of the street, too many (or too few) bulky clinical files, too many (or too few) surnames beginning with a particular letter. The bias may be quite unknown either in kind or degree. But it is no less likely to be there—it must be avoided by setting up rules of choice, to make that choice completely random and quite free of any element of personal choice.

In their simplest form the rules give everyone in the population an independent and *equal* choice of appearing in the sample. If the individual components of the population are already numbered serially, say from 1 to 790, then the required sample can be readily drawn with the aid of tables of random sampling numbers (see Appendix E). Starting, say, at the beginning of the table, we can read down the first three columns; the numbers of the items required (whether they be houses, files or people) are 136, 568, 193, 416, 156, 16, 89, etc. Any number outside the range (such as 812, which would otherwise have been the third item) is ignored. Similarly, if any number appears for a second time, it is ignored. The process continues until a sample of the required size has been drawn. The tables have been so constructed that every number has had an independent and equal chance of appearing and thus the sample is free from bias.

It does not inevitably follow that the sample is a 'good' sample, in the sense that it is a representative cross-section of the population. The play of chance itself must, of course, sometimes produce an unusual and, therefore, unrepresentative picture. If the sample is large (some hundreds) it is not likely to be seriously distorted; if it is very small (20 or less) it could easily be grossly in error. The solution to the problem must lie principally in a larger sample, but it can also sometimes be partially found by the device of *stratification*.

The stratified sample

Given sufficient knowledge of the population to be sampled we may divide it into well-defined subgroups or *strata* and then draw our sample from each of these strata separately. Within each stratum the choice is still entirely random, but automatically we have ensured that the final total sample includes the right proportion of each of the strata. For example, in sampling a population of children to measure their heights and weights we might first divide it into boys and girls and, within each sex, into the age groups 5–8, 9–12 and 13–15 years. Thc numbers in each group might be as shown in Table 2.1. If we needed a sample of 175, we might then sample from each group in proportion to its size, as shown in Table 2.2, where the number in the first group is calculated as $156 \times 175/1664$ rounded to the nearest whole number and similarly with the other groups. The total sample then contains, as nearly as may be, the correct proportions of boys and girls of the different age groups.

However, as already noted, the worth of a sample is related to its absolute size, and it will be seen that the samples of 5 in the 13–15 age groups are giving us very little information compared with the samples of over 60 in the 9–12 age group. So it may be better to go one stage further and use different sampling fractions in the different stata. If, for example, we were to take the sample sizes shown in Table 2.3, we should get better

Table 2.1 Number of schoolchildren in a population.

Years of age	*Boys*	*Girls*	*Total*
5–8	156	148	304
9–12	624	635	1259
13–15	49	52	101
Total	829	835	1664

Table 2.2 Number of schoolchildren in a (proportional) sample of 175.

Years of age	*Boys*	*Girls*	*Total*
5–8	16	16	32
9–12	66	67	133
13–15	5	5	10
Total	87	88	175

Table 2.3 Number of schoolchildren in a (non-proportional) sample of 175.

Years of age	*Boys*	*Girls*	*Total*
5–8	31	30	61
9–12	31	32	63
13–15	25	26	51
Total	87	88	175

information overall, without increasing our total sample size. Within each stratum the choice is still random and the chance of appearing is equal. Between the strata the chance has been allowed to vary but its level is known for each component group, and this *must* be taken into account in any calculations that follow.

If we have found the mean height in each stratum, as the figures in Table 2.4, the overall mean of the population is estimated as

$$(156 \times 113.19 + 624 \times 138.16 + 49 \times 159.64 + 148 \times 115.60 + 635 \times 135.19 + 52 \times 157.50)/1664 = 134 \text{ cm}.$$

Table 2.4 Mean height (cm) of schoolchildren in the sample of Table 2.3.

Years of age	*Boys*	*Girls*
5–8	113.19	115.60
9–12	138.16	135.19
13–15	159.64	157.50

Multiplying the sample mean of each group by the *population* size of that group, and finally dividing by the total population size, has made the necessary adjustment. This is *fundamental*. A quite false result would be reached by merely averaging the measurements taken, without such adjustment.

It is worth noting that, although a final result should not be quoted to a greater accuracy than the data warrant, it is better during calculations to keep as many figures as can reasonably be handled. Thus the figures in Table 2.4 are shown to two places of decimals, because they are to be used in further calculations, while the eventual result is quoted merely to the nearest centimetre. It never does any harm during calculations to keep more figures than are really justified, whereas it can be very harmful to round off too soon.

If stratification is to be worth while it is clear that we must know, or have good grounds to suspect, that the strata differ appreciably from one another in the characteristic, or characteristics, in which we are interested, e.g. that men differ from women, that one age group differs from another, that doctors differ from lawyers. If the strata do not differ or, in other words, the population as a whole is relatively uniform, there is no point in dividing it into subgroups. There can be no gain in accuracy in such circumstances. It is, therefore, necessary to think closely before adopting the more involved technique. And clearly it is impossible to adopt it if the population to be sampled is not defined in the necessary detail.

Sampling by stages

Sometimes a strictly random sample may be very difficult indeed to draw and it may be more practicable to take the required sample in a series of stages (this is known as multi-stage sampling). Suppose, for example, we wished to learn the number of X-ray examinations made of all the patients entering hospital in a given week in England and Wales. It would be very difficult, if not impossible, to devise a scheme which would allow the total population of patients to be directly sampled. On the other hand it would be relatively simple to list the health districts of the whole country and randomly to draw a sample of these areas. Within this sample of districts all the hospitals could then be listed by name and a random sample of these be drawn. Within each of these hospitals, a sample of the patients entering in the given week could be chosen randomly for observation and recording. Thus by stages we have reached the required sample. If appropriate, stratification could be introduced at one or more stages, e.g. the areas could be sampled in broad regions and subdivided into urban and rural, the hospitals could be broadly classified and sampled according to their function, and the patients could be subdivided by their sex and age and then randomly selected.

It must be made clear, though, that such a scheme should be adopted only when the practical difficulties of true random sampling are insuperable. Those who take a random sample of 12 hospitals, and within each of them a random sample of 10 patients, often believe that they have achieved the equivalent of a sample of 120 at random from all hospitals. They have not. For many purposes the worth of their sample is very little more than 12 rather than 120.

Other methods of sampling

It will be seen that the use of random sampling numbers requires that the population involved be already numbered or, at least, be numbered when the occasion arises. If that is not the case, one method of sampling that is *usually* effective is to start from a random number and then systematically take every *n*th name (or file, etc.). In this way, suppose that from a list of 1000 clinical case records 125 are to be drawn for study. This is a fraction of 1 in 8, so for the starting point a number between 1 and 8 is randomly selected, say 3. Every 8th file from that point is then drawn—3, 11, 19, 27, 35, etc. This procedure is known as 'systematic' sampling.

It should be fully realised that in certain circumstances it can give a biased result. For example, every fiftieth house in a series of streets might conceivably produce a sample with too many corner houses and too few in the centre of the street. More generally, the population to be sampled may have some periodicity in its characteristics. The fixed interval method of sampling may then produce relatively too many high (or too many low) values according to where the interval happens to fall in relation to the periodicity.

Occasionally some other simple method may present itself. For instance, every man serving in the Royal Air Force has a service number allotted to him at entry. It would be proper to choose a 1 in 10 sample by selecting all men whose number ended in, say, an 8, or a 1 in 1000 sample by selecting all men whose number ended in, say, 345. On the other hand, to choose all the men whose surnames began with certain letters is open to grave objection. Suppose we take the letters M, J, W and O. In Great Britain the sample will certainly include unduly large numbers of the Scots (Mackintosh, etc.) the Welsh (Jones and Williams), and the Irish (O'Brien, etc.). Yet if we deliberately leave out the letters M, J, W and O we shall have too few of these nationals. The method is not a good one and should be rejected.

Another procedure that is likely to be quite satisfactory in many situations is to select persons born on specified dates in any month in any year, perhaps on the 9th, 16th and 27th, for example.

While such special methods may sometimes be necessary, they should *never* be employed where true random selection could have been. The difficulties of drawing a truly random sample are often exaggerated, and investigators fall into these other methods on the grounds that they feel that they will probably be just as good. So they may be in some cases, but it is not possible to demonstrate that they were in any particular case if challenged. Furthermore, with special methods we have constantly to reflect on whether the mode chosen will tend to bring one kind of person rather than another into the net, and thus produce a biased picture. Much thought needs to be given to that before embarking upon any unorthodox sampling scheme.

Care will be especially needed when the characteristic being measured varies in time. For instance, the prevalence of sickness, e.g. influenza, varies seasonally. We shall therefore reach a quite false answer if our sample observations do not cover the whole year. The feature of *time* often calls for more careful thought.

In conclusion, whenever a worker has adopted a sampling method to derive his observations and is presenting the results of his work, he should give the reader an exact account of how he went to work. He must state briefly but comprehensively the sampling techniques that he adopted and the degree to which he was successful in applying them (including the incidence of non-response discussed below). Without this information the reader cannot judge whether the sample is likely to be a valid one, i.e. representative and unbiased.

Sampling by computer

Many computer programs and packages contain random number generators. These are actually *pseudo*-random numbers, that is to say they are produced by a mathematical process that is not really random at all, but nevertheless gives the appearance of randomness. The best of these do the job quite adequately, but unfortunately there are some very inferior ones in existence also, and it is not easy for a non-specialist to tell the one from the other. All that can be done here is to give a warning that it is unwise to use computer (or pocket calculator) produced random numbers for anything important, unless you know that the particular generator used has been competently tested and found to be adequate.

One bad feature often met is a system that always produces the same sequence of numbers unless the user specifically asks otherwise. Even worse, systems are not unknown in which the same numbers are always produced and the user is not even given the choice of asking for others. Certainly it can be useful to be able to repeat a sequence and the facility to ask for that should be available; but the default mechanism, unless the user specifically asks, should be a sequence that cannot be predicted, and is different on each occasion of use.

These numbers usually come not in the form of an individual digit from 0 to 9, as in random number tables, but as a fraction somewhere between 0.0 and 1.0. Thus the first to appear might be 0.197779, the second 0.910679 and so on. It is best to use them as a whole, and not to work down the individual digits, because if the generator is less than perfect the early digits within each number are likely to be better behaved than the later ones.

In using such numbers to draw records at random from a computer file, it is easiest to take each individual in the population in turn, and decide whether to include that one in the sample or not, before passing on to the next. The method is shown in Table 2.5, where, for each individual, the next random number in sequence is compared with the ratio of the number still needed in the sample to the number remaining available. If the random number is smaller than this ratio, then the individual is included in the sample, otherwise not. It is evident that this process must end with precisely the required number selected for the sample. What is not immediately so obvious is that each individual will have been given the necessary equal and independent chance of being selected, but in fact it is so.

Table 2.5 Drawing a random sample of 500 from a population of 10 000 individuals. Individuals are included in the sample if Column (5) is less than Column (4).

Individual number	*Number remaining available*	*Number still needed in sample*	(3)/(2)	*Random number*	*Include in sample?*
(1)	(2)	(3)	(4)	(5)	(6)
1	10 000	500	0.050000	0.197779	No
2	9 999	500	0.050005	0.910679	No
3	9 998	500	0.050010	0.391938	No
4	9 997	500	0.050015	0.006844	Yes
5	9 996	499	0.049920	0.139526	No
6	9 995	499	0.049925	0.029086	Yes
7	9 994	498	0.049830	0.613859	No
.	.	.	.	.	.
.	.	.	.	.	.
.	.	.	.	.	.

Non-response

One of the most difficult problems that will arise in working with a random sample is that of 'non-response'. Some of the persons included in the sample may refuse to be interviewed; some may be too ill; some, perhaps, cannot be traced; some of the children, in the example above, may be absent when we visit the school to take the measurements; even when dealing with a file of clinical case histories the information required may be missing from some proportion. Every missing 'individual' (person or item of information) detracts from the randomness of the sample. We do not know, and usually cannot know, that the individuals that we *can* include give a true picture of the total population. The absentees, whatever the cause of their non-response, may have different characteristics from those who are present. In other words the sample observed has thus become a biased sample and if the number of missing items is large it may be very seriously biased. It is for this reason that every possible effort must be made to gather into a drawn sample all those originally included in it. Indeed, one should remember that the best-laid sampling scheme is quite meaningless unless this effort is made. If there are missing individuals (and almost always there are some) then much thought must be given to them, as to whether their absence is likely to distort the sample in relation to the particular facts under study.

Such biases, introduced by non-response, or by dropping out, mean that mere numbers are not enough in themselves. It is common nowadays to see research proposals in which it has been calculated that a sample of 150, say, is needed, but because half the patients may drop out, it is proposed to start with 300 so that 150 will remain for analysis. It cannot be too strongly emphasised that this is incorrect; 150 remaining, after a 50% (and non-random) drop-out, are *not* the equivalent of a random 150 with no drop-outs. The bias is almost certain to be very considerable.

Sometimes, inevitably, the missing items may be numerous and it may be worth while drawing a random subsample upon which more intensive efforts can be made to draw in 100 per cent of the required individuals. The complete, or nearly complete, subsample can then be compared with the less satisfactory main sample to measure the amount of bias if any, that may exist in the latter. To give a specific example, in one inquiry into the earnings of doctors, before the advent of the National Health Service in Great Britain, nearly 6000 medically qualified men and women were approached for information. It was realised that the non-response rate would almost certainly be high—in fact it proved to be as much as 27 per cent. A small random subsample of 600 was, therefore, specially drawn. This much smaller number could then be more extensively studied from available records (e.g. the nature of their speciality) and more assiduously pursued for a reply. The results in this subsample strongly indicated that the more extensive income figures derived from the larger but incomplete group could not be seriously at fault.

Another procedure that may sometimes reduce the tendency of non-response lies in brevity and simplicity in one's requirements. Too many, and too difficult, questions do not encourage co-operation—particularly in the approach by questionnaire. There is always a desire to learn many things at the same time and by giving way to it one may end in learning nothing because of the resulting excessively high incidence of non-response. Once again a partial solution may sometimes be sought by using subsamples for different questions. Thus in studying in Great Britain the services given by general practitioners to their patients over a calendar year a sample of 6000 doctors was drawn. To reduce substantially the amount of work required of each, they were allocated randomly in subsamples of 500 to one month of the year. During that one month only, they were asked to keep a complete record of the number of attendances by and visits to patients. Thus for every month a fairly large sample of the total population of practitioners was available to give a measure of the services rendered in that month and a summation of the sample values would give the figure for the year. Further, each doctor was asked to carry out a relatively small task although an appreciable proportion (one-third) of the total number was used. In addition, information was sought on five other matters by randomly dividing the subsamples of 500 into five further subsamples of 100 each. One such group was asked to record the number of operations performed, a second group the number of injections given, a third the number of night visits paid, and so on.

Thus the demands on any individual were carefully restricted and the non-response rate proved to be very satisfactorily small (2–3 per cent). The original random sampling scheme was thereby maintained practically unimpaired. (For a full discussion of these sample surveys of the 'doctor's day and pay' see the *Journal of the Royal Statistical Society*, 1951, Series A, **114**, 1–36.)

The importance of randomness

In conclusion, the importance of this concept of random sampling could not be more clearly emphasised than in the illuminating comment once made by a newcomer to that field:

> The necessity of using a true random sample of the population in a survey of this nature is well known and needs no emphasis; nevertheless, it may be added that contact with such a sample provided a new experience. The actual practice of medicine is virtually confined to those members of the population who either are ill, or think they are ill, or are thought by somebody else to be ill, and these so amply fill up the working day that in the course of time one comes unconsciously to believe that they are typical of the whole. This is not the case. The use of a random sample brings to light the individuals who are ill and know they are ill, but have no intention of doing anything about it, as well as those who never have been ill, and probably never will be till their final illness. These would have been inaccessible to any method of approach but that of the random sample. Perhaps one of the deepest impressions left in my mind after conducting the survey is the fundamental importance of the random sample—unusual as it is in most medical work. It does not make for ease of working; all sorts of inaccessible personalities may be encountered, and it is more time-consuming; but the degree of self-selection imposed by the population on itself in regard to its approach to doctors inevitably gives anything other than a random sample a considerable bias. It has, however, one disadvantage in that the percentage of refusals may be high. (*The Social Medicine of Old Age*, The Report of an inquiry in Wolverhampton made in 1948 by Dr J.H. Sheldon, CBE, FRCP.)

Summary

In statistical work in the different fields of medicine we are constantly studying samples of larger populations. Sometimes we shall wish deliberately to draw such a sample from the population. Although in so doing we shall introduce a sampling error (which can be estimated), we shall nevertheless often gain in precision by the greater and more skilful attention that can be given to the collection of a smaller amount of data.

The sample should be drawn by some strictly random process that gives every individual in the parent population an equal chance (or known chance) of appearing in the sample. Random sampling numbers provide such a process. If groups within the population vary widely in their relevant characteristics, it may well be advantageous first to divide the population into those groups, or strata, and then to draw a sample from each stratum appropriately. Other schemes, such as sampling by stages, may sometimes be necessary, but they should not be adopted without the most careful thought. They are much less satisfactory.

Every effort must be made to keep non-response (or missing items) to a minimum. The most careful sampling scheme is of no value if a large proportion of the required data is not obtained.

Chapter 3

Collection of statistics: bias

In the previous chapter attention was devoted mainly to the situations in which a sample of observations could be deliberately drawn for study from some known population. The basic problems are then to define the population and to find an appropriate means of drawing a random sample from it. Often in medicine we are not in that situation at all. We have to accept whatever sample of observations (persons, records, etc.) may present itself in the natural course of daily events. *What is the nature of the sample* then becomes the crucial question. What are its characteristics? Are we entitled to argue from the particular to the general? For that is what we are invariably hoping to do. In seeking to advance knowledge we are, for example, not very interested in the fact that a particular and relatively small number of patients with a specific disease rapidly recovered when treated in some defined way. We are intensely interested in knowing whether that form of treatment is the method of choice for the generality of patients with that disease. We must then consider very carefully whether the sample is representative of all such patients and not in any way biased.

It is important to be clear on the meaning the statistician attaches to the word 'biased'. By a biased sample he denotes a sample that is not representative of the parent population of which it is part. The bias may have been deliberate, in which case the form of bias is known and the lack of comparability between the sample and the population is usually perfectly clear. For instance, if the treatment of some form of cancer by means of surgery were confined to those patients without mestasteses, then it is obvious that these patients are *not* a representative sample of all patients with this form of cancer, but are selected on the criterion of no detectable metastases. To compare their subsequent mortality experience with that of all patients is, therefore, a most doubtful procedure, for we are clearly not comparing like with like in all respects except with regard to surgical treatment. Even without that treatment the death rate of patients without metastases is likely to differ materially from the death rate of the general run of patients.

More often, however, the bias is not deliberate but is quite unforeseen or is unrealised. To say, therefore, that a research worker's figures relate to a biased sample of patients is not an aspersion on his scientific honesty; the statement implies only that owing to the method of collection of the figures, or to the limited field in which he was able to operate, it is quite impossible for his sample to be representative. It may be that with care the bias might have been avoided; often it is unavoidable. Its possible presence cannot be too carefully remembered or taken into account in interpreting all statistics. As, however, it is frequently overlooked by the authors of studies in medicine, this chapter is devoted to a series of examples.

Sex ratio at birth

As a simple illustration the frequency with which male and female births were recorded in the births column of *The Times* newspaper were taken over a period, the number of male births was 3304 and female births 3034, so that the sex ratio was 1089 males for each 1000 females. According to the Registrar-General's figures for England and Wales at that time, the sex ratio of births in the country as a whole rarely exceeded 1050 to 1000. It is clear that from the point of view of sex ratio the births recorded in *The Times* were unlikely to be representative of the births in the country as a whole. It is possible that first births are more frequently recorded in those columns than births of a later order, and such births have a different sex ratio; or that proud parents are more likely to record their sons than their daughters; or that the sex ratio differs between social classes. Whatever the explanation, with such a sample of births, if that were all that were available, one could not generalise about the population of the whole country with any security.

Hospital statistics

Turning to a more medical problem, hospital statistics of a disease can very rarely be regarded as representative of all cases of that disease. The patients are frequently drawn from particular areas which may have differing populations in age, sex and race; they may come disproportionately from particular social classes. Still more important, in many diseases only those patients who are seriously ill are likely to be taken to hospital. It is obvious that we cannot determine with any approach to accuracy the fatality rate of any disease, say measles at ages 0–5 years, if our statistics are based mainly upon the seriously ill—patients, for example, in whom a secondary pneumonia has developed—and ignore the mass of children whose symptoms are so slight that they can safely be treated in their own homes. Of *all* children with measles those in hospital would then form only a small and stringently selected group; our deductions from such a group are correspondingly limited, especially with regard to such factors as the incidence of complications and the rate of fatality or recovery. It is not too much to say that there is hardly any disease in which a hospital population must not initially be regarded with suspicion if it is desired to argue from the sample to the population of all patients. No such argument should be attempted without a preliminary and rigorous examination of the possible ways in which bias may have occurred.

The same difficulty arises with secular comparisons—e.g. when we wish to see whether the fatality rate from some disease has changed from one year to another. In each year the fatality rate is measured from the experience of the patients admitted to hospital, and in each year those patients are a small sample of all patients with the disease in question. It must be considered whether that sample has changed in type. In both years the sample may be a biased sample but the bias may not be identical. The kind of patient admitted may have changed. For example, in a group of American hospitals it was once reported that the fatality rate from appendicitis declined from 6 per cent in one year to 3.5 per cent four years later. Was that a 'real' decline, due to more efficient treatment maybe, or was there a concurrent change in the types of patients admitted? Examination of the basic figures showed that in these hospitals some 2500 patients were operated upon in the earlier year, while in the later year the number had risen to 3500, an increase of 40 per cent. It is impossible to believe that an increase of 40 per cent in four years was a real increase in the incidence of appendicitis. It is more likely that the desire for admission to these particular hospitals or some criterion of admission had changed, that some patients who were admitted in the later year would not have entered them in the earlier year. It is possible, therefore, that the type of entry has changed as well as the volume—perhaps that milder cases were admitted and operated upon in the later series which were not present in the earlier series. In the absence of positive evidence on that point the change in fatality cannot be accepted at its face value or as satisfactory evidence of the effect of a change in some other factor—e.g. the benefit of earlier admission to hospital in the later year. The following two questions must always be considered. Has there been a change in the population from which the samples are drawn at two dates—i.e. a change relevant to the question at issue? At each date was there the same probability that a particular type of patient would be included in the sample?

Day of treatment

In measurements of the value of some form of treatment, statistics of the type shown in Table 3.1 are frequently given.

It is possible that the level of this fatality rate at the different stages is seriously biased. Let us suppose, as is often the case, that the treatment is

Table 3.1

Day of disease upon which treatment was first given	*Fatality rate (per cent of treated patients)*
1	1.3
2	3.6
3	7.5
4	9.3
5	12.8
6 or later	16.4

given to those patients who are brought to hospital and that not *all* patients necessarily go to hospital. Then on the first day of disease a variety of patients will be taken to hospital, in some of whom, in the absence of the special treatment, the disease is destined to run a mild course, in others a severe course. The presence of a proportion of mild cases will ensure a relatively low fatality rate, even if the special treatment has no specific effect. But as time passes this proportion of mild cases in the hospital sample is likely to decline. By the time, say, the fourth day of disease is reached, a number of patients treated at home who were not seriously ill will have recovered or be on the way to recovery. Their removal to hospital is unnecessary. On the other hand, patients who have made a turn for the worse or whose condition has become serious are likely to be taken to hospital for immediate treatment. Thus on the later days of disease the sample removed to hospital for treatment may well contain an increasing proportion of persons seriously ill, it obviously being unnecessary to transfer those who are making an uninterrupted recovery. In other words the patients removed to hospital on the fourth day of disease are unlikely to be a random sample of all patients who have reached that day of the disease but may consist rather, perhaps mainly, of patients still seriously ill. Such a group will certainly have a relatively high fatality rate.

Another example of this statistical difficulty may be taken from some fatality rates recorded for appendicitis. It was reported that in a group of cases 2 per cent of those admitted to hospital within 24 hours of the onset of symptoms died compared with 10 per cent of those whose admission was delayed till after 72 hours. But it is likely that the group of patients admitted early is composed of a proportion of the seriously ill and a proportion that would do well whether admitted to hospital or not. On the other hand, those who are admitted after a delay of three days from onset are likely to be patients whose condition is serious, for clearly those whose condition has become quiescent are unlikely to be taken to hospital at that point of time. If such a sequence of events occurs, it is clear that the group of patients admitted early is not comparable with the group of patients admitted late. This may not be the whole explanation of the difference between the fatality rates; indeed it is not likely to be, for there are excellent reasons for the early treatment of appendicitis. But the bias outlined is a possible factor with statistics such as these, which makes it very difficult to measure accurately the *magnitude* of the advantage.

Post-mortem statistics

In an attempt to obtain more accurate data emphasis is often placed upon post-mortem statistics. One has, however, to remember that this increased accuracy is gained at the risk of using material that may be highly selective. It is rare for every death to be subjected to autopsy and those that are chosen are by no means chosen randomly. They are more likely to be chosen *because* the cause of death is obscure or because the case presents features of special interest. No measure of these or other selective factors is likely to be made easily.

In this field of autopsy statistics it is, too, important to note that the interpretation of the observed frequency of occurrence of *two* disorders in the same person is very difficult indeed. Thus it may appear that the number of persons found to have the two conditions is appreciably more than would be expected on a purely chance basis, i.e. based upon the known frequency of the occurrence of each condition separately. But the occurrence of two disorders in the same person not only make it more likely that he will seek hospital care, it may also make it more likely that he will die and thus come to autopsy. Such a possible selective mechanism must always be thought upon with much care in the specific study.

It should also be remembered that any feature observed at death is most unlikely to be representative of the living population. Outside the medical field, it is not unknown for attempts to be made to examine inequalities in society by investigating the amounts left in wills. Yet the difference between the wealth of *A* (who chose to save throughout life) and *B* (who chose to spend) will be a maximum at that point, and cannot, of itself, indicate any difference in opportunity.

Follow-up studies

Because of their incompleteness, follow-up studies of patients are often subject to selective influences. In basing conclusions upon the patients who *are* successfully followed up we are presuming that the results we record for such a group would be unchanged if we succeeded in tracing and adding in the lost-to-sight patients; in other words, we presume that the characteristic 'being followed up' is not correlated with the characteristic that we are measuring, e.g. survival. Is that likely to be true? On the one hand it might be easier to learn that a patient is dead—through the Registrar-General or other official records—than that he is alive and gone to Australia. On the other hand the living may answer inquiries more readily than the relatives of the dead. Once more all we can do is to consider the probabilities, and the possible biases in any specific situation or, of course, better still, avoid them at all costs by making the follow-up comprehensive.

Infant feeding

That, on *a priori* reasoning, there is a very strong case for the breast feeding of infants is obvious, but to secure a measure of the degree of its advantages has always been, owing to selective factors, extremely difficult.

Under actual conditions of life we cannot obtain two groups of infants, exclusively breast-fed and exclusively bottle-fed from birth, who will remain in their respective groups whether their progress is good or bad. If a baby on the breast is not thriving, its diet is likely to be changed to partial or complete artificial feeding. Thus, a purely breast-fed group would contain mainly babies who are doing well; those who are doing badly would often be transferred at different ages to the artificially fed group to the detriment of the latter in statistical comparisons. Even if one were to preclude any such additions to the artificially fed group, one would still have by the withdrawals from the breast-fed group a differentially selected group of breast-fed babies—unless the babies diverted to artificial feeding were deliberately retained in the breast-fed group in spite of the change in feeding; but such a group could not truly be designated 'breast-fed' and any satisfactory comparisons might well be impossible.

Such factors must be closely considered in interpreting results in this field.

Self-selection

An interesting real-life example of what may be termed *self-selection* is worth notice for it is characteristic of many a modern problem in occupational medicine. In its early work the Industrial Pulmonary Disease Committee of the Medical Research Council was concerned with the problem as to whether the working health and capacity of coal-miners were impaired by the inhalation of anthracite dust. To begin with, a study was made of the size and age constitution of the working population at the South Wales anthracite collieries. Underground workers had to be excluded from the investigation, since such workers, as well as being exposed to various concentrations of anthracite dust in the atmosphere, may in addition have been exposed to silica or other dust. If the health of these workers was found to be impaired it would be impossible to implicate the anthracite dust as the responsible agent. The impairment might equally well be due to exposure to stone dust containing silica, which, it is well known, can produce serious damage to health. In addition, it was considered necessary to exclude surface workers who had at any time worked underground, since the effects of exposure to silica dust will not necessarily be immediately apparent and also because impaired health may have been the reason for transference from underground to surface work. This was, in fact, known to be the reason in numerous cases, so that such workers would be a highly select group. Attention was therefore turned to workers who were exposed to anthracite dust on the surface and had *always* worked on the surface. Such workers, it was found, were employed on a relatively light task. Not only, as stated, was there a tendency to draft to it operatives who had previously worked underground and had for one reason or another become partially incapacitated, but in addition it was clear that a large number of boys were initially employed upon this work but rapidly moved away to other work. In the main these boys were drafted underground where the physical labour was heavier but the rate of pay superior.

The inevitable inference is that the healthy and strong individuals will transfer to underground work while those who remain on the surface are likely to be of under-average physique and health. In other words, there has been a form of self-selection. If the examination of such surface workers showed that they included a high proportion with impaired health or that they suffered an unduly high rate of sickness in comparison with some standard, this result could not with security be ascribed to the effects of dust inhalation. It might be considerably influenced by the fact that these surface workers were initially less healthy than a random sample of all surface workers. Such features of physical (and mental) selection must necessarily bear heavily upon the characteristics of persons entering, and remaining in, all specific occupations.

This investigation in South Wales also revealed an example of bias through volunteers being accepted for examination. For the reasons outlined above no inquiry was at that time made at the collieries and the field of study was transferred to dock-workers exposed to anthracite dust. Of 250 such workers it was arranged to examine, clinically and radiologically, a sample of 40 operatives—namely, 15 workers employed for 3–4 years and 25 older workers with 15–40 years' service. These two groups were selected at random from the complete list of operatives, to ensure, as far as possible with such small numbers, a representative sample. At the examination eleven of these men were absent and, to make good the deficiency in numbers, volunteers were secured in place of the absentees.

The results of the examinations suggested that there was a readiness to volunteer on the part of individuals who, on account of some known or suspected disability, desired to be medically examined. Such substitutions result in the sample ceasing to be random and it becomes unlikely to be representative of the population from which it was drawn. The stipulated *quantity* had been maintained but the *quality* had been lost.

It should be noted that if no substitutes had been accepted we still would not know that the 29 operatives who did attend were a cross-section of the originally chosen 40. Why did the 29 choose to come, why did the 11 stay away? If it was for any reasons connected directly or indirectly with their state of health, then the sample attending *must* give a biased picture. The neurotically inclined may more readily seek medical examination or they may more readily avoid it. The physically fit man may regard it as a waste of time; or, having nothing to fear he may be quick to submit. We do not know (or very rarely know) the motives at work and the greater, therefore, the departure from the original sample, the greater must be our doubts in interpreting the results we record.

Often we shall have knowledge of other characteristics of the individuals involved; perhaps we may know the ages and heights of the men concerned, for example. If so we can compare the two groups for these characteristics, see whether they are similar and make suitable adjustments to our calculations if not. But there always remains one characteristic that differs totally in the two groups—namely the very definition of the groups. *All* of one group attended, *all* of the other group stayed away, and without knowledge of the reasons there is no adjustment of the calculations that will allow for it.

Volunteers for treatment

For very similar reasons a sample which is composed of volunteers is not likely to be representative of the population at large. If, for example, the prevention of the common cold by a vaccine is offered to some specified group of persons, the volunteers are likely to belong mainly to that section of the group which suffers most frequently or severely from the complaint. They hope for some advantage from the treatment. Those who have been free from colds for a long time are unlikely to come forward. Those vaccinated are, therefore, a biased group, not comparable with the remainder of the population from which they were drawn (or, more strictly, drew themselves). In such a position the question must always arise: is the act of volunteering correlated with any factor which may influence the final results of the experiment? In the present example, if the vaccinated volunteers are mainly the common cold 'susceptibles' and the non-volunteers are mainly the 'resistants', then clearly the contrasts between their attack rates that we might make, to measure the vaccine, are quite meaningless.

Questionnaires

Inquiries carried out by means of questionnaires are above all those in which bias must be suspected. In all such inquiries, replies to the questions put—even to the simplest question—are received from only a proportion of the individuals to whom the form is sent. There can never be the slightest certainty that the individuals who choose to reply are a representative sample of all the individuals approached; indeed very often it is extremely unlikely that they are representative. For a simple example, one may take a careful enquiry once made by the Editor of *The Lancet* into the present-day openings of medical practice. To measure the success which recent graduates had achieved in their profession a questionnaire of three relatively simple questions was addressed to the 1490 men and women who in a specified year registered their names with the General Medical Council, namely (1) What branch of medicine have you taken up? (2) What led you to this choice? and (3) What was your approximate income from professional work last year? To overcome objections to providing such personal information no clue to the identity of the correspondent was required. Yet of the individuals approached only 44 per cent replied. Are these persons a representative sample of the 1490? It is possible, as is clearly pointed out in the report, that there might be a tendency for those who have been successful in their profession to be more eager to register their success than for those who have failed to register their failure. Alternatively, the latter might under the veil of secrecy be glad of the opportunity of stating frankly the drawbacks of the profession. Those who have turned to other professions might tend not to reply, under the impression that the inquiry cannot concern them. Successful and busy individuals might be unwilling to give time to the inquiry. It is impossible to determine whether any such factors are operative in the determination to answer or not to answer. The difficulty is inherent in all inquiries carried out by this method and must never be ignored.

It almost invariably *is* ignored by the daily press, which will report that 70 per cent of some group think, for example, this or that about atomic warfare, the National Health Service, or the President of the United States, and will overlook the fact that the 70 per cent is based upon the proportion (and possibly small proportion) of the group who chose to answer—and almost invariably it will never report the percentage who chose to answer. The figures are utterly misleading.

It is possible sometimes, however, to see whether the final sample is or is not biased in certain known respects. For example, suppose the population to be approached consists of all the persons on the medical register at a given time. For each of these persons we may know such characteristics as sex, age at qualification, degrees or other qualifications obtained, type of medical work upon which the person is engaged—general practice, public health, etc. Only 50 per cent of the total population, let us suppose, answer the questionnaire addressed to them all. In the statistical analysis of the available answers we can at least see, and it is of course essential to do so, whether the sample is representative of the total population in relation to the *known* characteristics of the latter. If 50 per cent of the men and 50 per cent of the women answered, then the sample obtained is not biased in its sex ratio; but if 60 per cent of the men and only 25 per cent of the women answered, then a bias has been introduced, for the ratio of males and females in the sample is different from the real ratio in the population. We must make allowance for that fact in analysing the results and cannot merely use the sample as it stands. Similarly, we may see whether older and younger persons answered to an equal degree and whether those engaged in different types of work were equally forthcoming. By such means we can then determine whether or not certain classes of persons have tended to answer more or less readily than others, and thus know whether or not our sample is biased in these known respects and, if necessary, make allowances for it. While such a check is highly important, indeed essential, it cannot be entirely conclusive. Even if the sample *is* representative in the known respects, we cannot be sure that those who chose to answer were in other respects representative of the total. For instance, 50 per cent of men and 50 per cent of women may answer, but in *each* group those who answer may mainly consist of those who feel more deeply upon the questions addressed to them, or be those (if any) who like filling in forms. In other words, the sample is correct in its sex proportions, but for neither sex do we know that the sample is such that it will accurately express the views of the total men and women originally approached.

In reporting these, or similar, inquiries a statement should always be made of the number of missing questionnaires or items. Clearly if 90 per cent of the required data were obtained, more reliance can be placed upon the results than if the proportion were only 45 per cent, and the reader should be in a position to judge.

The type of simple correction one can sometimes make for a known bias

Table 3.2

	Male	*Female*	*Total*
Number of persons to whom questionnaires were sent	10 000	2 000	12 000
Number of persons who answered	6 000	500	6 500
Mean income reported by those who answered	£20 000	£16 000	£19 692

can be demonstrated arithmetically from the hypothetical figures in Table 3.2. Using the sample as it stands, we see that the mean income reported by the 6000 men who answered was £20 000, and of the 500 women who answered was £16 000. The mean income of all persons in the sample is, therefore:

$$(6000 \times £20\,000 + 500 \times £16\,000)/6500,$$

which equals £19 692. But this figure is clearly too high, since men, on the average, earned more than women, and in the sample we have a ratio of 12 men to 1 woman, whereas in the total group the real ratio is only 5 men to 1 woman (due to the fact that 60 per cent of the men answered and only 25 per cent of the women). If we are prepared to believe that, for each sex, those who answered were a representative cross-section of the total approached, then the correct estimate of the average income of a person must be obtained by 'weighting' the observed mean incomes of the sexes by the correct numbers of persons of each sex. Thus we have $(10\,000 \times £20\,000 + 2000 \times £16\,000)/12\,000 = £19\,333$. In other words, we are accepting the sample figures as giving a true picture for each sex, but must combine them by using the known true proportion of men to women in the population sampled, in place of the untrue proportion given by the sample.

Forming groups by questionnaire

The difficulties with questionnaires, as detailed above, apply whenever we wish to estimate, from the answers, the proportions of the population with certain characteristics. Those difficulties are much less important if the object of the exercise is to form groups for a prospective inquiry.

For example, a questionnaire asking a group of people to specify their smoking habits is unlikely, if the proportion returned is small, to tell you accurately how many in the population are very heavy smokers, how many non-smokers, etc. But if the aim is to form a group of heavy smokers and a group of non-smokers, to compare their subsequent mortality over the years, this bias is of no importance.

House sampling

Finally an interesting and still highly topical example of bias in taking a sample of houses is suggested in a report on the historic influenza pandemic of 1918–19 (Ministry of Health's Reports on Public Health and Medical Subjects, No. 4). To obtain facts as to the incidence and fatality from influenza in that great epidemic a house-to-house inquiry was undertaken in five areas of a large city, information being obtained *so far as possible* at every fifth house. However, houses which were found closed at the time of visit were ignored. But houses in which there are young children are less often found closed and this would tend to affect the age-distribution of the population recorded in the sample. Compared with the population from which it was drawn the sample would be likely to contain an undue proportion of young children and a deficit in the number of adults. Any substitution of another house for the chosen one cannot correct for this bias in the sample, and such substitutions are to be avoided in sampling inquiries.

It will be noted that a bias of this type might be difficult to foresee. It is here that the statistician has some advantage, for his experience of such inquiries makes him familiar with the methods that are likely to ensure a random sample and those that are likely to lead to one that is unrepresentative of the population from which it is taken. Workers who are unfamiliar with sampling inquiries but wish to embark upon one may, therefore, find his advice of assistance.

To take an extreme example, if you take people walking down the street, and ask them whether they are permanently confined to bed, the answer will be that nobody is. The fallacy is obvious there and everyone will spot it at once. Precisely the same fallacy is too often ignored when it appears in more subtle guises.

Summary

If we wish to generalise from some sample group of observations—which is invariably the case in real life—we must possess a sample which is representative of the population to which it belongs. In taking samples deliberately, or in accepting samples that arise in the daily course of events, we must realise that bias may occur through the operation of various factors, leading to a sample which is not representative of the total population, but in which one type of member had more chance of appearance than another, whether that bias were due to deliberate choice or unconscious selection of the members incorporated in the sample. Self-selection of the members of a group is a common form of bias, e.g. in the physical or mental status of those who follow a certain occupation. Another exceedingly common form of bias lies in the absence of some of the required records, e.g. by individuals (uninterested, busy or lazy, whatever the reason may be) who do not reply to a questionnaire. In generalising from a sample, or in making comparisons between one sample and another, the possible presence of such bias must always be very closely considered.

Chapter 4

Collection of statistics: forms of record

In all scientific work we are involved in asking questions. In medicine, for example, we may seek to learn the effects of a specific treatment used upon patients with a specific disease. But whether we require the population concerned to reply themselves to a questionnaire, whether we seek the information by oral inquiry through ourselves or trained social workers, or whether we make a clinical examination or adopt some laboratory means of procedure, we are, in the last analysis, always asking questions. Accordingly, we need a form of record upon which to ask those questions and to record the answers. One of the first and, indeed, most decisive steps in any inquiry, therefore, is the construction of that form—*what* should be included and *how* it should be included. Each question must be given the closest thought to see whether it is clear and definite; what the possible answers are; whether the answers can be adequately, if not wholly accurately, obtained; how they can be analysed and put into a statistical table at the end of the inquiry or experiment. If the questions are incomplete, ill-conceived, or inadequately answered no statistical analysis, however erudite, can compensate for those defects or produce the answers that the worker had hoped to get. The time to remember that is not at the end of an investigation but at its beginning.

We must also remember two other things—on the one hand that the drafting of clear and unambiguous questions is an extremely difficult task and, on the other hand, that many people find the completion of any form an extremely difficult task. As an early report on the Census of England and Wales correctly emphasised, 'those who are conversant with forms and schedules, scarcely realise the difficulty which persons, not so conversant, find in filling them up correctly.'

Questions and answers

In formulating questions, or headings, for inclusion on a form there are a number of basic principles to bear in mind.

(1) To begin with, one should consider closely whether there is any *ambiguity in the question* and, consequently, in the answers received. A

very simple example can be found in that innocent question which appears on so many forms—age. Age last birthday? Or age nearest birthday? Generally, one might expect to be given the age last birthday. But at any moment of time, about one person in 12 is within one month of their *next* birthday and might therefore consider that that was the more appropriate age to give. Perhaps it does not matter. But that certainly does not mean that preliminary consideration should not be given to the question; *a decision should be taken that it does not matter*. In certain circumstances it certainly would matter. For instance, in a group of 100 children aged 5 years *last birthday*, the average age would be $5\frac{1}{2}$ years, or close to it (the individual ages would run from 5 to less than 6). In a group of 100 children aged 5 years to the *nearest birthday* the average age would be 5 years, or close to it (the individual ages would run from $4\frac{1}{2}$ to less than $5\frac{1}{2}$). The bodily measurements of the two groups would obviously differ appreciably since the average ages differ by 6 months. With a form on which 'age' is asked for we shall be unaware of what in fact is given. We should, therefore, specify 'age last birthday' (or, better still, ask for 'date of birth' and then make the required calculation ourselves).

Date of birth is also preferable to age because age changes as people grow older, but date of birth stays constant. Age at any given date, if wanted, can easily be calculated from it. In recording any dates, however, great care needs to be taken to define the manner of specifying them. In Britain it is usual to use six figures—two each for day, month, year; in the USA six figures are used similarly but in the order month, day, year. If there is any chance of confusion between these two systems, the form itself needs to state precisely what is wanted. It is also necessary to consider whether four digits might be wanted for the year: will the observations extend over more than 100 years?

Ambiguity in the question also arises if use is made of what may be called 'overlapping groups'. Thus, in a British inquiry into prematurity, the recipient of a form was asked to state the number of babies born within three birth weight groups (measured in pounds): (a) $2\frac{1}{2}$ to $3\frac{1}{2}$ lb (b) $3\frac{1}{2}$ to $4\frac{1}{2}$ lb, and (c) $4\frac{1}{2}$ to $5\frac{1}{2}$ lb. This does not make clear which is the correct group for a baby recorded as weighing exactly $3\frac{1}{2}$ or $4\frac{1}{2}$ lb. More precise definition of the group boundaries is required.

Better still is to record on the form the actual measurements taken, as precisely as possible. Groups can always be formed later, if wanted, from precise figures, whereas the reverse operation is impossible.

(2) As far as possible *every question should be self-explanatory* and not require the respondent to consult a separate sheet of instructions. The importance of this principle will, of course, vary with the circumstances. If a few highly trained persons are making the observations, conducting the clinical examinations, whatever it may be, then clearly they can be relied upon to turn to, and follow, detailed instructions. But if the form is to be completed by large numbers of less trained and less interested people, then experience shows that they cannot be relied upon to read and remember extensive footnotes or instructions. For example, in a trial of a vaccine against an infectious disease of childhood, details might be sought, for each

affected child, of any known exposure to another case. Such definitions of the possible varieties of exposure which are of interest must be given or some of the answers will undoubtedly be vague and uncertain. If possible the definitions should be incorporated in the question on the form itself. Thus for each case occurring the following alternatives of exposure might be specified *within the question*:
(a) within the child's own home
(b) at a day nursery
(c) at school
(d) elsewhere: specify place
(e) no known exposure.
The respondent is thus shown on the form itself the categories of information that are sought and can answer clearly and without undue trouble.

Whenever specifying alternative answers like this, it is essential to indicate whether more than one may be chosen (e.g. at home and at school) and if so to consider in advance how such multiple replies are to be handled in the analysis.

(3) Almost invariably *every question should require some answer*. Without that precaution it is often impossible to know for certain whether a person did not possess some characteristic or whether no information was in fact sought or obtained. For instance, the question at issue may be 'did the patient during pregnancy suffer an attack of rubella?' An answer 'yes' is a clear positive but no answer at all, or merely a dash (—), is by no means a clear negative. It *may* mean that, but it may, on the other hand, mean that the question was never asked or that no certain information was forthcoming. One does not know. The question should be given in such a form as invariably to require a clear answer, such as (a) yes; (b) no; (c) not known. Every question, therefore, needs some final category to make certain that some answer must be given—whether that final category be 'not known', 'no information', 'unspecified', or 'other'.

(4) *The Degree of accuracy to which measurements are required* should be specified. Should, for example, blood pressures be recorded to the nearest millimetre, to the nearest 5, or to the nearest 10? If no specification is given the recorders will inevitably vary amongst themselves in the accuracy to which they work. It may also be necessary sometimes to specify the nature of the measurement required—whether, for instance, body temperatures are to be taken orally or *per rectum*.

(5) It would seem needless to say that much thought should be given as to *whether it is likely that a particular question can be answered adequately by anyone at all*, yet much experience of forms suggests that the advice is not so needless. For example, very few people can give the certified cause of death of their parents or, even more, of other relations. It is very unlikely that people can remember accurately the minor illnesses (cold in the head, etc.) they have suffered in a previous 12 months. On the other hand it certainly does not follow that questions which obviously cannot be answered with complete precision are thereby rendered valueless in all circumstances. The amount of inaccuracy may be unimportant to the

problem at issue or it may still allow broad conclusions to be reached. Each situation must be carefully weighed on its merits. For instance, one can be sure that few people in their fifties or sixties can give an exact statement of their past smoking habits, the habits of a lifetime. It does not follow that they cannot give an answer which is sufficiently accurate to allow them to be classified into a few broad categories—heavy, moderate, light, or non-smoker—and for the frequency of some other characteristic, e.g. cancer of the lung, to be usefully examined within those broad categories. It should be remembered, too, that if the errors are unbiased, then the degree of association found between the two characteristics is likely to be *less* than the degree that exists in reality. (If in the incidence of cancer of the lung the *A*s really differ from the *B*s to a certain degree, that degree will be made less if through unbiased errors of memory, etc., we have included some *A*s with the *B*s and some *B*s with the *A*s. The contrast has been rendered less 'pure' and clear-cut.)

Under this heading there will also be occasions on which close attention must be paid to the problem of *observer error or variation*. Will doctors, or other workers, surveying the same patients classify them in the same way? For example, it is well known that experienced school medical officers will differ amongst themselves in the assessment of a child's nutritional condition. After a short interval of time, they will also differ from themselves. It is equally well known that experienced readers of X-ray films will differ over the interpretation of, or even over the presence or absence of, lesions in the chest. And according to the examiner involved, it has been shown that clinical histories will differ in the frequency with which somewhat ill-defined conditions, such as chronic bronchitis, are discovered. It is not really the observers who are at fault; it is the method. It is not precise enough to allow uniform and clear-cut decisions. It follows that close thought should be given to the problem of observer error or variation, whether it is likely to be an important feature and, if so, whether there are any means of reducing it.

(6) Any form of record which must be completed by many people should, to the utmost extent, be made *simple in wording and logical in the order of its questions*. The amount of work required of the respondent may often be reduced by putting the question in such a form that the answer demands only a cross or a tick or the ringing of one specified category (as in number (2) above). Such answers are also very easily tabulated—rather too easily sometimes. A respondent may, perhaps, tick 'yes' in answer to a question, but add in writing something like 'I have replied yes because that is literally true, but it may be misleading because . . . , etc.' All too often the 'yes' reply will then be entered into the computer file without the person in charge of the inquiry ever having seen the qualification or decided what to do about it. This is unfair to the respondent (even though he will be unaware of it), and also means that the results of the study may be worth less.

(7) Much attention should be given to *the number of questions*. Obviously this must vary widely with circumstances, but at the same time there are probably no circumstances in which the constructor of a form

should not ask himself of *every* question: Is this question essential? Can I obtain useful answers to it? Can I analyse them usefully at the end? Such a self-discipline is likely to reduce the size of any form. The temptation to collect information in case it might be useful should be resisted.

There are also circumstances, as described in Chapter 2, in which it may be wise to distribute a large number of questions over different samples and thus avoid too heavy a burden being placed upon any one respondent. On the other hand there are many circumstances in which it may be profitable deliberately to include questions for purposes of checking the nature of the response or to encourage an unbiased response. For instance, in collecting a sample of children aged 5 years to study, say, their previous attacks of infectious diseases, it might be useful to include a question on vaccination against measles. If the frequency of vaccination is known for the whole population from national statistics then, from its frequency in the sample, one may judge whether the sample is likely to be well chosen and representative.

In certain circumstances in approaching by questionnaire a population, marked or not marked by some characteristic, it is not unlikely that the marked persons will tend to reply and the unmarked not to reply. For example, patients who have undergone a blood transfusion are followed up by post some months later and asked, with appropriate questions, whether they have had symptoms of jaundice. Those who have had symptoms, the positives, may well be inclined to reply more readily than those who have not, the negatives, so that a false measure of the incidence of serum hepatitis is reached. It may therefore be profitable to put the question in such a form as to give everyone some reason to reply, e.g. by listing the symptoms of various common diseases (rheumatism, etc.) and asking the respondent to put an X appropriately, or by adding some very general question applicable to everyone, such as: 'Since leaving hospital have you been in good or poor health?' Except to encourage an unbiased response to the real question at issue, the nature of the extra query is, of course, unimportant.

It should be remembered that a form containing many questions may lead to less care on the part of the recipient; a shorter form, in other words, may promote greater accuracy of reply, as well as reducing the amount of non-response and the problems that that creates.

Form design for computer entry

Nowadays nearly all data are transferred to computer files as a first step, usually by typing the information in by hand from the completed questionnaires (or, occasionally, by 'optical mark reading' equipment that can do the transfer automatically). The form should be designed to make this transfer as easy and error-free as possible.

The usual form of computer file is derived from the days when such data storage was done using punched cards, and consists of one or more rows of characters for each subject of the inquiry, the same number of rows for each, with not more than 80 characters per row. The characters used are mainly the digits 0–9, but there are clear cases where it makes sense also to use the letters A–Z, for example if the names of the subjects are to be included.

Figure 4.1 shows the top of each of two pages of a questionnaire and the equivalent part of the corresponding computer file. Points to note are:

(1) Each part of the questionnaire starts with a line number, and the subject number. This enables the meaning of each computer record to be precisely known. Such identification is essential;
(2) All information to be included in the computer file is contained in boxes on the form, and nothing else is. Thus, in this instance, names and addresses are on the questionnaire, but not to be put into the computer file, so they are unboxed;
(3) Although the address is not to be included the post code is, as an indication of the subject's geographical location, and can be indicated as shown. In Britain post codes consist of two, three or four characters followed by a space and then another three characters. The space can be indicated on the questionnaire as shown but, not being in a box, is not to be entered into the computer. If the first half of the code is less than four characters, one or two dashes should be entered rather than leaving boxes blank;
(4) Each box, or set of boxes, has small numbers printed outside indicating the column numbers of the information in the eventual computer file.

It is essential to realise that the meaning of any item of information in the computer file is derived from the column in which it appears. To have the right information in the wrong columns is much worse than useless. To ensure that this does not happen, one character (and never more than one) must be entered in each box of the form. It is not unknown for someone to enter '10', for example, in a single box. This is disastrous, for if two digits instead of one are put into the computer file, all the rest of the information in that row will be pushed across one column. Such a form has to be returned to the person who filled it in for correction before it can be used.

Sometimes it will be necessary to derive the information for such a form from some other form. The fewer the number of such transfers from one form to another the better; every transfer gives an extra chance for copying errors to occur. Where initial observations are made numerically, it is better to design the form so that those initial observations are themselves entered into the computer records. Any grouping, or transformation, of the observations can be done with greater certainty inside the computer later.

Where forms are to be filled in directly by members of the public, it is usually wise not to ask them to fill in such boxes themselves, but to position all boxes within a 'For official use only' section, for later transcription of the information into the required numerical characters by a trained clerk.

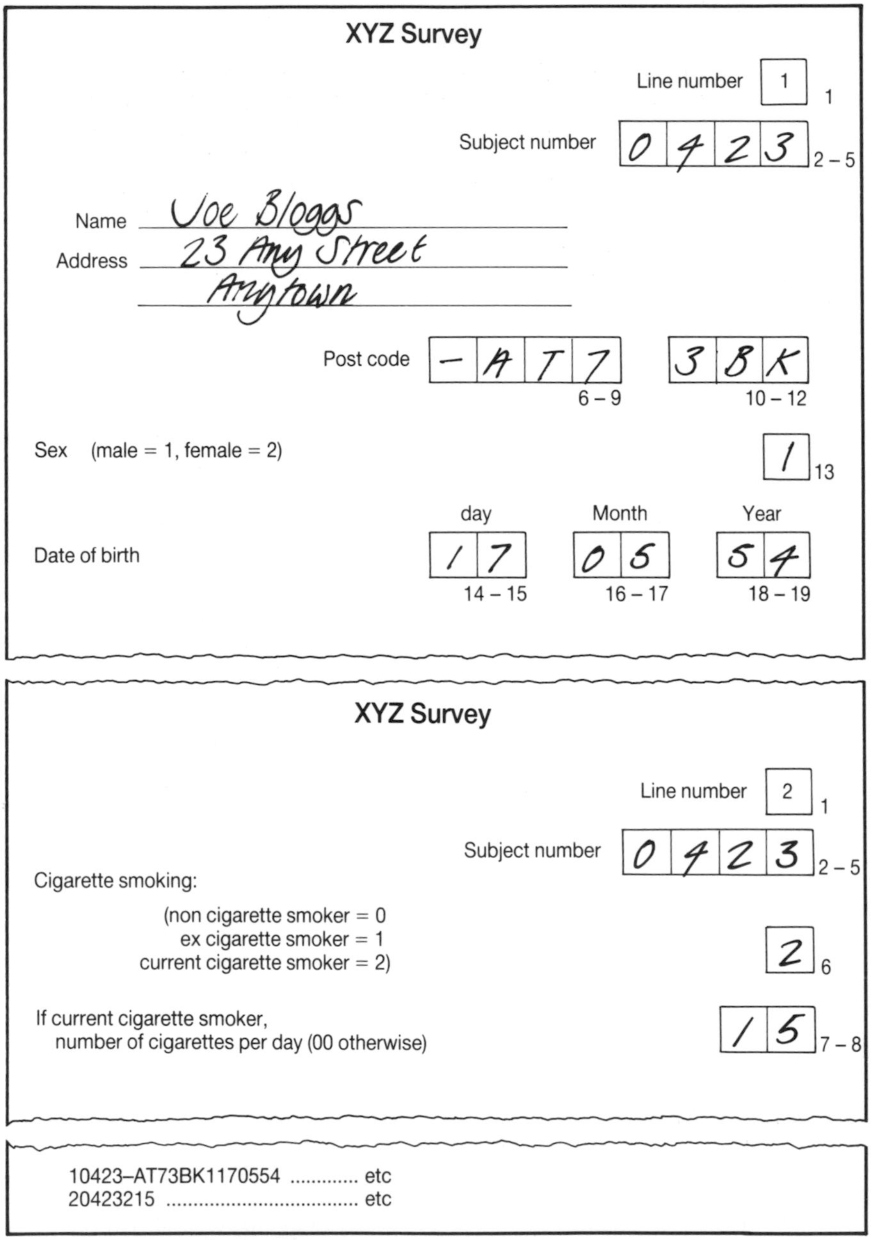

XYZ Survey

Line number [1] 1

Subject number [0 4 2 3] 2 – 5

Name: Joe Bloggs

Address: 23 Any Street, Anytown

Post code [– A T 7] 6 – 9 [3 8 K] 10 – 12

Sex (male = 1, female = 2) [1] 13

Date of birth: day [1 7] 14 – 15; Month [0 5] 16 – 17; Year [5 4] 18 – 19

XYZ Survey

Line number [2] 1

Subject number [0 4 2 3] 2 – 5

Cigarette smoking:

(non cigarette smoker = 0
ex cigarette smoker = 1
current cigarette smoker = 2) [2] 6

If current cigarette smoker,
number of cigarettes per day (00 otherwise) [1 5] 7 – 8

```
10423–AT73BK1170554 ............. etc
20423215 .................................. etc
```

Fig. 4.1 Two sections of a questionnaire and the corresponding sections of the resulting computer file.

Summary

One of the most decisive and difficult tasks in any inquiry is the construction of an appropriate form of record. Care must be taken to ensure that the questions are clear and unambiguous and, as far as possible, self-explanatory. Each question should require some answer and the standard of accuracy necessary for the purpose in hand should be considered. To ensure a high rate of return a form may need to be kept short. On the other hand, to ensure an unbiased return there may be occasions when extra questions are useful. So far as possible, forms should be designed to make transfer of the information to a computer file easy.

Chapter 5

Types of epidemiological inquiry

The pilot inquiry

In many inquiries—and particularly very large-scale or expensive inquiries—it will be extremely profitable to conduct a small pilot investigation. A small sample of the population can be drawn and approached—and that procedure itself will reveal any difficulties in the sampling method and in reaching the respondents. The responses themselves will give a measure of the non-response rate to be expected and throw light on questions which prove to have been ill-worded or ambiguous or which cannot be answered adequately at all. Revision can then be made before the inquiry itself is set in train. Thus it has been reported that a preliminary survey in the United States showed that a proposed new income tax form was 'incomprehensible to a substantial part of the public . . . As a result of this survey a new form was devised which everyone could understand and the Treasury gained millions of dollars from the increased revenue' (PEP Broadsheet on the *Social Use of Sample Surveys*, 1946). Apart, perhaps, from some legitimate doubts as to whether *any* income tax form could be devised which would be understood by everyone, the moral of the small pilot inquiry is clear.

The same principle may well apply in studying past records, e.g. clinical case notes. A small random sample will rapidly show whether the information required is available or whether there are too many gaps and omissions. It may also be made to show the scale of the proposed work in time and manpower.

Retrospective and prospective inquiries

Leaving aside the experimental approach (the strongest weapon in the scientist's armoury) there are, of course, in detail very many ways of investigation by observation. There are, however, two broad categories of approach, each with its own merits and defects, which are worthy of consideration—namely, the *retrospective* and the *prospective*.

In the *retrospective* inquiry the starting point is the affected person (e.g. the patient with cancer of the lung) and the investigation lies in the uncovering of features in his *history* which may have led to that condition (e.g. cigarette smoking, industrial hazards, air pollution, etc.). Does one (or more) of those features appear more frequently in the histories of affected persons than in the histories of an unaffected normal population? This, indeed, is the classical method of epidemiology which seeks to show that of individuals infected with typhoid fever, or cholera, most had consumed a particular supply of water while of those who were not attacked relatively few had done so. It is very rarely a question of all versus none since invariably some consumers are not attacked and frequently some non-consumers are attacked, e.g. through secondary infection. Thus, it is a comparison of *relative* frequencies and for this purpose not only the histories of the cases are required but the histories of some 'controls'. The choice of appropriate controls demands careful thought to ensure that the comparisons are valid. Usually the control group should be as similar as possible to the affected group except for the presence of the disease in question. Sometimes, indeed, it may be possible and valuable to pair each affected individual with a control individual of the same age, sex, etc., and thus deliberately equalise the groups in some features that will influence the comparison. Often, however, it will be highly profitable to seek more than one control group. If a whole series of control groups, e.g. of patients with different diseases, give much the same answer and only the one affected group differs, the evidence is clearly much stronger than if the affected group differs from merely one other group.

We must also bear in mind that the evidence is often based upon past records and is dependent upon the completeness and accuracy of those records.

The *prospective* method, on the other hand, starts with an unaffected sample of the population (e.g. without cancer of the lung), characterises each member of the sample by one or more features (e.g. smoking habits, occupation, place of residence) and then records the *future* occurrence of an event (the development of cancer of the lung) in relation to those features. Does the disease appear more frequently in some groups than others?

Case/control and cohort studies

These terms sometimes seem to be regarded as synonymous with retrospective and prospective studies respectively. It is indeed true that retrospective studies are usually performed in a case/control manner, as explained above, and that prospective studies usually adopt a cohort approach, where a cohort is taken to mean a group of subjects, the members of the group being precisely defined, which may sometimes diminish in size, through death for instance, but to which no new members may be added.

However, this correspondence is not universal and it does seem worth distinguishing the concepts. The case/control technique is by no means confined to the retrospective inquiry. It may well be used in prospective inquiries.

Moreover 'cohort' has a particular meaning in life-table usage, namely a group of the *same birth date* followed through life and it is regrettable that this meaning should be diluted.

If a doctor collects together the records of 50 patients with a specific disease that he saw over his lifetime of practice and examines from his records the trend of their illnesses, the inquiry is prospective in form, even though performed at a later date. While the 50 patients form a defined group, 'cohort' does not seem a good word to describe them.

The pros and cons

The advantages and disadvantages of the two approaches are these. With the prospective method the sample under study can usually be clearly defined and it is easier to consider whether it is likely to be representative of some population or is biased. It may be very difficult to identify similarly the nature of a retrospective sample or the nature of its very possible bias, e.g. are patients in hospital with a coronary infarct likely to be representative of all such patients in regard to some particular constitutional or environmental factor such as dietary habits? Are the controls, selected for comparison with them, representative of the general population (or some specific population) without a coronary infarct? What selective influences may bring the affected and the unaffected into observation? For instance are physicians more likely to elicit and record particular features in certain specified patients than in other patients being used as controls? Would they probe more closely into the smoking habits of a patient with cancer of the lung than of a patient with appendicitis? Those are difficult questions to answer. On the other hand, following up the same example, the difficulties of interpretation are much less if we take a random sample from a defined population, determine their dietary habits, and then record subsequent events. The nature of the sample is clearer. We may, however, note that in this prospective inquiry there may be refusals to co-operate, persons lost sight of and other forms of non-response, all of which will make the sample less certain in its nature.

Neither method, of course, can provide 'proof' of cause and effect. We are always seeking the *most reasonable* interpretation of an association. And usually we shall have to consider the most probable *order* of events in an association—do the dietary habits tend to lead to an infarct or does the person liable to an infarct tend to have certain dietary habits?

One great advantage of the prospective method is that, knowing the number of persons at risk in each group, the incidence rates of the events subsequently observed can be easily calculated and compared. Such a calculation is more difficult with the retrospective method. For example, we could observe pregnant women suffering an attack of rubella in the first, second, third and later months of pregnancy and subsequently observe the incidence of congenital defects in their babies and calculate its incidence in each group. On the other hand, starting with the defective babies, we might find retrospectively that all the mothers had had rubella whereas relatively few mothers of normal control babies had been attacked. To measure from such data the actual risks of a defective child might be impossible or, at least, very difficult.

The retrospective method, however, is likely to give an answer more speedily than the prospective with its prolonged follow-up and, in some circumstances, it is likely to be the only possible approach. For example, with a relatively rare condition like multiple sclerosis it might be quite impossible to categorise a sufficiently large population to give incidence rates from future occurrences within a reasonable span of time. It would be much less difficult to accumulate a large group of cases and retrospectively to explore their past.

The prospective method does not, however, always involve a subsequent waiting period. It can be applied so long as a population can be defined *at any specific time* and then its subsequent events noted, e.g. in records already accumulated. For instance, for the live births that took place in a given hospital over some earlier years it might be possible to determine from the available clinical records whether or not the mother had an X-ray of the abdomen. One might then determine by inquiry or other already available records, the health of the child 5 years later, applying the prospective method to existing records.

In some circumstances one might well choose to make a pilot retrospective inquiry before embarking upon a more arduous prospective investigation. Furthermore, if the same problem can be investigated by both retrospective and prospective methods, and if both give similar answers, then those answers are more certain, for although each can have difficulties of interpretation, those difficulties are unlikely to be the same in both methods.

In conclusion, though the prospective approach must usually be the 'method of choice', there can certainly be no *one* right way in which to make every investigation.

Summary

Pilot inquiries can be invaluable in revealing the difficulties and defects of a proposed large-scale investigation.

Many inquiries follow one of two forms of approach—the retrospective (looking backwards) and the prospective (looking fowards). The latter has much in its favour but usually takes much longer and, with rare events, may be impossible. There can be no one right or wrong way in all circumstances.

Chapter 6

Presentation of statistics

When observations or measurements have been made, or collected, the first object must be to express them in some simple form which will permit conclusions to be drawn, directly or by means of further calculations. The publication, for instance, of a long series of the responses of patients to a specific treatment is not particularly helpful (beyond providing material for interested people to work upon), for it is impossible to detect, from the unsorted mass of raw material, relationships between the various factors at issue. The worker must first consider the questions which he believes the material is capable of answering and then determine the form of presentation which brings out the true answers most clearly. For instance, let us suppose the worker has amassed a series of after-histories of patients treated for gastric ulcer and wishes to assess the value of the treatments given, using as a measure the amount of incapacitating illness suffered in subsequent years. There will be various factors, the influence of which it will be of interest to observe. Is the age or sex of the patient material to the upshot? Division of the data must be made into these categories and tables constructed to show how much subsequent illness was in fact suffered by each of these groups. Is the after-history affected by the type of treatment? A further tabulation is necessary to explore this point; and so on. The initial step must be to divide the observations into a relatively small number of groups, those in each group being considered alike in that characteristic for the purpose in hand.

To take another example, relevant to the remarkable and fascinating history of scarlet fever with its fluctuating virulence, Table 6.1 shows some past fatality rates (i.e. the proportion of patients with the disease who died) from scarlet fever in hospital in England many years ago; for this purpose children within each year of age up to 10 and in each five-year group from 10 to 20 are considered alike with respect to age. It is, of course, possible that by this grouping we are concealing real differences. The fatality rate at 0–6 months may differ from the fatality rate at 6–12 months, at 12–18 months it may differ from the rate at 18–24 months. To answer that question, further subdivision—if the number of cases justifies it—would be necessary. In its present form (accepting the figures of hospital cases at their face value) the grouping states that fatality declines nearly steadily with age, a conclusion which it would be impossible to draw from the

Table 6.1 The History of scarlet fever in the UK. Fatality rate of hospital cases in the years 1905–14.

Age last birthday (years)	*Number of cases*	*Number of deaths*	*Fatality rate per cent*
(1)	(2)	(3)	(4)
0–	46	18	39.1
1–	383	43	11.2
2–	881	50	5.7
3–	1169	60	5.1
4–	1372	36	2.6
5–	1403	24	1.7
6–	1271	22	1.7
7–	986	21	2.1
8–	864	6	0.7
9–	673	5	0.7
10–	1965	14	0.7
15–19	513	3	0.6

11 526 original unsorted and ungrouped records. The constuction of a *frequency distribution* is the first desideratum—i.e. a table showing the frequency with which there are present individuals with some defined characteristic or characteristics.

The frequency distribution

In constructing the frequency distribution from the original unsorted records, the first point to be settled is the number of classes or groups to be used. As one of the main objects of the resulting table is to make clear to the eye the general tenor of the records, too many groups are not desirable. Otherwise, with as many as, say, 50 groups, the tabulation will itself be difficult to read and may fail to reveal the salient features of the data. On the other hand, a very small number of groups may equally fail to bring out essential points. Also, in subsequent calculations made from the frequency distribution, we may sometimes need to suppose that all the observations in a group can be regarded as having the value of the middle of that group, e.g. if in a frequency distribution of ages at death there are 75 deaths at ages between 40 and 45 (i.e. 40 or over but less than 45) we shall presume that each can be taken as $42\frac{1}{2}$. In fact, of course, some will be less than $42\frac{1}{2}$, some more, but so long as the groups are not made unduly wide and the numbers of observations are not too few, no serious error is likely to arise; $42\frac{1}{2}$ will be the mean age of the 75 deaths, or quite near to it.

Of course, if the original observations are held on a computer file, and the computer is to be used for the analysis, those original observations should be used as they stand. The device of grouping, and assuming all observations in a group to be at its mid-point, applies only when making calculations by hand, or where figures in a grouped form are the only information available, and the original observations cannot be recovered.

Where grouping is to be used, 10 to 20 groups is, in general, usually an appropriate number to adopt. Also it is usually best to keep the class- or group-interval a constant size. For instance, in Table 6.1 the class-interval is 1 year of age up to age 10, and it is easy to see from the figures that the absolute number of cases per year of age rises rapidly to a maximum at age 5–6 and then declines. There is, however, an abrupt and large rise in the absolute number at age 10 merely because the class-interval has been changed from 1 year to 5 years—the mean number of cases per year of age would here be only 1965/5, or 393. This change of interval makes the basic figures (not the rates) more difficult to read and sometimes makes subsequent calculations more laborious. Generally, therefore, the class-interval should be kept constant. Also, as a general rule, the distribution should initially be drawn up on a fine basis, i.e. with a considerable number of groups, for if this basis proves too fine, owing to the numbers of observations being few, it is possible to double or treble the group-interval by combining the groups. If, on the other hand, the original grouping is made too broad, the subdivision of the groups is impossible without retabulating much of the material.

As an example of the construction of the frequency distribution we may use the 88 death rates in Table 6.2 which were taken from one of the

Table 6.2 The annual death rate per 1000 at ages 20–64 in each of 88 occupational groups (untabulated material).

(1)	(2)	(3)	(4)
7.5	10.3	7.7	6.8
8.2	10.1	12.8	7.1
6.2	10.0	8.7	6.6
8.9	11.1	5.5	8.8
7.8	6.5	8.6	8.8
5.4	12.5	9.6	10.7
9.4	7.8	11.9	10.8
9.9	6.5	10.4	6.0
10.9	8.7	7.8	7.9
10.8	9.3	7.6	7.3
7.4	12.4	12.1	19.3
9.7	10.6	4.6	9.3
11.6	9.1	14.0	8.9
12.6	9.7	8.1	10.1
5.0	9.3	11.4	3.9
10.2	6.2	10.6	6.0
9.2	10.3	11.6	6.9
12.0	6.6	10.4	9.0
9.9	7.4	8.1	9.4
7.3	8.6	4.6	8.8
7.3	7.7	6.6	11.4
8.4	9.4	12.8	10.9

Occupational Mortality Supplements of the Registrar-General of England and Wales (now the Office of Population Censuses and Surveys). The rates, as set out in four columns, have been copied merely in the order of occupations as adopted in the report and it is desired to tabulate them.

The first step is to find the upper and lower limits over which the tabulation must extend. The lowest rate is 3.9, the highest 19.3. We have therefore a range of 15.4. A class interval of 1 will give 16 groups and clearly will be convenient to handle. On this basis we may take the groups, or classes, as 3.5 to 4.4, 4.5 to 5.4, 5.5 to 6.4, and so on. If tabulating by hand, we may enter each rate by a stroke against the appropriate group. It is convenient to mark each fifth by a diagonal line as shown. Addition is then simple and errors are less likely to occur. (In making this tabulation the groups may be set out as above, 3.5 to 4.4, 4.5 to 5.4, etc. or as in Table 6.3, 3.5-, 4.5-, etc. It is most undesirable to have them in the form 3.5 to 4.5, 4.5 to 5.5, etc., since observations precisely on a dividing line, e.g. 4.5, will then sometimes be absent-mindedly put in one group and sometimes in the other.)

This method is satisfactory if the number of observations is not very large. For more than 100 observations, or thereabouts, it will always be found advantageous, if possible, to enter the observations into a computer file, after which any grouping, or other operations, can be done using such statistical software as is available. Grouping, by some means or other, will nearly always have to be done to produce suitable tables for a published report.

The final figures resulting from this construction of the frequency distribution are given in Table 6.4, from which it can be clearly seen that

Table 6.3 Process of tabulation of 88 death rates.

Death rate		*Score*
3.5–	𝍷	1
4.5–	𝍷𝍷𝍷𝍷	4
5.5–	𝍸	5
6.5–	𝍸 𝍸 𝍷𝍷𝍷	13
7.5–	𝍸 𝍸 𝍷𝍷	12
8.5–	𝍸 𝍸 𝍸 𝍷𝍷𝍷	18
9.5–	𝍸 𝍸 𝍷𝍷𝍷	13
10.5–	𝍸 𝍸	10
11.5–	𝍸 𝍷	6
12.5–	𝍷𝍷𝍷𝍷	4
13.5–	𝍷	1
14.5–		
15.5–		
16.5–		
17.5–		
18.5–	𝍷	1
19.5+		
Total		88

Table 6.4 The annual death rate per 1000 at ages 20–64 in 88 different occupational groups.

Death rate per 1000	*Number of occupational groups with given death rate*
3.5–	1
4.5–	4
5.5–	5
6.5–	13
7.5–	12
8.5–	18
9.5–	13
10.5–	10
11.5–	6
12.5–	4
13.5 and over	2*
Total	88

* 1 death rate of 14.0 and 1 death rate of 19.3.

the majority of the death rates lie between 6.5 and 11.5 per 1000 and that they are fairly symmetrically spread round the most frequent rate of 8.5–9.4. To avoid unduly lengthening the table by the inclusion of four groups with no entries against them, the final group has been termed 13.5 and over. As a rule this is an undesirable procedure unless the entries can also be precisely specified, as is done here in a footnote. Without that specification full information on the spread of the rates has not been given to the reader and he may be hampered if he wishes to make calculations from the distribution. A similar caution relates to the lowest group.

Statistical tables

Returning to Table 6.1, this may be used in illustration of certain basic principles in the presentation of statistical data.

(1) The contents of the table as a whole and the items in each separate column should be clearly and fully defined. For lack of sufficient headings, or even any headings at all, many published tables are quite unintelligible to the reader without a search for clues in the text (and not always then). For instance, if the heading given in Column (1) were merely 'age', it would not be clear whether the groups refer to years or months of life. The unit of measurement must be included.

(2) If the table includes rates, as in Column (4), the base on which they are measured must be clearly stated—e.g. death rate per cent, or per thousand, or per million, as the case may be (a very common omission in published tables). To know that the fatality rate is '20' is not helpful unless we know whether it is 20 in 100 patients who die (1 in 5) or 20 in 1000 (1 in 50).

(3) Whenever possible the frequency distributions should be given in full, as in Columns (2) and (3). These are the basic data from which

conclusions are being drawn and their presentation allows the reader to check the validity of the author's arguments. The publication merely of certain values descriptive of the frequency distribution—e.g. the arithmetic mean or average—severely handicaps other workers. For instance, the information that for certain groups of patients the mean age at death from cancer of the lung was 54.8 years and from cancer of the stomach 62.1 years is of very limited value in the absence of any knowledge of the distribution of ages at death in the two classes.

(4) Rates or proportions should not be given alone without any information as to the numbers of observations upon which they are based. In presenting experimental data, and indeed nearly all statistical data, this is a fundamental rule (which, however, is constantly broken). For example, the fatality rate from smallpox in England and Wales (ratio of registered deaths to notified cases) was 42.9 per cent in 1917 while in the following year it was only 3.2 per cent. This impressive difference becomes less convincing of a real change in virulence at that time when we note that in 1917 there were but 7 cases notified, of whom 3 died, and in 1918 only 63 of whom 2 died. (Though the low rate of 1918 *may* mark the presence of variola minor.) It is, however, the essence of science to disclose both the data upon which a conclusion is based and the methods by which the conclusion is attained. By giving only rates or proportions, and by omitting the actual numbers of observations or frequency distributions, we are excluding the basic data. In their absence we can draw no valid conclusion whatever from, say, a comparison of two or more percentages. The news 'media' are great offenders in this respect. They will glibly report that the influenza epidemic has risen five-fold in a week but rarely are we told whether the basic figures are 10 and 50 cases or 100 and 500.

It is sometimes stated that it is wrong to calculate a percentage when the number of observations is small, e.g. under 50. But *so long as the basic numbers are also given* it is difficult to see where the objection lies and how such a presentation can be misleading—except to those who disregard the basic figures and who, therefore, in all probability will be misled by *any* presentation. If, for example, we have 9 relapses in 23 patients in one group and 4 relapses in 15 patients in another, it is difficult for the mind to grasp the difference (if any). Some common basis would seem essential, and percentages are a convenient one. In this comparison they are, in fact, 39% and 27%. They will thus be seen, in view of the small numbers involved, to be not very different. The fundamental rules should be that the writer gives no percentages without adding the scale of events underlying them and the reader accepts no percentages without considering that scale. Inevitably with small numbers care will be needed in drawing conclusions; but that is true whatever the basis from which they are viewed.

(5) On a point of detail it is sometimes helpful in publishing results to use one decimal figure in percentages to draw the reader's attention to the fact that the figure is a percentage and not an absolute number. An alternative, especially useful in tables, is to give percentages (or rates) in italics and absolute numbers in bold type. This variation in type, too, often makes a large table simpler to grasp. As a general principle two or three

small tables are to be preferred to one large one. Often the latter *can* be read but its appearance may well lead to it going unread.

(6) Full particulars of any deliberate exclusions of observations from a collected series must be given, the reasons for and the criteria of exclusion being clearly defined. For example, if it be desired to measure the success of an operation for, say, cancer of some site, it might, from one aspect, be considered advisable to take as a measure the percentage of patients surviving at the end of five years, *excluding those who died from the operation itself*—i.e. the question asked is 'What is the survival rate of patients upon whom the operation is successfully carried out?' It is obvious that these figures are not comparable with those of observers who have included the operative mortality. If the exclusion that has been made in the first case is not clearly stated no one can necessarily deduce that there is a lack of comparability between the records of different observers, and misleading comparisons are likely to be made. Similarly one worker may include among the subsequent deaths only those due to cancer and exclude unrelated deaths, irrespective of their cause. Definition of the exclusions will prevent unjust comparisons.

Sometimes exclusions are inevitable—e.g. if in computing a survival rate some individuals have been lost sight of so that nothing is known of their fate. The number of such individuals must invariably be stated, and it must be considered whether the lack of knowledge extends to so many patients as to stultify conclusions.

Beyond these few rules it is very difficult, if not impossible, to lay down rules for the construction of tables. The whole issue is the arrangement of data in a concise and easily read form. In acquiring skill in the construction of tables probably the best way is to consider published tables critically with such questions as these in mind: What is the *purpose* of this table? What is it *supposed* to accomplish in the mind of the reader? Wherein does its failure of attainment fall? Study of the tables published by the professional statistician—e.g. in the reports of the Office of Population Censuses and Surveys—will materially assist the beginner.

Graphs

Even with the most lucid construction of tables such a method of presentation will often give difficulties to the reader, especially to the non-numerically minded reader. The presentation of the same material diagrammatically often proves a very considerable aid and has much to commend it if certain basic principles are not forgotten.

(1) The sole object of a diagram is to assist the intelligence to grasp the meaning of a series of numbers by means of the eye. If—as is unfortunately often the case—the eye itself is merely confused by a criss-cross of half a dozen, or even a dozen, lines, the sole object is defeated. The criterion must be that the eye can with reasonable ease follow the movements of the various lines on the diagram from point to point and thus observe what is

the change in the value of the ordinate (the vertical scale) for a given change in the value of the abscissa (the horizontal scale). The writer should always remember, too, that he is familiar with the data, and what may be obvious to him is not necessarily obvious to the reader. The object of the graph is to make it obvious, or at least as clear as possible, and simplicity is invariably the keynote.

(2) The second point to bear in mind in constructing *and in reading* graphs is that by the choice of scales the same numerical values can be made to appear very different to the eye. Figures 6.1 and 6.2 are an example. Both show the same data—namely the trend of the death rate in women from cancer of the breast in England and Wales between 1951 and 1981. In Fig. 6.1 the increase in mortality that has been recorded appears to have been exceedingly rapid and of serious magnitude while in Fig. 6.2 a slow and far less impressive rise is suggested. This difference is, of course, due to the differences in the vertical and horizontal scales and to the fact that in Fig. 6.1 the vertical scale does not start at zero (see below). In reading graphs, therefore, the scales must be carefully observed and the magnitude of the changes interpreted by a rough translation of the points into actual figures. In drawing graphs undue exaggeration or compression of the scales must be avoided.

It must be considered also whether a false impression is conveyed, as quite frequently happens, if the vertical scale does not start at zero but at some point appreciably above it. Wherever it can reasonably be done, a true zero should be used. Somctimes a journal editor may object to a graph such as Fig. 6.2, complaining about showing so much blank paper. The answer is that the blank paper is *not* being wasted, but is fulfilling the purpose of presenting, to the eye, the true picture of a steady but not dramatic increase. To redraw it as Fig. 6.1 is *wholly* to waste the paper, as it does not give an honest picture to the eye. It can be interpreted only by translating from the graph back to the underlying figures, and a table of those figures would give the information more accurately in less space. One always has to use intelligence, of course, to decide when a true zero would be helpful and when not. For instance, in Fig. 6.2, to go back to zero on the *horizontal* scale would be ludicrous and would not help anyone in any way. The basic rule should be that a true zero should always be used except when a conscious decision has been taken that there are advantages in not using one, and that the graph will not mislead as a result.

(3) Graphs should always be regarded as subsidiary aids to the intelligence and not as the *evidence* of associations or trends. That evidence must be largely drawn from the statistical tables themselves. It follows that graphs are an unsatisfactory *substitute* for statistical tables. A deaf ear should be turned to such editorial pleading as this: 'If we print the graphs would it not be possible to take the tables for granted? Having given a sample of the process by which you arrive at the graph is it necessary in each case to reproduce the steps?' The retort to this request is that statistical tables are *not* a step to the diagram, they are the basic data. Without these basic data the reader cannot adequately consider the validity of the author's deductions, and he cannot make any further analysis of the

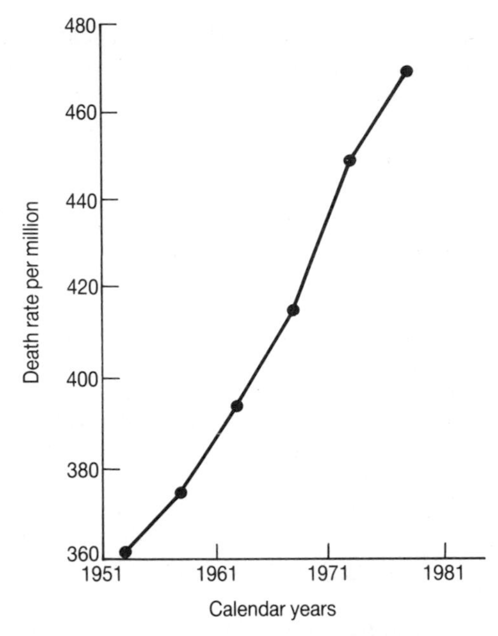

Fig. 6.1 Death rates per million women from cancer of the breast. England and Wales 1951–81. Example of a misleading graph.

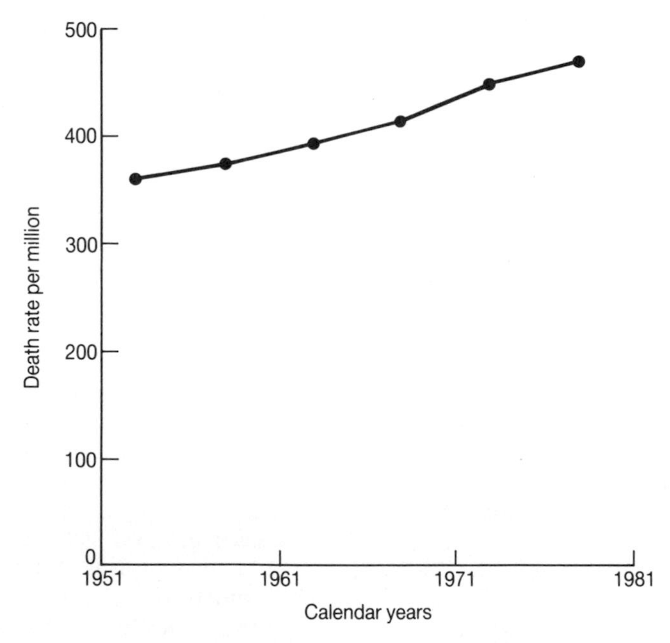

Fig 6.2 Death rates per million women from cancer of the breast. England and Wales 1951–81. The same data as in Figure 6.1, honestly presented.

data, if he should wish, without laboriously and inaccurately endeavouring to translate the diagram back into the statistics from which it was originally constructed (and few tasks are more irritating). There are, of course, some occasions when the statistical data are not worth setting out in detail and a graph may be sufficient, but careful thought is advisable before that procedure is followed.

(4) The problem of scale illustrated in Figs. 6.1 and 6.2 is also an important factor in the comparison of trend lines. Thus Fig. 6.3 shows the trend of the infant mortality rate (i.e. the death rate within the first year of life) and of the death rate of young children (i.e. within the following four years) in England and Wales between 1946 and 1981. Unless the scale of the ordinate (the vertical scale) is carefully considered, the inference drawn from this graph might well be that over this period the infant mortality rate declined relatively more than the death rate in young children. Actually the reverse is the case—relatively the death rates at ages 1–4 years declined slightly more than the infant mortality rate. In 1976–80 the infant mortality rate was 36 per cent of the rate recorded in 1946–50 while the death rate at ages 1–4 years was only 30 per cent of its earlier level. *Absolutely*, infant mortality shows the greater improvement (from 35 per 1000 to 13 per 1000 compared with 1.8 and 0.53 in the death rate at ages 1–4); but *relatively* the death rate of young children shows the advantage. If it is the *relative* degree of improvement that is at issue, Fig. 6.3 is insufficient. For this purpose the rates in each quinquennium may be

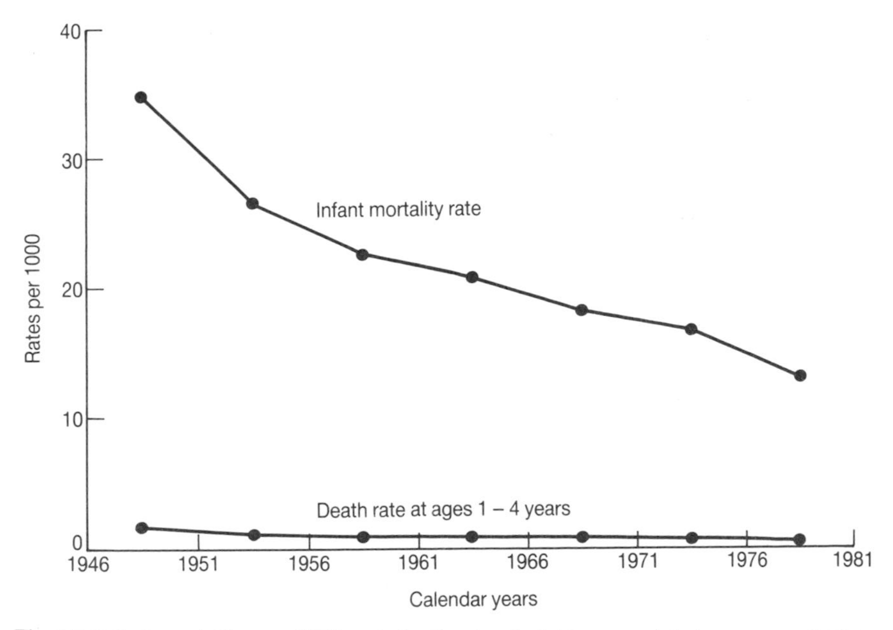

Fig 6.3 Infant mortality per 1000 and death rate of children aged 1–4 years per 1000 in England and Wales from 1946–81.

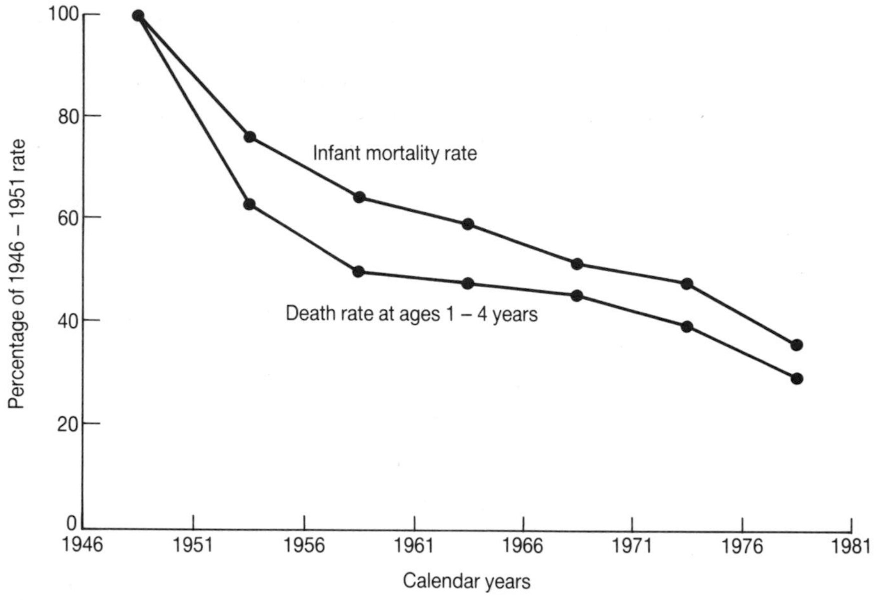

Fig 6.4 The rates in each 5-year period expressed as percentages of the corresponding rates in 1946–51.

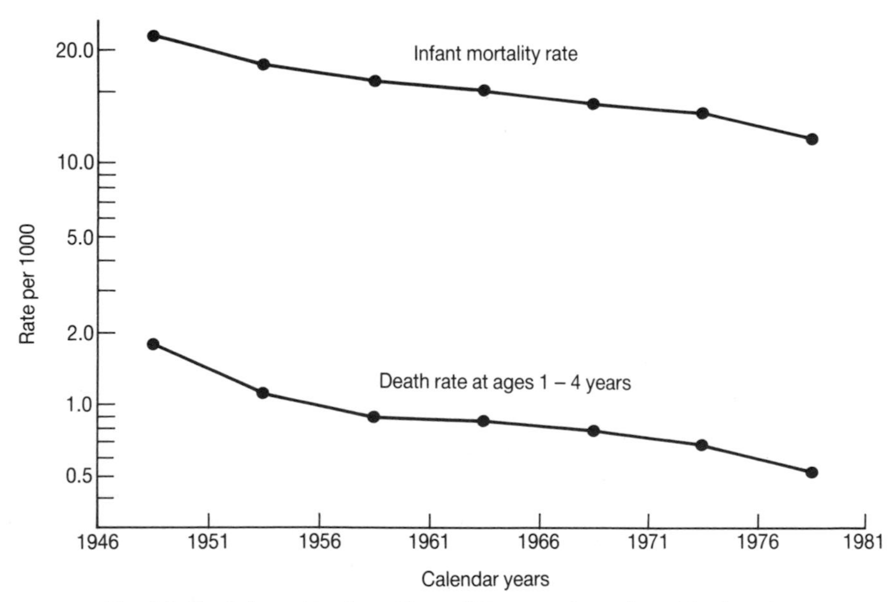

Fig. 6.5 The information from Figure 6.3 repeated on a logarithmic scale.

converted into percentages based upon the rates in 1946–50 as shown in Fig. 6.4.

For any figures such as these, where the *relative* increase or decrease is more relevant than the *absolute* increase or decrease, the use of a graph on logarithmic graph paper should be considered. The same figures as in Fig. 6.3 are shown again in Fig. 6.5 drawn on a logarithmic scale, in which equal distances (on the vertical scale) indicate *multiplying* by a given quantity, instead of *adding* a given quantity as on an ordinary graph. The two lines are now virtually parallel, showing (as Fig. 6.4 did) that there is little difference in the two relative rates of decline.

It should be noted that the use of a true zero is impossible on a logarithmic graph. No matter how far downwards you extend the scale, zero is never reached. Nor, of course, can negative values be shown—but for the sort of data for which a logarithmic scale is suitable, negative values are usually impossible anyway.

(5) It is a *sine qua non* with graphs, as with tables, that they form self-contained units, the contents of which can be grasped without reference to the text. For this purpose inclusive and clearly stated headings must be given, the meaning of the various lines indicated, and a statement made against the ordinate and abscissa of the characteristics to which these scales refer, including the units employed.

Frequency diagrams

Many types of diagrams have been evolved to bring out the main features of statistical data. In representing frequency distributions diagrammatically the *histogram* is most commonly used. In this, the base line denotes the characteristic which is being measured and the vertical scale reveals the frequency with which it occurs. In Table 6.1 the numbers of deaths from scarlet fever in a certain study were as follows:

Age last birthday (years)	0–	1–	2–	3–	4–	5–	6–	7–	8–	9–	10–	15–19
Number of deaths	18	43	50	60	36	24	22	21	6	5	14	3

A histogram of these figures is shown in Fig. 6.6. The frequency is represented by an area corresponding to the number of observations. Thus in the present example a point is placed against 18 on the vertical scale against both age 0 and age 1 on the base line. A rectangle is then drawn to show this frequency. Similarly a point is placed against 43 both above age 1 and age 2, and this rectangle is completed. The area of each rectangle is thus proportional to the number of deaths in the year of age concerned. To maintain a correct area when the scale of age grouping changes, we must divide the recorded deaths by the new unit of grouping, i.e. the 14 and the 3 in the two final groups must be divided by 5, giving, on average, 2.8 and

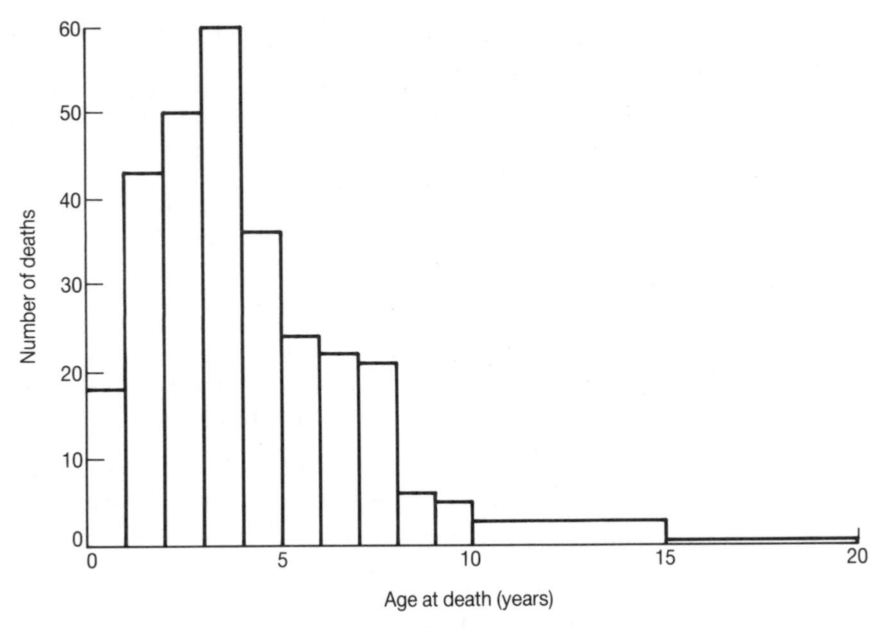

Fig 6.6 A histogram showing the frequency of some deaths from scarlet fever in relation to age.

0.6 deaths per year of age. These figures then relate to the whole of the 5-year age group and the rectangles will extend over the ages 10–15 and 15–20. In other words, in the absence of more detailed data, we plot the figures as if there were 2.8 deaths at ages 10–11, 2.8 at ages 11–12, 2.8 at ages 12–13, and so on; 0.6 at ages 15–16, 0.6 at ages 16–17, 0.6 at ages 17–18, and so on. If thought desirable the actual number of deaths reported can be written inside, or just above, the rectangles.

Another simple form of diagram is the *bar chart*, which can be used to show pictorially the absolute, or relative, frequency of events, e.g. the number of deaths due to specified causes or the percentage of patients with a particular disease showing certain symptoms, as illustrated in Fig. 6.7.

In a preliminary exploration of the degree of association between two characteristics, a *scatter diagram* can be very illuminating (such as shown in Chapter 18). Unfortunately we often need to visualise the interrelationship of more than two variables. Nobody yet has derived a meaningful way of presenting a scatter diagram in four or more dimensions.

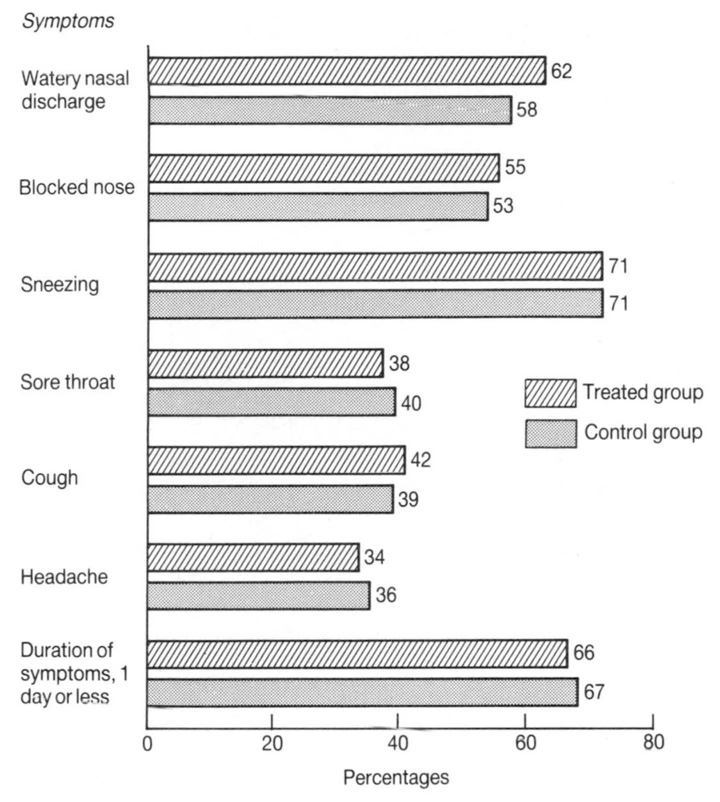

Fig. 6.7 A bar chart showing the percentage of patients with defined presenting symptoms of the common cold in a treated group and a control group.

Over-elaboration

It has recently become the fashion in some circles to produce graphs that are so over-elaborate as to take away nearly all (or even all) their usefulness. The daily press constantly gives examples, with diagrams that are given a three-dimensional appearance, even though the information is only two-dimensional, that are distorted to turn them into a picture of an object on which the graph is drawn, etc. With the help of computer software this sort of thing has become too easy. Presumably those responsible, and their editors, believe such efforts to be worth while, but it need hardly be said that all such gimmicks should be completely avoided in any scientific statistical work.

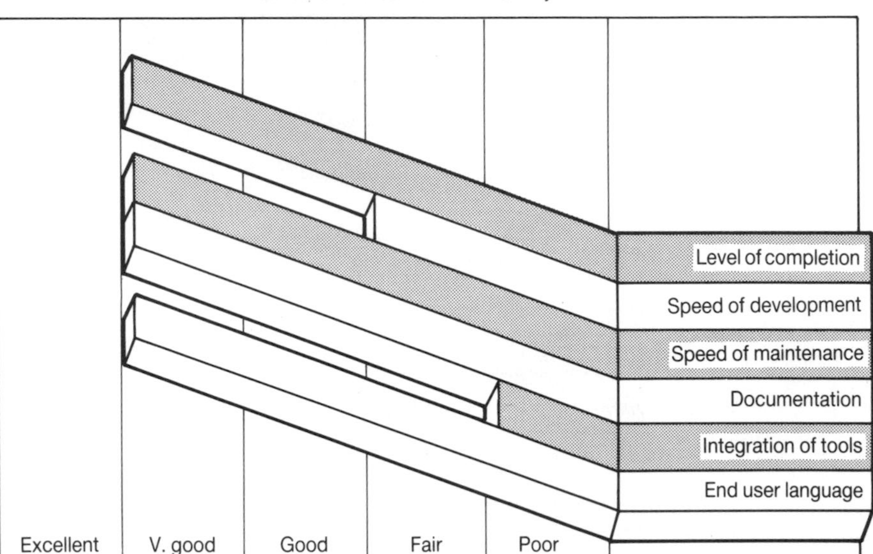

Fig. 6.8 An example of misuse of computer graphics.

For example, Fig. 6.8 comes from a weekly newspaper dealing with computing matters. After studying it, the conclusion seems to be that it is saying either

Level of completion	Excellent
Speed of development	Good
Speed of maintenance	Excellent
Documentation	Excellent
Integration of tools	Fair
End user language	Excellent

or possibly

Level of completion	Very good
Speed of development	Fair
Speed of maintenance	Very good
Documentation	Very good
Integration of tools	Poor
End user language	Very good

There is no way of telling which is intended. In either case, a short table such as given above would be much easier to understand. The graphical format, sloping lines etc., are pure baloney.

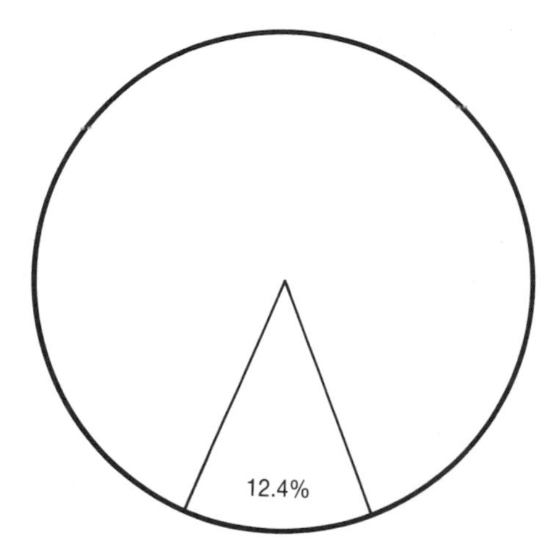

Fig. 6.9 A pie chart.

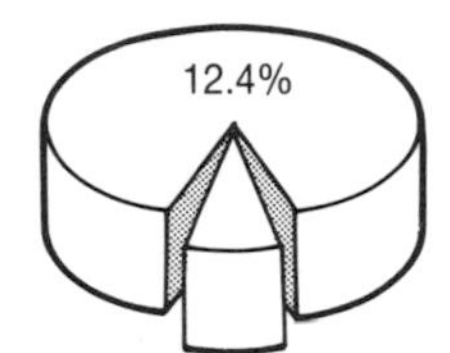

Fig 6.10 Misuse of the pie chart idea.

Should it be thought that such a warning is not needed here—that nobody would think of adopting such methods in any serious publication—perhaps reference may be made to the Medical Research Council's *Corporate Strategy: Update on Developments 1989–1990*. Here we find the percentages of investment to be applied to various fields of research. If anyone finds that a figure of 12.4 per cent, say, is not easy to visualise and needs a diagram such as Fig. 6.9 to illustrate it, no harm is done by that. In fact, though, the document gives Fig. 6.10 instead, in which the meaning is distorted by the perspective of the view, and a meaningless third dimension is introduced. Furthermore the 'slice of cake' illustrated clearly does not fit what has been cut from the 'cake', and one is left wondering where the missing investment may have gone!

Summary

For the comprehension of a series of figures tabulation is essential; a diagram (*in addition to tables but usually not in place of them*) is often of considerable aid both for publication and, still more, for a preliminary study of the features of the data. In publication, however, it should always be remembered that quite simple data are perfectly clear in a table and therefore a graph is merely a waste of space. Alternatively with very complicated data a diagram may be equally of no assistance and a waste of space. In medical literature the amount of space wasted by useless or unreadable diagrams is quite astonishing.

Both tables and graphs must be entirely self-explanatory without reference to the text. As far as possible the original observations should always be reproduced (in tabulated form showing the actual numbers belonging to each group) and not given only in the form of the percentages falling in each group. The exclusion of observations from the tabulated series on any grounds whatever must be stated, the criterion upon which exclusion was determined clearly set out, and the number of such exclusions stated. Conclusions should be drawn from graphs only with extreme caution and only after careful consideration of the scales adopted.

Chapter 7

The average

When a series of observations has been tabulated, i.e. put in the form of a frequency distribution, the next step is the calculation of certain values which may be used as descriptive of the characteristics of that distribution. These values will enable comparisons to be made between one series of observations and another. The two principal characteristics of the distribution which it is invariably necessary to place on a quantitative basis are (1) its average value and (2) the degree of scatter of the observations round that average value. The former, the average value, is a measure of the *position* of the distribution, the point around which the individual values are dispersed. For example, the average incubation period of one infectious illness may be 7 days and of another 11 days. Though individual values may overlap, the two distributions have different central positions and therefore differ in this characteristic of location. In practice it is constantly necessary to discuss and compare such measures. A simple instance would be the observation that people following one occupation lose *on the average* 5 days a year per person from illness; in another occupation they lose 10 days. The two distributions differ in their position and we are led to seek the reasons for such a difference and to see whether it is remediable.

The three terms, *average*, *mean*, *arithmetic mean*, are customarily taken as interchangeable, indicating the sum of all the observations divided by the number of observations. There are, however, two other 'averages' that are sometimes of value, the *geometric mean* and the *median*.

If a distribution has a short left-hand tail and a much longer right-hand tail, the arithmetic mean may be much higher than the usual experience because of a few unduly large values. For instance, in discussing something such as the length of an illness we may be interested not so much in the mean, or average, length as in the *usual* length. A few very long cases cannot be balanced at the other extreme because a length of zero or less is impossible. The fact that these few unduly long cases exist is important, but to get a feel for the general experience the geometric mean may be a better measure.

The meaning of a geometric mean is most simply seen by comparing it with an arithmetic mean in a very simple example. Given the four observations 1.2, 2.4, 9.6 and 19.2, the arithmetic mean is 8.1 because

$$1.2 + 2.4 + 9.6 + 19.2 = 32.4$$
and
$$8.1 + 8.1 + 8.1 + \ 8.1 = 32.4$$

whereas the geometric mean is 4.8 because

$$1.2 \times 2.4 \times 9.6 \times 19.2 = 530.8416$$
and
$$4.8 \times 4.8 \times 4.8 \times \ 4.8 = 530.8416$$

The median of a series of observations is the value of the central or middle observation when all the observations are listed in order from lowest to highest. In other words, half the observations lie below the median and half the observations lie above it. The median therefore divides the distribution into two halves, and it defines the position of the distribution in that way.

It is much less affected than the mean by a few exceptional observations. To take again the example of the length of an illness, if most of the patients in question are ill for between 10 and 15 days the median will not be appreciably altered by the addition of 2 patients ill for 3 and 6 months, they merely represent two more cases lying above the middle point, and how much above is immaterial. On the other hand, the mean might be increased by their addition to 20 days and be, therefore, a poor measure of the general location of the distribution. In such circumstances to quote the values of both the arithmetic mean and the median is usually advisable.

Calculation of the mean

With a short series of observations the calculation of the mean is quite simply made. Let us take the figures in Table 7.1 relating to the recorded age of onset of disease (age last birthday) for a group of 27 patients suffering from some specific illness.

The arithmetic mean is the sum of all the values divided by their number. For the above example this is 39 + 50 + 26 + 45 + etc., giving a total of 1320. The average of the ages as given is, therefore, 1320/27, or 48.9 years. If we call each individual observation x, it is usual to denote the mean as $\bar{x}$

Table 7.1

Age at last birthday of patient at onset of disease (in years)		
39	40	47
50	51	44
26	66	48
45	63	59
47	55	42
71	36	54
51	57	47
33	41	53
40	61	54

(pronounced 'x-bar'). Then the symbolic formula is

$$\bar{x} = \frac{\text{Sum } x}{n}$$

where n is the sample size.

Nowadays few people would carry out this calculation by hand. They would enter the 27 figures into a calculator, or a suitable computer program, very often without further thought. Having been saved the labour of the calculation, there should be more time available for thinking, but experience seems to suggest that such thinking is often absent.

The 48.9 years, as calculated, is indeed the observed average of the *age last birthday*, but surely what we are wanting is the average of the actual ages of onset and these will (on the average) be half a year greater, for someone aged 40 last birthday may be anywhere between 40 and 41—$40\frac{1}{2}$ is the best estimate we can make. It would be possible to add $\frac{1}{2}$ to each age before finding the average, but it is easier merely to add $\frac{1}{2}$ to the average afterwards, which must give the same result. Thus the average age is derived as 49.4 years.

Where the mean of a large number of observations is required, and no calculator or computer is available, the methods set out in Appendix A may be helpful. For just a few observations, mental calculation is often simplest, and a point to remember is that use of an 'arbitrary origin' can simplify the operations. Thus we might have observations of the number of cases of an infectious disease in each week of the year in each of 3 years (Table 7.2) and require the average annual number in each week.

In week 1 each total has 120 as a common feature and it is necessary to add only $6+1+8=15$, and dividing by 3 and adding the 120 back in, the mean is 125. In week 2, again using 120 as a base line, we have $12+6+0=18$ and the mean is 126. In week 3 we may take 160 as base line and we have $3-1+1=3$ and the mean is 161. In week 4 we may take 180 as base line and have $2+11+10=23$ and the mean is 187.7. With experience these small differences from a base line can be accurately noted and mentally added without difficulty.

Table 7.2

Year	*Week 1*	*Week 2*	*Week 3*	*Week 4*	*etc.*
1974	126	132	163	182	
1975	121	126	159	191	
1976	128	120	161	190	

Interpretation of the mean

One sometimes sees the mean misinterpreted as a value that is typical, in the sense that most people would exhibit that value, and fun is made of the fact that the average Member of Parliament has 2.3 children (or whatever the figure may be). In fact such an example shows not the absurdity of the figure but the absurdity of the person making fun of it. All that the figure says is that if we divide the total number of MPs' children by the number of MPs, the result is 2.3 children per MP. This can be useful knowledge in comparing family sizes, of one group and of another, but there is no suggestion that any *individual* has 2.3 children.

In statistical work, we often use the words 'expected value' or 'expectation' to indicate the mean that would occur among some group on the basis of a given hypothesis. Suppose, for example, we have one group of 40 British men, and another of 30 American men, of whom 7 die altogether. The hypothesis might be that, except for sampling variability, the death rate among the British and among the Americans would be the same, so we should 'expect' $40 \times (7/70) = 4$ deaths among the 40 British and $30 \times (7/70) = 3$ among the Americans. But such expected values are expected only on the average, in the long run, rather than to be precisely observed on any particular occasion, and the use of 'expected' in statistics is always intended in this way.

If we had had 39 British and 31 Americans, the expected values would have been

$$39 \times (7/70) = 3.9 \text{ deaths}$$

and $$31 \times (7/70) = 3.1 \text{ deaths.}$$

The fact that fractions of a death are impossible does not invalidate these figures, any more than fractions of a child being impossible invalidates 2.3 children per MP.

Similarly the word 'expectation' is much misunderstood and much misused. That the 'expectation of life' in a certain population is 60 years, does not mean that most people in that population die aged 60 (or even that any of them do). It is only a particular sort of average. We shall examine this in more detail later (see Chapter 21).

Calculation of the geometric mean

The simplest way to find a geometric mean is to take the logarithms of the observations, calculate the arithmetic mean of those values, and finally take the antilogarithm of the result.

Using the same 27 values, of age last birthday, we should proceed as in Table 7.3. The sum of the logarithms is found to be 45.46603 their average to be $45.46603/27 = 1.68393$, of which the antilogarithm is 48.3, so the

Table 7.3

Estimate of age	*Logarithm*
39.5	1.59660
50.5	1.70329
26.5	1.42325
etc.	etc.

geometric mean is 48.3 years, quite close to the arithmetic mean in this instance.

Note that $\frac{1}{2}$ was added to each age before taking the logarithms. Merely to add $\frac{1}{2}$ to the answer, as was done with the arithmetic mean, would not have got the right answer.

Calculation of the median

Writing now, the values in order of magnitude from lowest to highest we have the following list: 26, 33, 36, 39, 40, 40, 41, 42, 44, 45, 47, 47, 47, (48), 50, 51, 51, 53, 54, 54, 55, 57, 59, 61, 63, 66, 71. The median being the central value will be the 14th observation, there being 13 lower values than this and 13 higher values. It is, therefore, 48 years, and half the patients had lower ages of onset than this and half had higher.

A simple method of determining which value is required for the median is to divide by 2 the number of observations plus 1, or $(n+1)/2$; in the present instance $(27+1)/2 = 14$, and the 14th value is the median. If there were 171 values we should calculate $(171+1)/2 = 86$, so the 86th value is the median, there being then 85 lower values, the median, 85 higher values = 171 in all.

In these instances when the total number of observations is an odd number there is no difficulty in finding the median as defined—the central value with an equal number of observations smaller and greater than itself. Often, however, the definition cannot be strictly fulfilled, namely when the total number of observations is an even number. If in the above series an additional patient had been observed with an age of onset of 73 there would have been 28 observations in all. There could be no central value. In such a situation it is usual to take the mean of the *two central values* as the median. Thus we should have: 26, 33, 36, 39, 40, 40, 41, 42, 44, 45, 47, 47, 47, (48, 50), 51, 51, 53, 54, 54, 55, 57, 59, 61, 63, 66, 71, 73. The two central values are 48 and 50, with 13 values lying on either side of them, and the median is taken as $(48+50)/2 = 49$ years.

The method of finding which are the required observations will, again, be to divide by 2 the number of observations plus 1; and the median will be the average of the values immediately above and below. Thus with 28 observations we have $(28+1)/2 = 14.5$ and the 14th and 15th values are required.

Difficulties with the median

This customary extension of the definition of the median presents no difficulty and is a reasonable procedure. Often, however, in a short series of observations the definition cannot be completely fulfilled. Thus we might have as our observations of ages of onset the following values: 26, 33, 36, 39, 40, 40, 41, 42, 44, 45, 47, 48, 48, (48), 48, 51, 51, 53, 54, 54, 55, 57, 59, 61, 63, 66, 71. With 27 observations the middle one, the 14th, can of course be found. But it will be seen that there are *not* 13 smaller

Table 7.4 Two hypothetical frequency distributions.

A		B	
Height of a group of children (cm)	*Number of children*	*Number of children in a family*	*Number of families observed*
127–	96	0	96
129–	120	1	120
131–	145	2	145
133–	83	3	83
135–	71	4	71
137–	32	5	32
139–141	18	6+	18
Total	565	*Total*	565

observations than this value and *not* 13 larger, for three of them are equal to the 14th value. In strict terms of the definition the median cannot be found in this short series. Even with a large number of observations it may be impossible to find a median value if the characteristic under discussion changes discontinuously. Take, for example, the frequency distributions in Table 7.4.

In the left-hand distribution, A, we have measurements of the heights of a group of 565 children and in the right-hand distribution, B, the number of children observed in 565 families.

The median height will be that of the 283rd child when the observations are listed in order. To a considerable extent they are already in order, in the frequency distribution. Adding up the observations from the lower end of the distribution $96 + 120 = 216$, and we therefore need 67 more beyond this point to reach the 283rd. The median value, accordingly, lies in the group 131–133 cm and merely by putting the 145 observations in that group in exact height order the required 67th can be found. Variation in stature is continuous and thus a median value can reasonably be calculated.

On the other hand, the number of children in a family cannot vary continuously but must proceed by unit steps. The 283rd family must again be the 'middle' one and it must have two children, but there can be no central value for family size which divides the distribution into two halves, half the families having fewer children and half having more children. There is no real median value. Sometimes, however, it may be reasonable to extend the definition and to accept for the median the value which divides the distribution in such a way that half the observations are *less than or equal to* that value, half are *greater than or equal to* it.

The weighted average

Let us suppose the fatality rates shown in Table 7.5 are observed. It would be wrong to compute the general fatality rate at all ages by merely taking

Table 7.5

Age last birthday (years)	*Fatality rate (per cent)*	*Number of patients*
Under 20	47.5	40
20–39	15.0	120
40–59	22.4	250
60 and over	51.1	90

the average of these four rates, i.e. (47.5 + 15.0 + 22.4 + 51.1)/4. The rate at all ages will depend upon the number of patients who fall ill at each age, as shown in the third column. To reach the rate at all ages the separate age-rates must be 'weighted' by the number of observations in each group. Thus we have (47.5 × 40) + (15.0 × 120) + (22.4 × 250) + (51.1 × 90) divided by 500, the total number of patients, which equals 27.8 per cent. By such weighting we are in effect calculating the total number of deaths that took place in the total 500 patients. These deaths divided by the number of patients gives the required rate, and the unweighted average of the rates will not produce it unless either the number of patients at each age is the same or the fatality rate remains the same at each age. In general it is, therefore, incorrect to take an unweighted average of rates or of a series of means.

Summary

The general position of a frequency distribution on some scale is measured by an average. There are three averages in common use: (1) the arithmetic mean (2) the geometric mean and (3) the median. The arithmetic mean, usually termed the mean or average, is the sum of all the observations divided by their number. The geometric mean is similarly defined, but on a multiplicative instead of an additive scale. The median is the central value when all the values are listed in order from the lowest to the highest. In calculating the mean of a series of sub-means the 'weights' attached to the latter must be taken into account.

Chapter 8

The variability of observations

In the previous chapter we have calculated and discussed average values, but the very fact that we thought it necessary to calculate an average, to define the general position of a distribution, introduces the idea of *variation* of the individual values round that average. For if there were no such variation, if, in other words, all the observations had the same value, then there would be no point in calculating an average. In introducing and using an average, usually the arithmetic mean, we therefore ignore—for the time being—that variability of the observations. It follows that, taken alone, the mean is of limited value, for it can give no information regarding the variability with which the observations are scattered around itself, and that variability and its degree are important characteristics of the frequency distribution.

As an example Table 8.1 shows the frequency distributions of some

Table 8.1 Frequency distribution of some recorded deaths of women according to age from (1) diseases of the Fallopian tube, and (2) abortion.

Age last birthday (years)	*Diseases of the Fallopian tube*	*Abortion*
0–	1	—
5–	—	—
10–	1	—
15–	7	—
20–	12	6
25–	35	21
30–	42	22
35–	33	19
40–	24	26
45–	27	5
50–	10	—
55–	6	—
60–	5	—
65–	1	—
70–74	2	—
Totals	206	99

recorded ages at death from two causes of death amongst women. The mean, or average, age at death, does *not* differ greatly between the two, being 37.2 years for the deaths registered as due to diseases of the Fallopian tube and 35.2 years for those attributed to abortion. But both the table and the diagram based upon it (Fig. 8.1) show that the difference in the variability, or scatter, of the observations round their respective means is very considerable. With diseases of the Fallopian tube the deaths are spread over the age-groups 0–4 to 70–74, while deaths from abortion range only between 20–24 and 45–49.

As a further description of the frequency distribution, we clearly need a measure of its degree of variability round the average. A measure commonly employed in medical (and other) papers is the *range*, as quoted above—i.e. the distance between the smallest and greatest observations. Though this measure is often of interest, it is not very satisfactory as a description of the general variability, since it is based upon only the two extreme observations and ignores the distribution of all the observations within those limits—e.g. the remainder may be more evenly spread out over the distance between the mean and the outlying values in one distribution than in another. Also the occurrence of the rare outlying values will depend upon the number of observations made. The greater the number of observations the more likely it is that the rare value will appear amongst them. As a result differences between the ranges recorded in two similar investigations may arise solely from a differing total number of observations. They will give a distorted view of the variability found in the

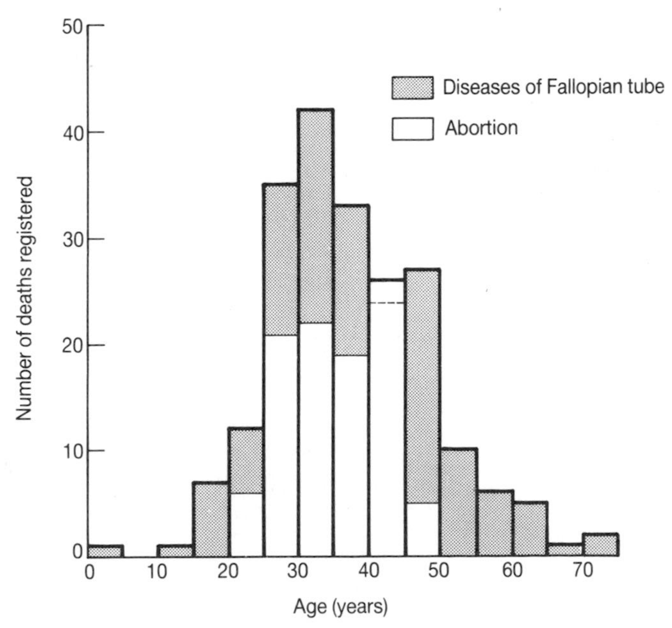

Fig. 8.1 Frequency distributions of some recorded deaths of women from two causes.

two inquiries. Thus, in publishing observations, it is certainly insufficient to give only the mean and the range; as previously pointed out, the frequency distribution itself should be given when possible—even if no further calculations are made from it.

Deviations, the variance and the standard deviation

We can think of the scatter of the observations in terms of the *deviation* of each from the mean of the distribution, that is

$$\text{deviation} = \text{observation} - \text{mean}$$

and, to measure the scatter, we need some sort of average of how much deviation each observation shows. However we cannot just average the deviations as such. If we have the 5 observations 1.3, 3.4, 6.7, 8.5 and 8.9, for example, their mean is 5.76, and the average of the deviations would be found as in Table 8.2. That the total, and thus the average, of the deviations is zero is not an accident in this particular case but follows inevitably from the definition of the mean.

It would be possible to take a measure of scatter that averaged the absolute values of the deviations (that is to say ignoring the minus signs on the negative ones) but this causes mathematical awkwardness and, in general, a preferable measure is derived from using another method of making all the values positive, namely by squaring them. Continuing with the same example, we have Table 8.3.

Table 8.2

Observation	*Deviation*
1.3	1.3 − 5.76 = −4.46
3.4	3.4 − 5.76 = −2.36
6.7	6.7 − 5.76 = 0.94
8.5	8.5 − 5.76 = 2.74
8.9	8.9 − 5.76 = 3.14
Total	0.00

Table 8.3

Deviation	*Deviation squared*
−4.46	(−4.46) × (−4.46) = 19.8916
−2.36	(−2.36) × (−2.36) = 5.5696
0.94	0.94 × 0.94 = 0.8836
2.74	2.74 × 2.74 = 7.5076
3.14	3.14 × 3.14 = 9.8596
Total	43.7120

To find an average of these squared deviations it might be thought that we should now divide by the sample size n (5 in this instance). In fact, however, we divide by $n-1$ instead, to get 43.7120/4 = 10.928. The reason is that, in general, we are not interested in the scatter in the sample as such, but in an estimate of the scatter in the population from which the sample was drawn. If we knew the true value of the mean in that population, we should find the deviations by using that population mean. We should then indeed have n independent values, and would use n as the divisor. In practice, though, we do not know that population mean, but have to use the sample mean as an estimate of it, which introduces a bias. This can be seen from the fact that if we know the values of $n-1$ deviations from the sample mean, we can say what the nth one must be, since their total is always zero. Thus although we have n observations, their deviations about their own mean have only $n-1$ *degrees of freedom*, so $n-1$ is the appropriate divisor.

An alternative way of looking at it is to note that one observation gives us no information at all about the amount of scatter. A second observation gives us the first piece of information about scatter. So it is always one unit in arrears.

The value we have derived, 10.928, the sum of squares of deviations divided by their degrees of freedom, is a valid measure of scatter, much used by statisticians and called the *variance*. It is a measure that must be used with care, however, because its units are peculiar. The squaring of deviations has also squared the units of measurement. If, for example, we have measured heights in metres, we should expect a measurement of scatter of those heights also to be in metres, not in square metres, representing an area rather than a length; but at least square metres do mean something. If, however, we are investigating the numbers of nurses employed by various health authorities and needed a measure of variability around the average, to report that variability in terms of 'square nurses' would lead to even more doubt about statisticians' sanity than exists at present.

To restore the original units we must take the square root of the variance

$$\sqrt{10.928} = 3.306$$

This is called the *standard deviation* and has the same units as the original observations. The units must, of course, always be attached to a standard deviation, just as much as to a mean or to an individual observation; it should not be reported as a bare figure.

Turning to the meaning of the result, a large standard deviation shows that the frequency distribution is widely spread out from the mean, while a small standard deviation shows that it lies closely concentrated about the mean with little variability between one observation and another. For example, the standard deviation of the widely spread age distribution of deaths attributed to diseases of the Fallopian tube (see Table 8.1). is 11.3 years, while of the more concentrated age distribution of deaths attributed to abortion it is only 6.8 years. The frequency distributions themselves clearly show this considerable difference in variability. The standard deviations have the advantage of summarising this difference by measuring the variability of each distribution in a single figure; they also enable us to test, as will be seen subsequently, whether the observed differences between two such means and between two such degrees of variability are more than would be likely to have arisen by chance.

In making a comparison of one standard deviation with another it must, however, be remembered that this criterion of variability is measured in the same units as the original observations. The mean height of a group of schoolchildren may be 48 inches and the standard deviation 6 inches; if the observations were recorded in centimetres instead of in inches, then the mean would be 122 cm and the standard deviation 15.2 cm. It follows that it is not possible by a comparison of the standard deviations to say, for instance, that weight is a more variable characteristic than height; the two characteristics are not measured in the same units and the selection of these units—e.g. inches or centimetres, pounds or kilograms—must affect the comparison. In fact, it is no more helpful to compare these standard deviations than it is to compare the mean height with the mean weight. Further, a standard deviation of 10 round a mean of 40 must indicate a relatively greater degree of scatter than a standard deviation of 10 round a mean of 400, even though the units of measurement are the same.

The coefficient of variation

To overcome these difficulties of the comparison of the variabilities of frequency distributions measured in different units or with widely differing means, the *coefficient of variation* may occasionally be useful. This coefficient is the standard deviation of the distribution expressed as a percentage of the mean of the distribution—i.e. coefficient of variation = (standard deviation/mean) × 100. If the standard deviation is 10 round a mean of 40, then the coefficient of variation is 25 per cent; if the standard deviation is 10 and the mean is 400, it is 2.5 per cent. The original unit of measurement is immaterial for this coefficient, since it enters into both the numerator and the denominator of the fraction. For instance, with a mean height of 48 inches and a standard deviation of 6 inches the coefficient of variation is $(6/48) \times 100 = 12.5$ per cent. If the unit of measurement is a centimetre instead of an inch, the mean height becomes 122 cm, the standard deviation is 15.2 cm and the coefficient of variation is $(15.2/122) \times 100 = 12.5$ per cent again.

Quartiles

Just as we may sometimes choose to measure position by using the median instead of the mean, we can measure scatter by a similar technique. The *quartiles* are defined as the three values that divide the distribution into four equal parts. Taking again the values used in the previous chapter which, arranged in order, were 26, 33, 36, 39, 40, 40, (41), 42, 44, 45, 47,

47, 47, (48), 50, 51, 51, 53, 54, 54, (55), 57, 59, 61, 63, 66, 71, the quartile values are shown in brackets as 41, 48 and 55 years, the middle quartile being simply the median of course. The scatter can be quoted in terms of the difference between the upper quartile and the lower quartile, called the *inter-quartile range*, which here is $55 - 41 = 14$ years.

In many distributions the inter-quartile range will be found to be about 1.4 standard deviations—a relationship well illustrated by the example figures, which have a standard deviation of 10.32 years.

Other quantiles

Similarly distributions can be divided into other portions of equal frequency. For example, the quintiles are the four values that divide a distribution into five equal parts, the deciles the nine values that divide it into ten equal parts and so on. The general name for such values is quantiles.

Unfortunately there is a tendency among some users of statistics (and even among some statisticians who should know better) to misuse these words. Having divided a distribution into five equal parts they refer to the five quintiles, when what they mean is the five fifths, of which the four quintiles are the boundary values. Such slovenly misuse of words is deplorable and causes trouble.

The importance of variability

These measures of variability are just as important characteristics of a series of observations as the measures of position—i.e. the average round which the series is centred. As was said by Udny Yule, one of the leading British statisticians of the early years of the twentieth century, the important step is to

> get out of the habit of thinking in terms of the average, and think in terms of the frequency distribution. Unless and until he [the investigator] does this, his conclusions will always be liable to fallacy. If someone states merely that the average of something is so-and-so, it should always be the first mental question of the reader: 'This is all very well, but what is the frequency distribution likely to be? How much are the observations likely to be scattered round that average? And are they likely to be more scattered in the one direction than the other, or symmetrically round the average?' To raise questions of this kind is at least to enforce the limits of the reader's knowledge, and not only to render him more cautious in drawing conclusions, but possibly also to suggest the need for further work.

Examples of variability

The practical application of some of these measures of variability may be illustrated by the figures in Table 8.4, which are taken from a statistical study of blood pressure in healthy adult males. In the original the full frequency distributions are also set out. The variability of these physiological measurements, which is apparently compatible with good health at the time of measurement, is striking. It led the authors to conclude that we must hesitate to regard as abnormal any isolated measurements in otherwise apparently fit individuals. Some of the measurements they found are definitely within the limits usually regarded as pathological, and study is necessary to determine whether such large deviations from the 'normal' have any unfavourable prognostic significance. It is clear that the mean value alone is a very insecure guide to 'normality' (see Chapter 27).

As a further example of the importance of taking note of the variability of observations, the incubation period of a disease may be considered. If the day of exposure to infection is known for a number of persons we can construct a frequency distribution of the durations of time elapsing between exposure to infection and onset of disease as observed clinically. If these durations cover a relatively wide range, say 10–18 days with an average of 13 days, it is obvious that observation or isolation of those who have been exposed to infection for the *average* duration would give no high degree of security. For security we need to know the proportion of persons who develop the disease on the fourteenth, fifteenth, etc., day after exposure; if these proportions are high—i.e. the standard deviation of the distribution is relatively large—isolation must be maintained considerably beyond the *average* incubation time. In such a case the importance of variability is indeed obvious; but there is a tendency for workers to overlook the fact that in *any* series of observations the variability, large or small, is a highly important characteristic.

For the beginner, who at first finds the standard deviation a somewhat intangible quantity, it is useful to remember that six times the standard deviation usually includes practically all the observations. Thus, in Table 8.4, the standard deviation of the diastolic blood pressures is 9.39 mm, and six times this, or 56 mm, should include very nearly all the observations. In

Table 8.4 The blood pressure in 566 healthy adult males. Means, standard deviations, coefficients of variation, and the range of measurement.

	Mean	*Standard deviation*	*Coefficient of variation (per cent)*	*Range*
Age (years)	23.2	4.02	17.33	18–40
Heart rate (beats per minute)	77.3	12.83	16.60	46–129
Systolic BP (mm)	128.8	13.05	10.13	97–168
Diastolic BP (mm)	79.7	9.39	11.78	46–108
Pulse pressure (mm)	49.1	11.14	22.69	24–82

fact, the observations lie within 108 − 46 = 62 mm. If the distribution is symmetrical, the mean plus 3 times the standard deviation should give approximately the upper limit of the observations, and the mean minus 3 times the standard deviation should similarly give their lower limit. Thus for the diastolic blood pressure we have 79.7 + 28.2 and 79.7 − 28.2, or a range of, approximately, 52 to 108 mm, very close to the observed range of 46 to 108 mm. This rule also serves, it will be seen, as a check upon the calculation of a standard deviation—not, of course, to show that a small error has been made but whether some serious mistake has led to a standard deviation which is quite unreasonable. It must not, however, be expected to hold with a few observations only.

Calculation of the standard deviation

Methods of calculation are detailed in Appendix B. However, it is necessary to give a warning here against misuse of a common method. If we call the standard deviation s, the formula is

$$s = \sqrt{\frac{\text{Sum } (x - \bar{x})^2}{n - 1}}$$

Let us concentrate on just the sum of squares of deviations, since the final division by $n - 1$ and the taking of the square root cause no trouble.

It can be shown algebraically that

$$\text{Sum } (x - \bar{x})^2 = \text{Sum } (x^2) - (\text{Sum } x)^2/n$$

and many people have been taught to use this second formulation instead of the first one because, for calculating by hand, it is easier. For example, taking the five observations used earlier in the chapter (1.3, 3.4, 6.7, 8.5 and 8.9) we found the sum of squares of deviations to be 43.712. Using the alternative formula we should proceed as in Table 8.5. The result is exactly correct, without having had to find deviations from the mean. However, this equivalence of the two holds only in terms of pure mathematics, where every number is known to infinite precision. For practical calculation,

Table 8.5

	x	x^2
	1.3	1.69
	3.4	11.56
	6.7	44.89
	8.5	72.25
	8.9	79.21
Total	28.8	209.60

Sum (x^2) − (Sum $x)^2/n$ = 209.60 − $(28.8)^2/5$ = 43.712

numbers have to be rounded to a finite number of significant figures; many computers hold numbers only to about seven significant figures unless specially instructed otherwise.

Suppose our sample of five, instead of being 1.3, 3.4, 6.7, 8.5 and 8.9 had been 1001.3, 1003.4, 1006.7, 1008.5 and 1008.9. This represents merely a change of mean, the variability remaining exactly as before, so the sum of squares of deviations should remain as 43.712. If we use the second formulation and work with complete accuracy we get the figures in Table 8.6. If, however, the numbers are rounded, correct to 7 significant figures, 5 057 809.60 will become 5 057 810 and 5 057 765.888 will become 5 057 766, giving 5 057 810 − 5 057 766 = 44, correct to only 2 figures. In even more extreme cases, there may be no figures correct at all.

This trouble occurs only when the coefficient of variation is very small. If making calculations by hand, it will be noticed if it occurs, and can be corrected. Unfortunately, the dangerous formulation has been written into many computer programs and pocket calculator operations. Usually it does no harm, but on the odd occasion inaccurate answers are found without any warning. Unless one is very sure that, in the particular usage, a small coefficient of variation will never occur, the second formulation should not be used when programming a computer to find standard deviations.

Table 8.6

	x	x^2
	1 001.3	1 002 601.69
	1 003.4	1 006 811.56
	1 006.7	1 013 444.89
	1 008.5	1 017 072.25
	1 008.9	1 017 879.21
Total	5 028.8	5 057 809.60

$$\text{Sum}\,(x^2) - (\text{Sum}\,x)^2/n = 5\,057\,809.60 - (5028.8)^2/5$$
$$= 5\,057\,809.60 - 5\,057\,765.888$$
$$= 43.712$$

Summary

As descriptions of the frequency distribution of a series of observations certain values are necessary, the most important of which are, usually, the mean and standard deviation. The mean alone is rarely, if ever, sufficient. In statistical work it is necessary to think in terms of the frequency distribution as a whole, taking into account the central position round which it is spread (the mean), the variability it displays round that central position (the standard deviation), and the symmetry or lack of symmetry with which it is spread round the central position. The important step is to think not only of the average but also of the scatter of the observations around it.

Chapter 9

Shapes of frequency distribution

An important classification of frequency distributions is into the *discontinuous* and *continuous* types. To repeat an earlier example, the number of children in a family is discontinuous 0, 1, 2, etc. but not 1.3 or 2.7, whereas measurement of heights of children can be thought of as continuous, any value within a given range being possible. In practice, all distributions are really discontinuous, because we can measure things, and record the results, only to a finite degree of precision. Nevertheless it is helpful to think in terms of distributions being continuous, and to represent the observed values, which often give a 'lumpy' histogram merely by the play of chance, by a smooth curve which we assume better to represent the underlying distribution in the population.

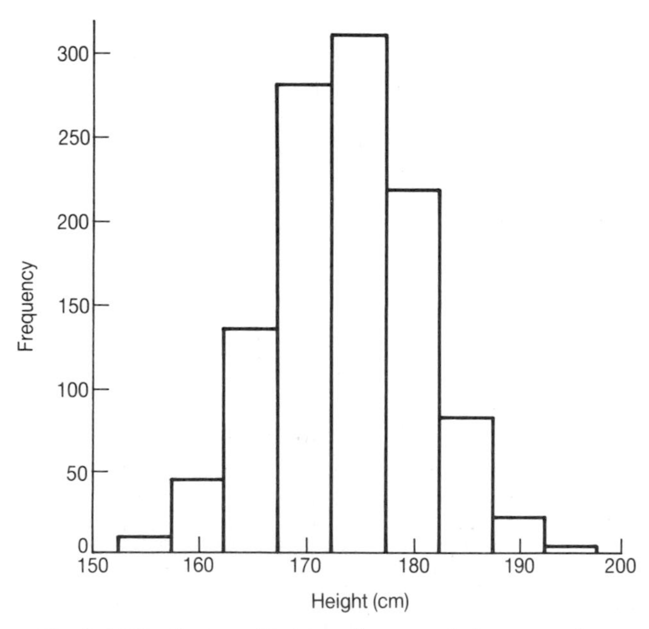

Fig. 9.1 Distribution of heights of some male factory workers.

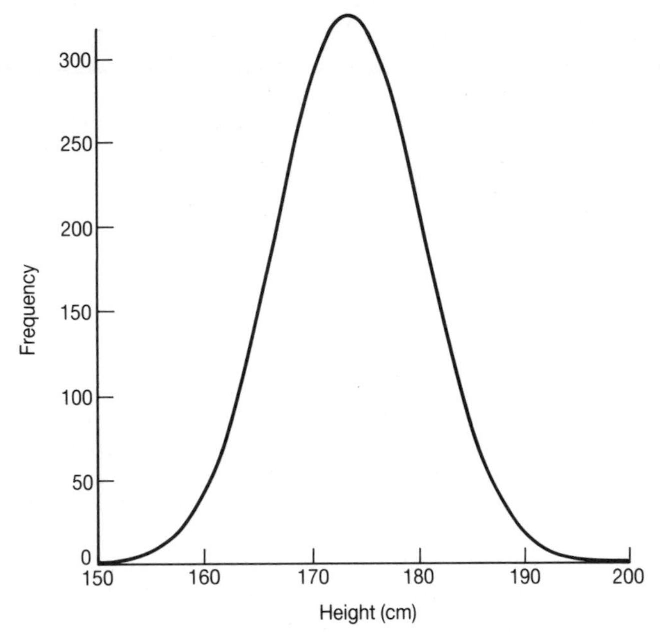

Fig. 9.2 A normal curve corresponding to the heights in Figure 9.1.

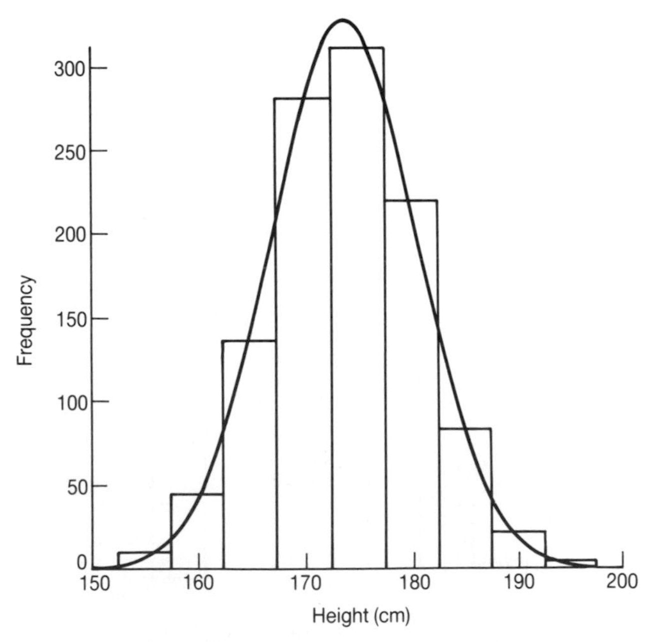

Fig. 9.3 Combination of Figures 9.1 and 9.2.

As with a histogram, a continuous curve used to represent a frequency distribution is interpreted by equating the frequency between any pair of values to the corresponding area under the curve. For example, Fig. 9.1 shows a histogram of observations of the heights of 1111 men included in a survey at a number of factories, while Fig. 9.2 shows a continuous curve representing the same data. Figure 9.3, putting the two together, shows that the curve represents the actual distribution very well. The curve has the advantage that, if its mathematical form is known, the quoting of the values of just a few constants gives a brief description of the entire distribution.

The normal distribution

The curve shown in Fig. 9.2 is a particular example of a 'normal' distribution, which has very great importance in statistical theory, being fundamental to many of the methods discussed in later chapters. The name 'normal' is a little unfortunate, sometimes leading people to think that it is the normal in the sense that all measurable characteristics occurring in nature, e.g. of men, animals or plants, should conform to it. This is not so. While many distributions do show, approximately, such a shape, it should not be thought that those that do not are in any way abnormal.

It is sometimes thought that any distribution that has a single peak and is symmetrical is a normal distribution. This is a mistake; the curve has a specific mathematical shape, varying only in mean and standard deviation from one normal distribution to another.

Figure 9.4 shows three normal distributions with different standard

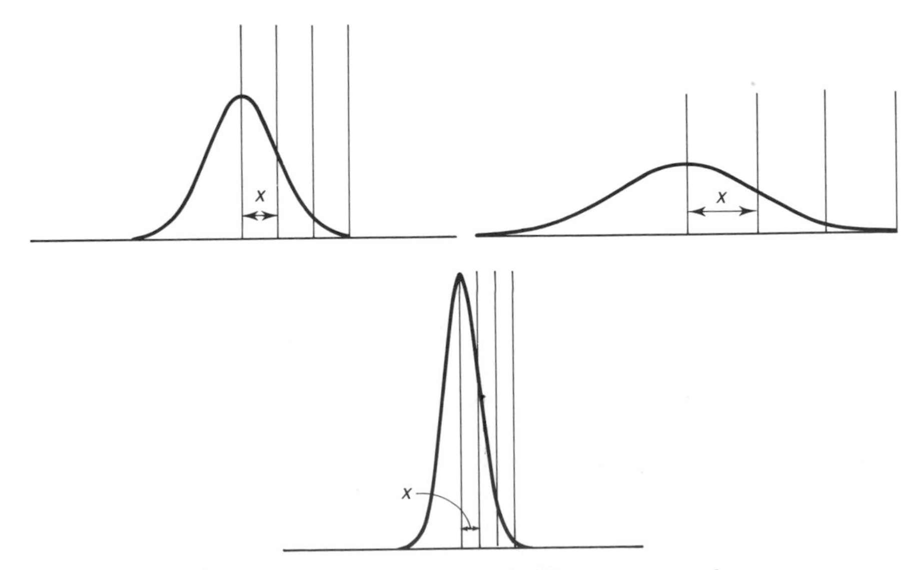

Fig. 9.4 Three normal distributions with different amounts of scatter.

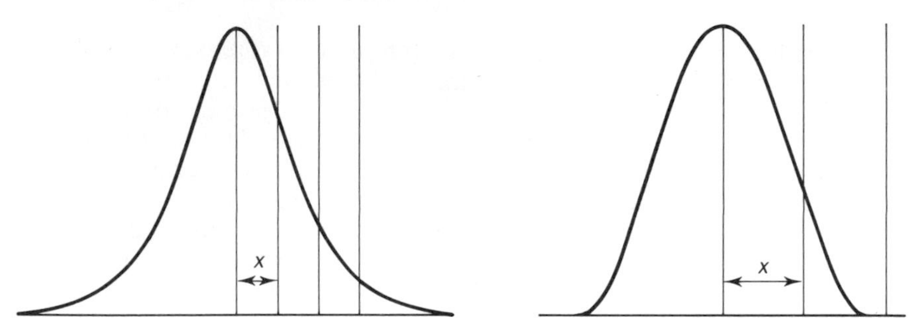

Fig. 9.5 Two symmetrical, but non-normal, distributions.

deviations. If we call the horizontal distance, from the mean to the place on the curve where its slope is steepest, x, we see that in each case virtually the whole distribution just fits within $mean \pm 3x$. Adopting the same procedure with the two curves in Fig. 9.5 it is seen that in one case the extent of the curve is much less than $mean \pm 3x$ and in the other case much greater—although single peaked and symmetrical, neither of these therefore is a normal distribution.

Theoretically speaking, a normal distribution actually extends to infinity in either direction, never quite reaching the base line, but only about one observation in 370 falls outside the $mean \pm 3x$ limits as described. The value of x as defined above is, in the case of a normal distribution, precisely one standard deviation, which allows a useful mental picture of the standard deviation and how it measures the amount of scatter of observations, but this is not true of other distributions.

Fact and fiction

It might be claimed that the original observations represent the fact of the actual heights of the 1111 men previously mentioned, whereas the normal curve fitted to that distribution is a fictitious approximation. It is not as simple as that though, because the distribution in Fig. 9.1 shows the observations grouped into 5 cm groups. The original observations were made, supposedly, to the nearest centimetre, and if we plot a histogram of them as recorded, we get the curiously lumpy effect shown in Fig 9.6.

It looks as though those who recorded the heights had a strong preference for numbers that end in 0, 3, 5 and 8, rather than 1, 2, 4, 6, 7 and 9. Indeed almost exactly half of the recorded heights end in those digits rather than the 40 per cent of them that would be expected to do so on the average. In Chapter 12 we shall find that this almost certainly indicates a 'real' effect, rather than being merely a result of chance. In those circumstances it may well be that it is the actual observations that are

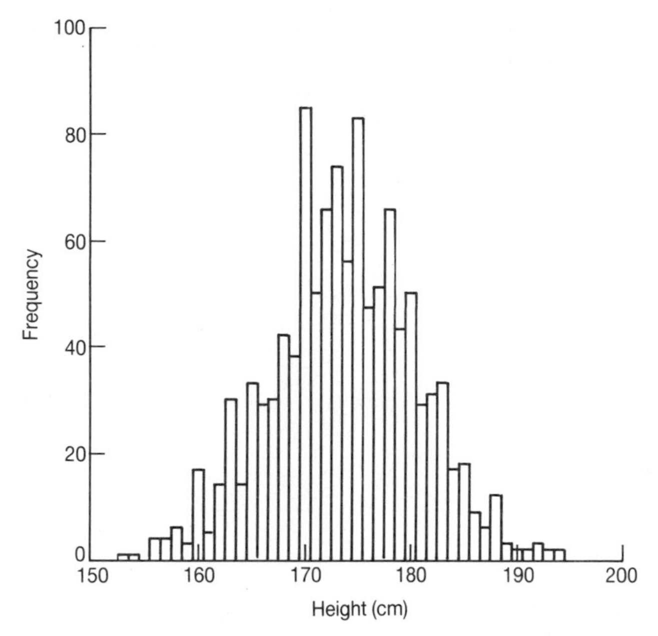

Fig. 9.6 Distribution of heights of some male factory workers as originally reported, showing digit-preference.

fiction if taken to 1 cm precision, and that the fitted distribution better represents the underlying facts of the actual heights of the men involved.

Further examination of normal curve

Calculation from the original data shows the mean of the height distribution to be 173.7 cm and the standard deviation to be 6.8 cm. Examining the frequency distribution, we can see how many men have heights that differ from the mean by not more than 1 standard deviation, i.e. lie between $173.7-6.8=166.9$ cm and $173.7+6.8=180.5$ cm. There are $385+396=781$ of them, i.e. 70.3 per cent of the total 1111 men (see Table 9.1).

Similarly we find that 94.4 per cent of the men have heights within two standard deviations either side of the mean and that 99.9 per cent have heights within three standard deviations of the mean.

This is a useful way of looking at a frequency distribution, namely to see how many of the observations lie within a given distance of the mean, not in terms of the actual units of measurement but in terms of multiples of the standard deviation.

Returning to the ideal normal frequency distribution, as derived mathematically, its characteristics are: (1) the mean and median coincide; (2) the

Table 9.1 Heights of 1111 men showing characteristics of a normal distribution.

	Height (cm)	*Number of men of given height*		*Percentage of men within given distance of the mean*		
				within 1 s.d.	*within 2 s.d.*	*within 3 s.d.*
	153	1	0.1%			
mean − 3 s.d.						
	154–160	35	3.2%			
mean − 2 s.d.						
	161–166	125	11.3%			
mean − 1 s.d.						
	167–173	385	34.7%			
mean				70.3%	94.4%	99.9%
	174–180	396	35.6%			
mean + 1 s.d.						
	181–187	143	12.9%			
mean + 2 s.d.						
	188–194	26	2.3%			
mean + 3 s.d.						
	Total	1111				

curve is perfectly symmetrical round the mean; and (3) we can calculate *theoretically* how many of the observations will lie in the interval between the mean itself and the mean plus or minus *any* multiple of the standard deviation. This calculation gives the following results:

Proportion of observations that lie within ± 1 times the SD from the mean	68.27 per cent
Proportion of observations that lie within ± 2 times the SD from the mean	95.45 per cent
Proportion of observations that lie within ± 3 times the SD from the mean	99.73 per cent.

These figures are quite closely matched by the real example.

It will be seen that with a measurement that follows a normal distribution nearly one-third of the values observed will differ from the mean value by more than once the standard deviation, only about 5 per cent will differ from the mean by more than twice the standard deviation, and only some 3 in 1000 will differ from the mean by more than 3 times the standard deviation. In a normal distribution, in other words, values that differ from the mean by more than twice the standard deviation are fairly rare, for only about 1 in 20 observations will do so; values that differ from the mean by more than 3 times the standard deviation are very rare, for only about 1 in 370 will do so.

Skew distributions

Although, as already described, not all symmetrical distributions are normal, the most obvious departure from normality that frequently arises is skewness—that is to say a lack of symmetry. Of the 1111 factory workers mentioned earlier, 413 were selected as working on processes that subjected them to atmospheric pollution. An estimate of the degree of pollution for each of these men was made, and the resulting distribution is shown in Fig. 9.7. It is seen to be very skew, most of the men experiencing values close to zero, but with a few having much higher values.

A distribution such as this can often be made more nearly symmetrical, by considering the logarithms of the observations instead of the observations themselves. The result of taking logarithms in this instance is shown in Fig. 9.8, where it is seen that a normal curve is now not too bad a fit. Performing such a *logarithmic transformation* of the data has the advantage that those statistical methods that work best for normally distributed data can now be used, the eventual results being translated back, of course, to their meaning for the original data.

There are other, similar, mathematical transformations of data that can be used to convert other shapes of distribution to normality, but the logarithmic transformation is the one most frequently used.

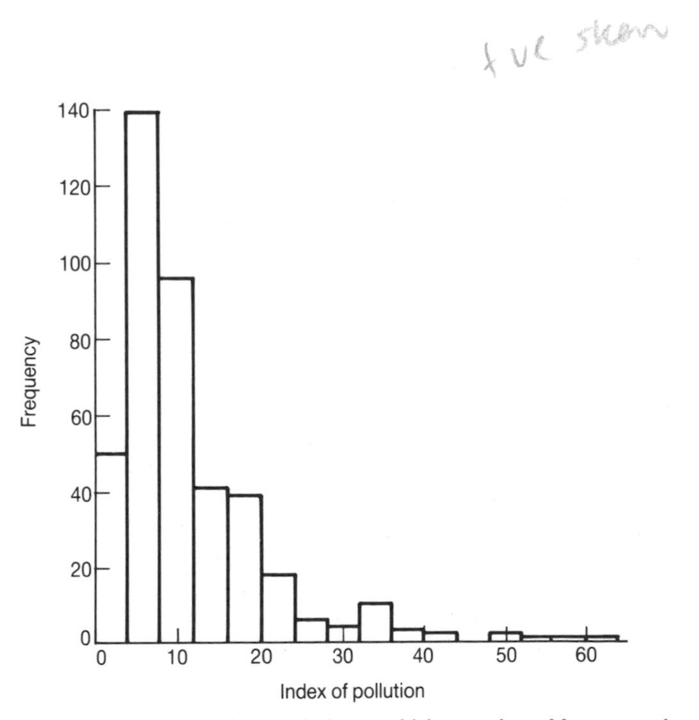

Fig. 9.7 Example of a skew distribution: pollution to which a number of factory workers were exposed.

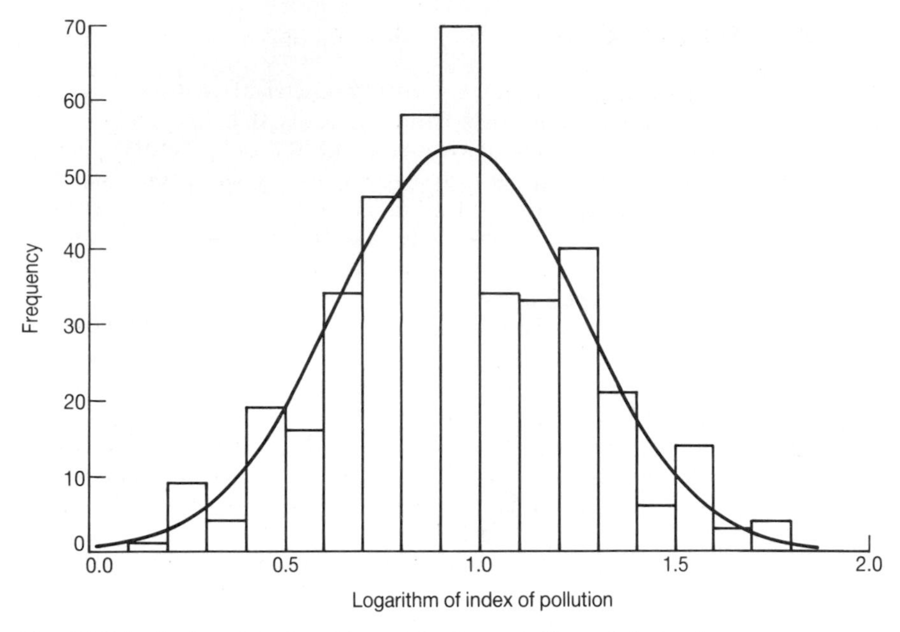

Fig. 9.8 Distribution of the logarithms of the figures in Figure 9.7, showing more symmetry.

Approximating discontinuous distributions with continuous ones

Returning to the heights of the 1111 men, we found that a normal distribution was a good approximation. Yet we must take care in using a continuous curve to represent discontinuous observations; although we can conceive of height as being continuously variable, these measurements were supposed to be taken only to the nearest centimetre.

Suppose then we wished to use a normal curve, with mean 173.7 cm, and standard deviation 6.8 cm, to estimate the number of men who would have been recorded as having a height of 182 cm or greater. First we have to find how far from the mean 182 cm is, in units of the standard deviation, but instead of using the figure 182 for the purpose we must use 181.5, because we are trying to distinguish between 182 or greater on the one hand, and 181 or less on the other. Within a continuous curve 181.5 will be the point of division between these two groups. Using 181.5 in this way, instead of the 182 in the original question, is known as a *continuity correction* and we always have to ask ourselves whether such a correction is needed when using a continuous distribution to approximate a discontinuous one.

In the current instance we calculate

$$(181.5 - 173.7)/6.8 = 1.15$$

It is worth noting that the numerator and denominator of this fraction are both measured in centimetres, so the units cancel out and the result, 1.15, is a pure number without units attached. Using a table, or a suitable computer program, for the normal curve, 1.15 standard deviations from the mean is found to be exceeded by 12.5 per cent of observations. 12.5 per cent of 1111 is 139 to the nearest whole number, almost the same as the 140 actually observed in the original distribution.

Summary

Frequency distributions may be continuous or discontinuous, symmetrical or skew. A particular shape of continuous, symmetrical, distribution is known as the 'normal' distribution, and it has very important properties for statistical analysis. Other shapes of distribution can often be transformed mathematically to approximate normality. If using a continuous curve to represent a discontinuous distribution, a continuity correction may be required.

Chapter 10

Problems of sampling: averages

The observations to which the application of statistical methods is particularly necessary are those which are influenced by numerous causes, the object being to disentangle that multiple causation. Furthermore the observations utilised are nearly always only a sample of all the possible observations that might have been made. For instance, the frequency distribution of the stature of Englishmen—e.g. the number of Englishmen of different heights—is not based upon measurements of all Englishmen but only upon some sample of them. The question that immediately arises is how far is the sample representative of the population from which it was drawn, and bound up with that question, to what extent may the values calculated from the sample—e.g. the mean and standard deviation—be regarded as precise estimates of the values in the population sampled? If the mean height of 1000 men is 169 cm with a standard deviation of 7 cm, may we assert that the values of the mean and standard deviation of all the men of whom these 1000 form a sample are not likely to differ appreciably from 169 and 7? This problem is fundamental to all statistical work and reasoning; a clear conception of its importance is necessary if errors of interpretation are to be avoided, while a knowledge of the statistical techniques in determining errors of sampling will allow conclusions to be drawn with a greater degree of security.

Elimination of bias

Consideration must first be given, as previously noted, to the presence of bias in the sample. If owing to the method of collection of the observations, those observations cannot possibly be a representative sample of the total population, then clearly the values calculated from the sample cannot be regarded as valid estimates of the population values, and no statistical technique can allow for that kind of error. That problem was discussed in Chapter 3. In the present discussion we shall presume that the sample is unbiased and devote attention entirely to the problem of the variability which will be found to occur from one sample to another in such values as means, standard deviations, and proportions, due solely to what are sometimes known as the 'errors of sampling'. This is not in fact a very good description. We are not concerned with *error* in the sense of 'mistake' but in the sense of 'wandering' i.e. the variability that must occur through the play of *chance*.

The mean

Let us suppose that we are taking observations of counts from a radioactive marker, each count being over a period of 1 minute, where it is known that the marker concerned shows an average count of 3 per minute. Each observation must, of course, be an integer 0, 1, 2, 3, etc., with a population mean of 3.

The observations to be used were not in fact found by using a radioactive substance but by means of a computer simulation of the process. From that simulation samples of 5 observations were taken, each observation being a 1 minute count. To what extent will these means in the small samples diverge from the real mean in the population, 3.0?

In Table 10.1 are set out 100 such samples of 5 observations each. We see that each observation consists of a number in the range 0–9. (There is no reason why a value of 10 or more should not occur, but it can be shown that, from this distribution, the probability that any observation exceeds 9 is only just over 1 in 1000, so it is not surprising that we have not observed such an occurrence in 500 tries.)

For each sample of 5 the mean was found as shown in the table, and we can make a frequency distribution of these means. As measured by the range, we can notice at once that the means are less variable than the individual observations: instead of varying from 0 to 9, they vary only from 1.4 to 5.0. As might be expected with samples of only 5, there will be instances, due to the play of chance, in which the observed mean is rather different from the population mean. On the other hand, these extreme values of the observed mean are relatively rare, and a large number of the means in the samples lie fairly close to the population mean (3.0), as shown in Table 10.2; 49 per cent of them between 2.5 and 3.5.

The exercise was repeated, using samples of 10, 20 and 50 instead of only 5, and the results are shown in the later columns of Table 10.2. We notice how, with larger samples, the means cluster more and more closely around the population mean, 61 per cent of them between 2.5 and 3.5 for 10, 85 per cent for 20, and 95 per cent for 50. At the foot of the table we have the mean of each distribution of means, which is always quite close to 3.0. We also have the observed standard deviation of each distribution of means, getting smaller and smaller as the sample size increases.

Two factors in precision

These results show, what is indeed intuitively obvious, that the precision of an average depends, at least in part, upon *the size of the sample*. The larger the random sample we take the more precisely are we likely to reproduce

Table 10.1 100 samples, each of 5 counts from a radioactive marker (simulated figures).

Sample number	1	2	3	4	5	6	7	8	9	10	11	12	13	14	15	16	17	18	19	20	21	22	23	24	25
	6	3	2	2	3	2	3	4	2	2	1	3	2	3	2	0	2	3	2	4	7	2	3	3	1
	1	2	1	2	3	3	7	1	3	2	1	2	2	2	2	4	5	3	1	2	4	2	4	4	4
	1	1	3	6	2	2	1	2	3	0	3	5	4	3	0	1	7	1	2	3	4	2	1	3	3
	2	2	2	4	1	9	1	3	0	5	3	1	3	3	3	2	6	4	4	3	2	1	4	4	3
	4	4	0	8	0	3	2	3	3	2	7	1	5	4	2	2	2	1	0	3	0	2	6	3	1
Mean of each sample	2.8	2.4	1.6	4.4	1.8	3.8	2.8	2.6	2.2	2.2	3.0	2.4	3.2	3.0	1.8	1.8	4.4	2.4	1.8	3.0	3.4	1.8	3.6	3.4	2.4

Sample number	26	27	28	29	30	31	32	33	34	35	36	37	38	39	40	41	42	43	44	45	46	47	48	49	50
	2	3	2	3	2	6	3	4	3	3	4	4	4	5	6	6	1	5	5	4	2	1	1	3	5
	2	4	0	3	2	4	2	4	3	5	2	2	3	2	2	0	4	4	2	2	3	1	2	1	1
	1	7	2	4	3	1	5	3	4	0	4	2	1	5	3	2	2	4	0	6	3	2	2	5	3
	2	3	3	6	5	0	2	2	3	3	2	3	3	5	3	3	4	4	3	3	3	1	2	2	4
	1	2	0	3	2	2	5	2	1	5	4	0	5	4	2	3	2	3	5	3	2	3	2	4	3
Mean of each sample	1.6	3.8	1.4	3.8	2.8	2.6	3.4	3.0	2.8	3.2	3.2	2.2	3.2	4.2	3.2	2.8	2.8	4.0	3.0	3.6	2.6	1.6	1.8	3.0	3.2

Sample number	51	52	53	54	55	56	57	58	59	60	61	62	63	64	65	66	67	68	69	70	71	72	73	74	75
	4	3	1	4	2	0	4	0	3	5	5	4	3	1	2	5	3	3	5	4	2	2	4	2	5
	4	5	5	5	4	2	2	5	2	0	2	4	4	3	0	4	5	4	3	4	1	2	1	7	4
	2	0	2	3	1	3	0	5	1	7	6	5	3	0	8	2	7	3	2	1	1	1	4	1	3
	3	3	1	3	4	8	7	3	1	1	2	4	2	3	6	4	4	2	2	7	6	2	2	1	3
	2	1	5	4	9	1	4	3	3	4	1	3	2	1	3	3	5	1	2	2	8	3	3	4	1
Mean of each sample	3.0	2.4	2.8	3.8	4.0	2.8	3.4	3.2	2.0	3.4	3.2	4.0	2.8	1.6	3.8	3.6	4.8	2.6	2.8	3.6	3.6	2.0	2.8	3.0	3.2

Sample number	76	77	78	79	80	81	82	83	84	85	86	87	88	89	90	91	92	93	94	95	96	97	98	99	100
	1	5	3	3	3	2	5	7	6	6	2	4	2	5	3	3	1	3	6	2	2	4	3	2	2
	1	2	3	3	2	3	6	2	3	1	6	4	6	2	2	0	2	3	2	2	2	3	1	3	3
	4	1	3	1	4	1	2	3	6	2	2	2	3	4	4	3	2	4	3	5	2	5	3	3	5
	4	6	0	1	4	4	6	1	3	5	4	1	2	1	0	1	1	4	6	7	5	4	3	2	2
	2	3	9	0	1	3	6	4	1	2	2	3	3	7	1	3	3	2	1	2	3	6	5	2	2
Mean of each sample	2.4	3.4	3.6	1.6	2.8	2.6	5.0	3.4	3.8	3.2	3.2	2.8	3.2	3.8	2.0	2.0	1.8	3.2	3.6	3.6	2.8	4.4	3.0	2.4	2.8

Table 10.2 Distributions of means in samples of different sizes.

	Observed frequencies			
Value of mean in sample	*Samples of 5*	*Samples of 10*	*Samples of 20*	*Samples of 50*
Less than 0.5				
0.5–	1			
1.5–	26	16	7	2
2.5–	49	61	85	95
3.5–	22	21	8	3
4.5–	2	2		
5.5+				
Total number of means	100	100	100	100
Total observations	500	1000	2000	5000
Grand mean	2.952	3.036	2.988	2.989
Standard deviation of means	0.770	0.598	0.362	0.257

the characteristics of the population from which it is drawn. (The statistical use of the word *population* should be noted, as meaning a collection of values from which a sample has been taken. It does not imply that the figures relate to people.) The size of the sample, however, is not the only factor which influences the precision of the values calculated from it. A little thought will show that it must also depend upon the *variability of the observations in the population*. If every individual in the population could have only one value then clearly, whatever the size of the sample, the mean value reached would be the same as the true value. If on the other hand the individuals had greater variability the means of samples could, and would, have more variability too. The precision of a value calculated from a sample depends, therefore, upon two considerations:

(1) the size of the sample;
(2) the variability of the characteristic within the population from which the sample is taken.

The statistician's aim is to pass from these simple rules to more precise formulae, which will enable him to estimate, with a certain degree of confidence, the value of the mean, etc., in the population and also to avoid drawing conclusions from differences between means or between proportions when, in fact, these differences might easily have arisen by chance.

It is worth emphasising again that the precision does *not* depend upon the population size, or, in particular, on the ratio of the sample size to the population size. Indeed in the present example the population size, the number of observations (of a radioactive count) that could be made is virtually infinite, being limited only by our observing powers, so any of our samples are virtually 0 per cent of the population, but that fact does not stop them from telling us that the population mean is close to 3.0.

Measuring the variability of means

As a first step we may return to Table 10.2 and measure the variability shown by the means in the samples of different sizes. We have illustrated that variability by drawing attention to the range of the means, the extent to which the means are concentrated round the centre point, and in the last line of Table 10.2, the standard deviations of the frequency distributions.

The standard deviation, or scatter, of the means round the grand mean of each of the total 100 samples becomes, as is obvious from the frequency distributions, progressively smaller as the size of the sample increases. It is clear, however, that the standard deviation does not vary *directly* with the size of the sample; for instance, increasing the sample from 5 to 50—i.e. by ten times—does not reduce the scatter of the means by ten times. The scatter is, in fact, reduced not in the ratio of 5 to 50 but of $\sqrt{5}$ to $\sqrt{50}$—i.e. not ten times but 3.16 times (for $\sqrt{5} = 2.24$ and $\sqrt{50} = 7.07$ and $7.07/2.24 = 3.16$). This rule is very closely fulfilled by the values of Table 10.3; the standard deviation for samples of 5 is 0.770, and this value is 3.0 times the standard deviation, 0.257, with samples of 50. The first more precise rule, therefore, is that *the accuracy of the mean computed from a sample does not vary directly with the size of the sample but with the square root of the size of the sample.* In other words, if the sample is increased a hundredfold the precision of the mean is increased not a hundredfold but tenfold.

As the next step we may observe in samples of different sizes how frequently means will occur at different distances from the true mean. For instance it was pointed out above that with samples of 5 individuals 49 per cent of the means lay within half a unit of the true mean of the population. The grand mean of these 100 samples, 2.952 is not quite identical to the true mean of the population, 3.0, as, of course, the total 500 observations are themselves only a sample; it comes very close to it as the total observations are increased—it is 2.989 with 100 samples of 50. Instead, therefore, of measuring the number lying within half a unit, or one unit, of the grand mean, we may see how many lie within the boundary lines 'grand mean plus the value of the standard deviation' and 'grand mean minus the value of the standard deviation'—i.e. $2.952 + 0.770 = 3.722$ and $2.952 - 0.770 = 2.182$. It is found that 67 of the 100 means in Table 10.1 lie

Table 10.3 Values computed from the frequency distribution of means given in Table 10.2.

Number of individuals in each sample	*The mean, or average, of the 100 means*	*The standard deviation of the 100 means*	*(The population standard deviation)/(square root of size of sample)*
5	2.952	0.770	0.775
10	3.036	0.598	0.548
20	2.988	0.362	0.387
50	2.989	0.257	0.245

between those limits. If we extend our limits to 'grand mean plus or minus *twice* the standard deviation'—i.e. $2.952 + 1.540 = 4.492$ and $2.952 - 1.540 = 1.412$ we find 97 of the 100 means within those limits.

These figures of 67 and 97 per cent within one and two standard deviations, respectively, are remarkably close to the theoretical figures of 68.27 and 95.45 per cent of a normal distribution (see Chapter 9). Even for samples as small as five observations each, and starting from a noticeably skew distribution, the distribution of means is approaching normality. Similar results will be reached if these methods are applied to the larger samples. Our conclusions are therefore:

(1) If we take a series of samples from a population, then the means of those samples will not all be equal to the true mean of the population but will be scattered around it.
(2) We can measure that scatter by the standard deviation shown by the means of samples; means differing from the true mean by more than twice this standard deviation, above or below the true mean, will be only infrequently observed.

The means show an approximately 'normal' distribution

To be rather more precise, it can be proved that the means of the samples will be distributed round the mean of the population approximately in the

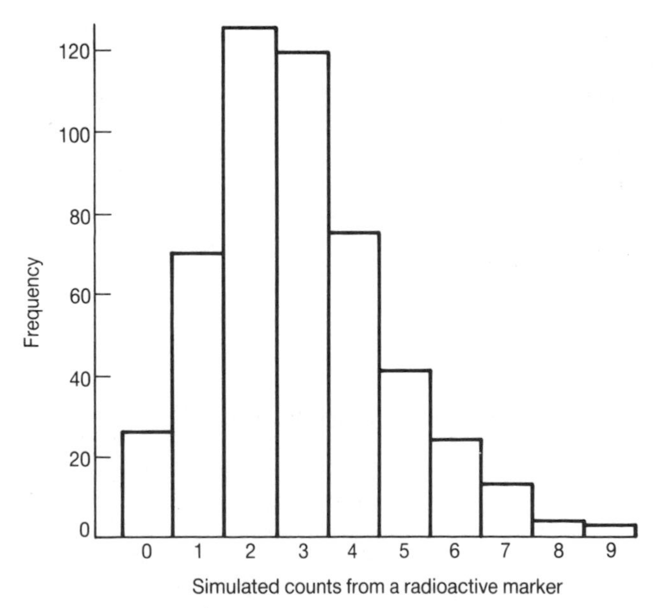

Fig. 10.1 Histogram of the observations in Table 10.1, showing skewness.

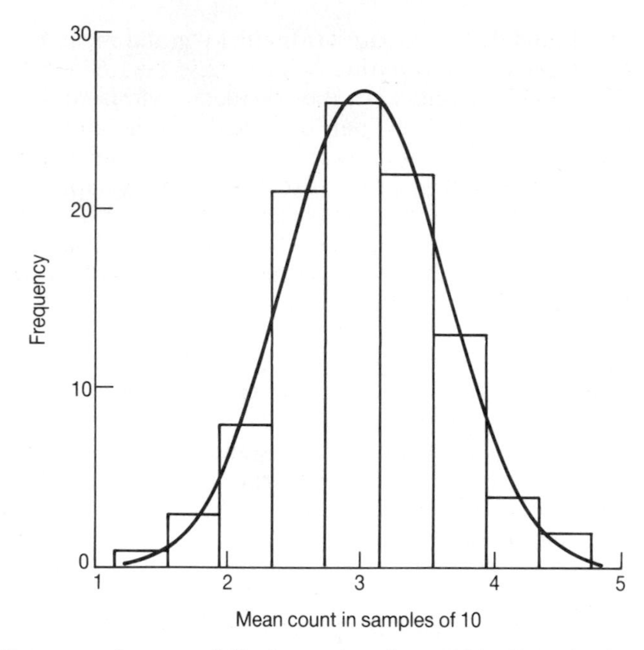

Fig. 10.2 Histogram of means of 10 observations from Table 10.1, showing approach to normal curve.

shape of a normal curve. In other words, it can be calculated how many of them will lie, in the long run (if we take enough samples), within certain distances of the real mean, those distances being measured as multiples of the standard deviation. This becomes more and more true the larger the sample size. Figure 10.1 shows a histogram of the original observations of Table 10.1, a noticeably skew distribution with an abrupt termination at its lower end, yet Fig 10.2 shows a histogram of the means of samples of size 10 from that same parent population as observed, and a superimposed normal curve, demonstrating a remarkably good fit.

It follows that, provided we have a reasonably large sample, and an estimate of the standard deviation that would be observed in the distribution of means of samples of that size, we can make inferences about the population mean by assuming the normal distribution shape.

Deducing the standard deviation

In practice, however, we do not have an observed value for this standard deviation of the means, for we do not usually take repeated samples. We take a single sample, say of patients with diabetes, and we calculate a single mean, say of their body weight. Even if we did have 10 samples of 5, say, we should combine them into one sample of 50, and the difficulty would arise again of wishing to know the variability associated with one mean.

Our problem is this: how precise is that mean—i.e. how much would it be likely to vary if we did take another, equally random, sample of patients? What *would* be the standard deviation of the means *if* we took repeated samples? It can be shown mathematically that the standard deviation of means of samples is equal to *the standard deviation of the individuals in the population sampled* divided by the square root of the number of individuals included in the sample. These values are in the final column of Table 10.3 and it will be seen that they agree very closely with the standard deviations calculated from the 100 means themselves (they do not agree exactly because 100 samples are insufficient in number to give complete accuracy).

Suppose we had made the observations given in Table 10.1, not now considered as 100 samples of 5 but as a single sample of 500 observations. We have calculated our observed mean as 2.952 and our observed standard deviation as 1.739. We can then argue that if the true standard deviation in the population were 1.739 and we took repeated samples of 500 observations, their means would vary around the population mean in a normal distribution (approximately) with a standard deviation of $1.739/\sqrt{500} = 0.078$. If the population mean were greater than $2.952 + 2\,(0.078) = 3.108$ our observed mean would be unlikely to have occurred; similarly if the population mean were less than $2.952 - 2(0.078) = 2.892$ our observed mean would be unlikely to have occurred. So the population mean probably lies somewhere within the limits 2.892 and 3.108.

As this is an artificial example, and we started off knowing the population mean to be 3.0, we can see that, in this instance, the derived limits are correct. The true mean does lie between them. Suppose, however, that we had had a hypothesis, before the observations were made, that the distribution from which they were to be drawn had a mean of 2.7, we could now say that our result gave a *significant difference* from the hypothesis, meaning that if the hypothesis were correct we should have observed an improbable event, so we prefer to believe the hypothesis to be incorrect.

The value used above, 0.078 or in general $s/\sqrt{n}$, where s is the observed standard deviation and n is the sample size, is known as the *standard error* of the mean, the different term standard error being used instead of standard deviation to indicate that we have not actually observed a distribution of means (as we did earlier in this chapter) and observed its variability, but have calculated an estimate of what that standard deviation of means would have been had we observed it. It is useful to have a different term to indicate this. Unfortunately the square of a standard error is called a variance, just as is the square of a standard deviation, without any distinction. It is necessary to be on one's guard therefore in determining the meaning of 'variance' according to context.

It is important to use the complete phrase 'standard error *of the mean*', since other calculated values also have their standard errors, measuring the precision with which they are known. In publications standard errors should be named as such, or by the abbreviation S.E. or s.e., *not* merely by the sign ±, which can be seriously misleading.

So far these ideas have been considered only in an intuitive way. They will be considered more precisely in the next chapter. We should note immediately, however, that special care is necessary with small samples, for two reasons: (1) the sample standard deviation of a small sample may differ considerably from the population standard deviation; (2) the distribution of means from small samples will be less accurately represented by a normal distribution than is the case with large samples.

Summary

In medical statistical work we are, nearly always, using samples of observations taken from large populations. The values calculated from these samples will be subject to the laws of chance—e.g. the means, standard deviations, and proportions will vary from sample to sample. It follows that arguments based upon the values of a single sample must take into account the inherent variability of these values. It is idle to generalise from a sample value if this value is likely to differ materially from the true value in the population sampled. To determine how far a sample value is likely to differ from the true value, a standard error of the sample value is calculated. The standard error of a mean is dependent upon two factors—the size of the sample, or number of individuals included in it, and the variability of the measurements in the individuals in the population from which the sample is taken. This standard error is calculated by dividing the standard deviation of the individuals in the sample by the square root of the number of individuals in the sample. The mean of the population from which the sample is taken is unlikely to differ from the value found in the sample by more than plus or minus twice this standard error.

Chapter 11

Confidence intervals and significance tests

In the last chapter we introduced the idea that if we have a sample mean and its standard error, then the true mean of the population sampled probably lies somewhere within limits set at the observed mean plus and minus twice its standard error, i.e. $\bar{x} \pm 2s/\sqrt{n}$. We can make this procedure more precise in two ways. The first is that instead of taking 2 standard errors and saying 'probably' we can take a value that ties up with a chosen probability level. If, as is most often done, we choose 0.95 as the probability level, this corresponds to 1.96 standard deviations rather than 2 on a normal curve. Secondly we can make allowance for the fact that the standard deviation obtained from the sample is only an estimated value, not the true value in the population. This allowance is made by referring not to the normal curve, but to a rather differently shaped curve known as the t distribution, or often as *Student's t*. (It was first derived by W.S Gosset, who published statistical papers using the pseudonym 'Student'.)

Values of this may be looked up in tables (such as Table II at the back of the book) or found from suitable computer programs. Unlike the normal curve, which has only one value corresponding to any given probability, the t value varies according to its associated *degrees of freedom*, namely the degrees of freedom used in estimating the standard deviation, $n-1$. In the particular example, with a sample mean of 2.952, a sample standard deviation of 1.739, and a sample size of 500, if we require limits giving 95 per cent confidence we look in Table II under the $P=0.05$ heading (i.e. $1.0-0.95$) and find for 499 degrees of freedom that $t=1.97$. So the formula is

$$2.952 \pm 1.97 \times 1.739/\sqrt{500}$$

i.e. 2.80 and 3.11

These are called 95 per cent *confidence limits* for the mean, while the range from 2.80 to 3.11 is called the 95 per cent *confidence interval*.

Suppose, however, that we had had only the first 25 observations of Table 10.1 instead of all 500 values. Calculation shows a sample mean of 2.60, and a sample standard deviation of 1.915. A 95 per cent confidence interval for the population mean would then be

$$2.60 \pm 2.06 \times 1.915/\sqrt{25}$$

i.e. 1.81 and 3.39

The interval, though still including the true value, is much wider than before partly because we have to use 2.06 instead of 1.97 for the value of t (derived from 24 degrees of freedom instead of 499) but mainly because of a much larger standard error (using $\sqrt{25}$ instead of $\sqrt{500}$ in its denominator).

If we had only those 25 observations, we could then say, with 95 per cent confidence, that the true mean lay somewhere between 1.81 and 3.39, meaning that 95 per cent of the time, or 19 occasions out of 20 in the long run, we should be correct in making such a statement, but that 5 per cent of the time, or 1 occasion out of 20, we should be incorrect. If that proportion correct is thought inadequate it is perfectly possible to get a higher proportion—there is nothing magic about the 95 per cent figure that is commonly used—by choosing the appropriate value of t for 99 per cent, or 99.9 per cent, or any other figure. The trouble is that, the greater the confidence required, the wider the limits become until, when we are virtually certain of being right, the limits are so wide as not to tell us anything much. The only way to get greater confidence without widening the limits is to have a larger sample size.

Confidence and probability

In the discussion above we have been talking of 95 per cent confidence. Is this the same thing as a 95 per cent probability? Not quite. The distinction lies in the fact that if we repeat the experiment, it will not be the confidence limits that stay the same while the true mean varies, but the true mean that stays the same, while the limits will vary (both in their positioning and in their width) from one trial to another.

Figure 11.1 shows, again for the data in Table 10.1, the confidence limits derived from each of the 20 successive sets of 25 observations. It will be seen how the limits vary from one sample to another, for a fixed population mean, but nearly always including that fixed value within them. In fact, with 20 95 per cent confidence intervals, we should expect, on the average, that 19 of them would include the true value and 1 would not. That is exactly what we observe here (and there was no cheating in deriving the observations), but of course it is not to be expected that 19 right and 1 wrong will occur from every set of 20 consecutive occasions. Each individual interval has 1 chance in 20 of being wrong, and how many are wrong in any given set of 20 will itself vary according to the laws of probability.

It might seem to be a distinction without a difference, and merely a change of wording to represent the same facts, whether we say that there is a 95 per cent probability that the true mean lies within the limits, or a 95

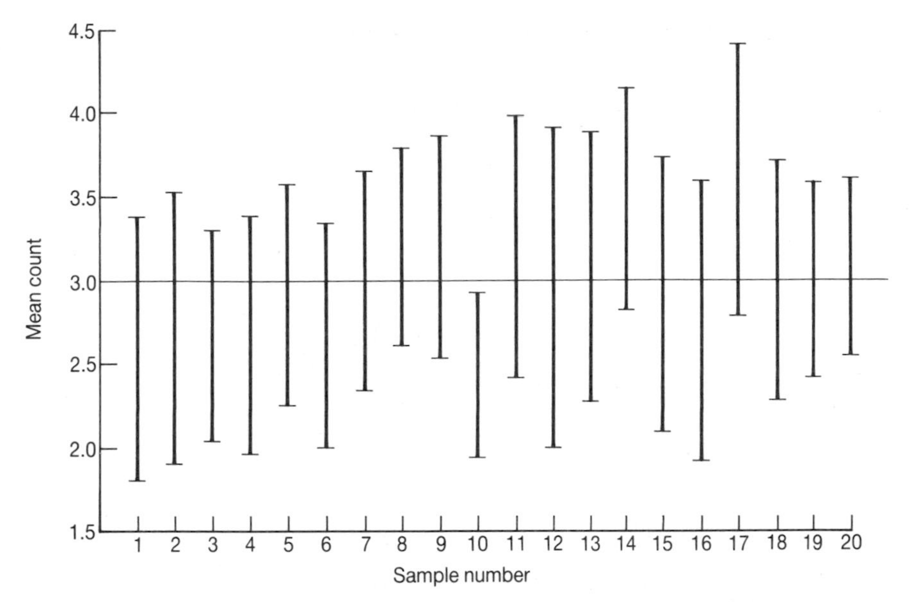

Fig. 11.1 Twenty 95% confidence intervals for a mean, where the true value is known to be 3.0.

per cent probability that the limits come out so as to enclose the true mean. At the simplest level that is indeed so, but in more complicated situations paradoxes can be produced if great care is not taken with the precise wording.

In saying, therefore, that the true mean lies within the limits with 95 per cent confidence (instead of 95 per cent probability) we are using coded language that means 'Yes. I do know that paradoxes can arise if I am not careful. This wording takes care of that.' For practical purposes, at the level of this book, regarding probability and confidence as the same thing will do no harm.

Deriving confidence limits

It is important to recall just how confidence limits are derived, because an incorrect argument is sometimes employed, using a diagram such as that shown in Fig. 11.2, which seems to imply some sort of probability distribution of the population mean.

The correct argument is as shown in Fig. 11.3, saying that if the population mean were any less than the lower limit, there would be less than a $2\frac{1}{2}$ per cent probability of observing a sample mean as great as that actually observed. Similarly if the population mean were any greater than the upper limit, there would be less than a $2\frac{1}{2}$ per cent probability of observing a sample mean as small as that actually observed. So either the

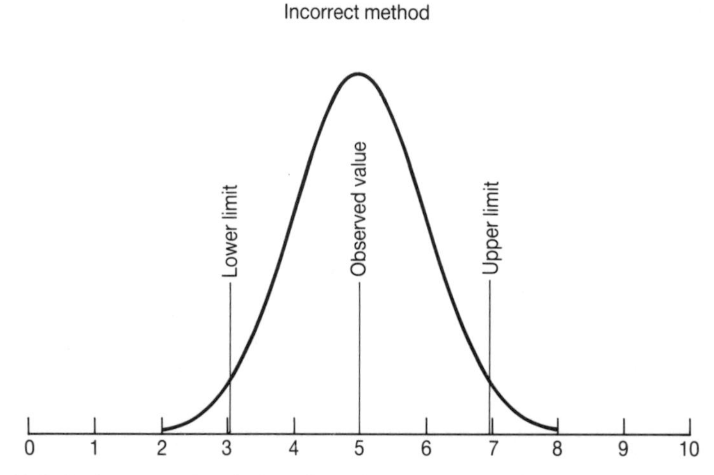

Fig. 11.2 An incorrect description of the derivation of confidence limits for a mean.

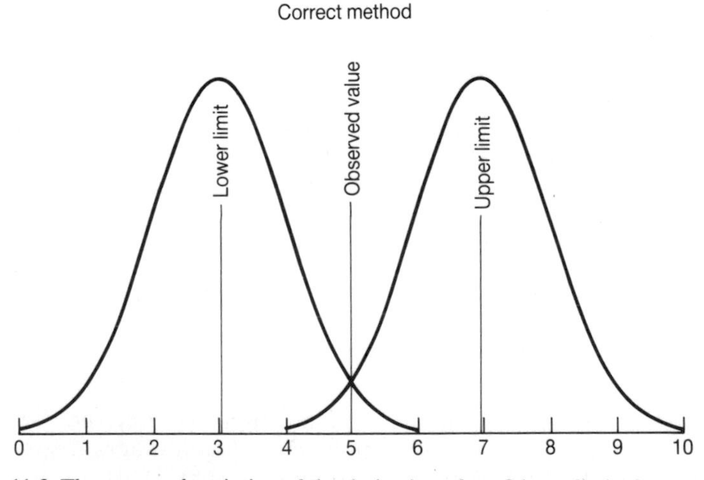

Fig. 11.3 The correct description of the derivation of confidence limits for a mean.

population mean lies between those limits or an event of fairly low probability has occurred.

Exactly the same numerical values could be derived from the incorrect argument in Fig. 11.2, so does it matter? Yes. At present we are dealing with limits for the mean, using either normal or *t* distributions, and these are symmetrical. If finding limits for, say, a population standard deviation which has a skew distribution, the values derived from the incorrect argument would be seriously at fault.

Significance tests

A similar argument to that used above for a confidence interval, saying that *either* the population mean lies within the derived limits *or* an unusual event has occurred, may also be employed when we have a hypothesis, formed before the observations were taken, that we wish to test.

Suppose, for example, that (as mentioned briefly in Chapter 10) our hypothesis was that the radioactive marker of Table 10.1 would have an average count of 2.7 per minute, and we then observed the 500 counts given there. We could argue as follows: we have observed a mean of 2.952, with an associated standard error of 0.0778. If the true mean were really 2.7, our sample mean would be $(2.952-2.7)/0.0778 = 3.24$ standard errors away from it. Very few observations from a normal distribution are as far from their mean as that. (Strictly we ought to be using a *t*-distribution, but with 499 degrees of freedom this is negligibly different from a normal curve.) Indeed Table I shows that the probability of an observation so far into the tail is nearly as low as 0.001, and fuller normal curve tables show it to be only 0.0012, or about 1 in 800. So either we have been very unlucky and this is the 1 case in 800 to give so extreme a result by chance, or our hypothesis must be wrong. Under those circumstances we prefer to believe the hypothesis to be wrong, and say that the sample value is *significantly different* from the hypothetical value, at a probability level given by $P = 0.0012$. (A number of journals nowadays insist on printing p instead of P in this context. This results from an agreement between their editors taken without statistical advice. We shall continue to use P here, and hope that the editors reverse their decision eventually.)

It has become traditional to regard a probability of 1 chance in 20 as the borderline between *significant* and *not significant*, and in the past one would often see statistical results marked $P<0.05$, or $P<0.01$ for highly significant results, or just *NS* meaning 'not significant'. ($<$ is the mathematical sign for 'is less than'.) In days when such probability values had to be derived from printed tables, better information than that was often too difficult to derive. With modern computing machinery we can usually derive P values more precisely, and they should then be so quoted. It is pointless to use the less informative $P<0.01$ when one could have used the more informative $P = 0.0012$, say. Similarly, *NS* is a very poor substitute for $P = 0.06$ (not quite there but looking worth further enquiries), or $P = 0.80$ (more discrepant values than the observed one would be expected 4 times out of 5 if the hypothesis were true, so there is no reason whatever to disbelieve it on these observations).

The null hypothesis

It might be argued that we vary rarely have any prior hypotheses of what a mean (or any other) value should be. This is not in fact so, because we are often comparing two sets of results, perhaps one set of patients given a specific new treatment, and another set not given that treatment, or perhaps, using patients as their own controls, comparing results when the new treatment is given with results when it is withheld.

We wish to discover whether the treatment has any effect and (as with the law courts assuming innocence until guilt is demonstrated) we start with the hypothesis that it has no effect. This initial assumption is known as the *null hypothesis*, and we ask whether the distribution of differences has a mean that differs significantly from zero. If the test shows a *not* significant result then the null hypothesis remains tenable, i.e. it may well be true that the treatment has no effect. On the other hand, a *significant* test result demonstrates that it is unlikely that the null hypothesis is true. We prefer to pin our faith on there being a true effect of the treatment.

For example Table 11.1 shows the systolic blood pressures of 9 patients before and after being given a specific drug. Taking the difference in values for each patient, it is evident that we should expect those differences to come from a distribution with a mean of zero if that drug really had no effect on blood pressure. Using the techniques specified above, we find the mean observed difference to be −11.44 mm and its standard error to be 4.14 mm, so $t = -11.44/4.14 = -2.76$. The probability corresponding to a given t value does not depend upon the sign of t, so we can look up 2.76 in Table II against 8 degrees of freedom and find that P is somewhere between 0.05 and 0.02 (written as $0.02<P<0.05$). More precise tables or a suitable computer program would show $P = 0.025$. So if the drug really had no effect, there would be only 1 chance in 40 of observing so large an

Table 11.1 The systolic blood pressure (mm) of 9 patients before and after treatment.

Before treatment	*After treatment*	*Difference*
132	136	4
160	130	−30
145	128	−17
114	114	0
125	115	−10
128	117	−11
154	125	−29
134	136	2
123	111	−12

Mean = −11.44 mm
Standard deviation = 12.43 mm
Standard error of mean = $12.43/\sqrt{9} = 4.14$ mm
t (8 degrees of freedom) = $-11.44/4.14 = -2.76$

absolute difference; we infer that *other things being equal*, it *appears likely* that the drug lowered the average blood pressure.

The italicised words must be emphasised. It must be recognised that we are weighing probabilities, never, as is sometimes suggested by non-statistical authors of medico-statistical papers, reaching 'mathematical proof'. A difference between the observed and expected values *may* be a 'real' difference (in the sense that the treatment was effective) even though it is not twice the standard error; but the calculation shows that the hypothesis that the difference has occurred by chance is equally valid. If, on the other hand, the difference between the observed and expected values is, say, four times the standard error, this does not 'prove' that it is a 'real' difference; it may still be the result of chance. But the calculation shows that the hypothesis that it is due to chance is unlikely to be true, for such a chance difference is a rare event. The advantage of the calculation is that the investigator is thus enabled to estimate critically the value of his results; he may be prevented from wasting his time by developing some elaborate argument on a difference between two averages (or proportions) which is no greater than a difference that might easily be obtained on drawing two random samples from one and the same record.

Finally, presuming that the difference recorded between the observed and expected values is more than would be expected from the play of chance, then we must consider carefully whether it is due to the factor we have in mind—e.g. the special treatment—or to some other factor which affected the results. For example, suppose that all the pre-drug blood pressures were taken in the evening, and the post-drug blood pressures next morning. We may think that we are observing a drug effect when, really, we are observing only a time-of-day effect. Such extraneous effects always have to be guarded against most carefully, which shows the necessity for careful planning.

Statistical significance and clinical importance

The word 'significance' sometimes misleads people into thinking that a significant effect is necessarily an important one and, contrariwise, that the absence of a significant effect demonstrates the absence of anything important to be found. In the hope of overcoming such beliefs the phrase 'statistical significance' is sometimes used, but it does not seem to help. The phrase is itself misused by journalists and others with no appreciation of its meaning.

In Fig. 11.4 we see possible results of trials to try to detect whether each of 5 specific drugs reduces blood pressure, in circumstances where an average reduction of 5 mm would be considered important, but a lesser reduction is not worth bothering with.

For each drug 95 per cent confidence limits are shown for the mean difference detected in its trial. If we relied solely upon significance tests, it would be reported that none of drugs A, C and D showed a significant difference, whereas B and E each gave $P<0.01$ perhaps. The assumption,

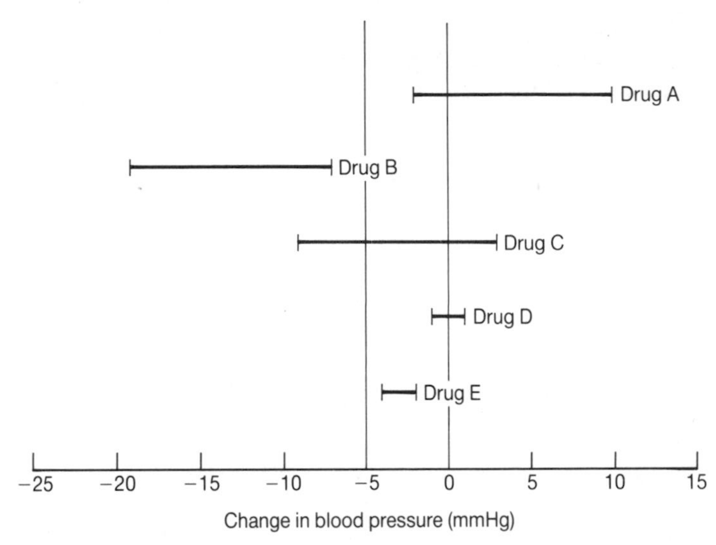

Fig. 11.4 Statistical significance versus clinical importance: hypothetical results of change in blood pressure from five drug trials (95% confidence limits).

that *B* and *E* are therefore those of clinical importance, is shown to be incorrect by study of the confidence limits, which give far more information than significance tests. We see that the results for *D* and *E* are much more precise than those for *A*, *B* and *C*; perhaps they used larger sample sizes, or were more accurately performed. *A* and *D* are not significantly different from zero, but they are different from 5 mm or more—they are probably not worth trying further.

However *E* is also not worth trying further (unless in a different dose). It almost certainly has a real effect, but this effect is a long way short of the 5 mm required for clinical importance. Drug *B*, however, is both statistically significant and clinically important. Its limits show its effect to be almost certainly greater than 5 mm. Drug *C* cannot be dismissed as merely 'not significant'. The trial has not shown it to have an effect, but nor has it ruled out the possibility of an important effect. It could be even better than *B*. Without more work, using a larger sample, we cannot tell.

One-tailed and two-tailed tests

So far we have tacitly assumed it to be correct, in judging significance, to calculate the probability of the observed result or a more extreme one in either direction, taking the portion of a normal curve, for example, picked out in Fig. 11.5 and saying that there is only a small chance that an observation falls into the shaded area. But why that particular shaded area? If we took any other equal area and shaded that instead the argument would still be true that if the null hypothesis were correct there

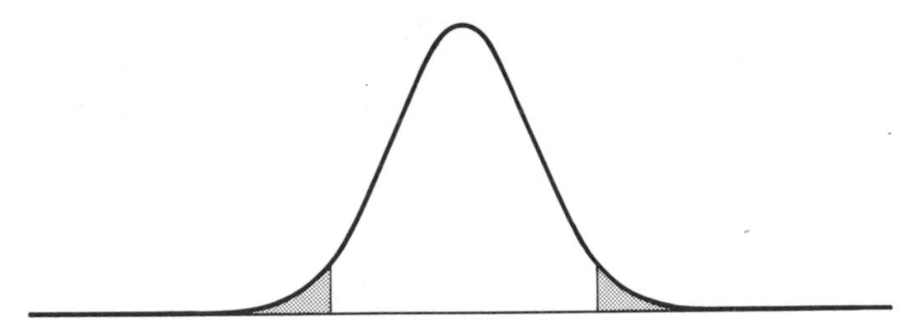

Fig. 11.5 Normal curve with 'significant' tail areas indicated.

would be only a small chance of such an observation.

The answer to this question lies in realising that two types of error may occur. The false positive, saying that we have found a real difference even when the null hypothesis is actually true, sometimes called a Type I error, is guarded against by the significance test. The false negative, failing to demonstrate a difference even though the null hypothesis is not actually true, sometimes called a Type II error, also needs to be taken into account. We need to ask—if the null hypothesis is not true, what *alternative hypothesis* is true instead, and take such a shaded area as will tend to minimise the chance of a Type II error, for a given chance of a Type I error.

The usual alternative hypothesis is that, if the null hypothesis, that the mean has a particular given value, is untrue then the actual mean may be either greater than or less than that given value. If we wish to discover such a difference in whichever direction it occurs, the shaded area of Fig. 11.5 is the most efficient. Having made our observation of a mean, say, and calculated a value of t from it, the P value is derived as the area beyond that t value plus an equal area at the other end of the distribution. This is called a two-tailed test and tables of t are so constructed as to apply it automatically, which is why, when we found $t = -2.76$ in the example above, it was correct to find P corresponding to $t = 2.76$ instead.

Occasionally, however, it may be sensible to adopt an alternative hypothesis that says that if the null hypothesis is not true then the true mean will be greater than the null value; or similarly we might specify 'less' instead of 'greater'. In other words we are choosing in advance to take notice of a change in one direction (provided it is large enough) but to ignore a change in the other direction. If we do that we can halve the P value. For example, in our example above, if we had specified an alternative hypothesis that, if the drug had any effect, it would lower the blood pressure not raise it, then having found $t = -2.76$ we could take $P = 0.0125$ instead of $P = 0.025$. This is called a one-tailed test. It is evident that the direction of difference must be specified in advance of seeing the results in such a case. To see which way the result comes out and then use a one-tailed test in that direction would double the chance of a Type I error, and is clearly impermissible.

The general rule should be to use two-tailed tests, except in special circumstances where particular justification for a one-tailed test can be found. One example of such circumstances arises if a second trial is performed to see whether a result from a first trial is confirmed or refuted. On the first trial a two-tailed test should be used, but on the second trial the direction of change that is being sought can be specified as the same as that found in the first trial, because a change in the other direction would not confirm the result but refute it. Of course, if two trials showed strongly contradictory results, one would wish to investigate why, not merely ignore the fact, but simply to answer the question of whether the second result confirms the first, a one-tailed test is correct.

Tail areas of discontinuous distributions

For a discontinuous distribution we can measure a single tailed probability easily enough. Considering Distribution A in Table 11.2, the single-tailed P-value corresponding to an observation of 3, and using the upper tail, is $0.24 + 0.06 = 0.30$, while for Distribution B it is $0.23 + 0.05 = 0.28$. There is some difficulty, however, about a two-tailed probability. It is easy enough for Distribution A because it is symmetrical; we can take the one-tailed value and double it, to incorporate an equal probability in the other tail.

Distribution B, however, is asymmetrical and there is no equal area in the other tail. The best solution would appear to be to take as many terms as possible from the other tail, subject to the condition that their total

Table 11.2 Two discontinuous frequency distributions—symmetrical and asymmetrical.

		Distribution A		
		P-values		
Value	*Probability*	*1-tailed*		*2-tailed*
0	0.06	0.06	1.00	0.12
1	0.24	0.30	0.94	0.60
2	0.40	0.70	0.70	1.00
3	0.24	0.94	0.30	0.60
4	0.06	1.00	0.06	0.12
		Distribution B		
		P-values		
Value	*Probability*	*1-tailed*		*2-tailed*
0	0.07	0.07	1.00	0.12
1	0.25	0.32	0.93	0.60
2	0.40	0.72	0.68	1.00
3	0.23	0.95	0.28	0.35
4	0.05	1.00	0.05	0.05

probability does not exceed that of the observed tail. Thus for an observation of 3 we take $0.23+0.05=0.28$ for the first tail and add in 0.07 from the second tail to give $0.28+0.07=0.35$. We stop there because to add in the next term, 0.25, would give a second tail $(0.07+0.25)$ which is greater than 0.28.

Rather strangely at first sight, this leads to the result than an observation of 4 (from the B distribution) has a two-tailed P-value identical to its one-tailed P-value, both are 0.05. This is correct however—if the question is 'What is the probability of an observation as far into the tails, in either direction, as 4?' the answer is that the other tail contributes nothing, because it contains nothing so discrepant in probability terms.

It must be admitted, however, that not all statisticians agree that this is the right approach. Some support merely doubling the one-tailed value, irrespective of asymmetry, and others have other strange approaches. The matter remains controversial.

Summary

Having observed a mean, or any other value that may be calculated from observations of a sample, we can give limits within which the corresponding value in the population lies, with a known degree of confidence. It is conventional to use 95 per cent confidence limits but other degrees of confidence may be employed if desired.

Significance tests may be employed where it is desired to test whether a population mean (or other value) differs from a predetermined hypothetical value. The test involves weighing probabilities and can never amount to proof. This test can give no information as to the *origin* of a difference, beyond saying whether or not chance is a plausible explanation.

Statistical significance should not be confused with clinical importance. A very precise experiment can show an effect to be almost certainly present, even though its size is too small to matter, whereas a less precise experiment may demonstrate the *possibility* of an effect of importance without being convincing that there is really any effect at all.

Two-tailed significance tests are preferable to one-tailed ones, except in special circumstances where a one-tailed test can be justified.

Chapter 12

Problems of sampling: proportions

In practical statistical work a value which is of particular importance, owing to the frequency with which it has to be used, is the *proportion*. For example, from a sample of patients with some specific disease we calculate the proportion who die. Let us suppose that from past experience, *covering a very large body of material*, we know that the fatality rate of such patients is 20 per cent (the actual figure, from the point of view of the development of the argument, is immaterial). We take, over a chosen period of time, a randomly selected group of thirty patients and treat them with some drug. In reality, of course, the ethical questions arising from such a course of action are of great importance. They will be considered in Chapter 23. For the purposes of this chapter we shall assume that we, and an appropriate ethics committee, are satisfied that the trial is justifiable. Then, presuming that our sample is a truly representative sample of all such patients—e.g. in age and in severity—we should observe, if the treatment is valueless, about 6 deaths (it may be noted that we are also presuming that there has been no secular change in the fatality rate from the disease). We may observe precisely 6 deaths, or owing to the play of chance we may observe more or less than that number. Suppose we observe only 3 deaths; is that an event that is likely or unlikely to occur by chance with a sample of 30 patients? If such an event is quite likely to occur by chance, then we must conclude that the drug *may* be of value, but so far as we have gone, we must regard the evidence as insufficient and the case unproven. Before we can draw conclusions safely we must increase the size of our sample. If, on the other hand, such an event is very unlikely to occur by chance, we may reasonably conclude that the drug is of value (that is, of course, having satisfied ourselves that our sample of patients is comparable with those observed in the past in all respects except that of the treatment). Before we can answer the problem as to what is a likely or an unlikely event we must determine the frequency distribution to be expected when observing a proportion in samples of a given size taken from the appropriate population. Presuming the treatment is of no help, then the fatality rate we should observe on a very large sample is 20 per cent (or nearly that). How far is the rate likely to differ from that figure in smaller samples?

Sample of three

Instead of starting with the relatively difficult problem of a sample of 30, as actually observed, let us first get our ideas in order by examining a simpler case, namely a sample of only 3 patients. Four events are then possible in our sample: (1) all three may recover; (2) two may recover and one die; (3) one may recover and two die; (4) all three may die.

We shall assume that the fate of each patient is independent of the fate of any other, that is to say that each patient's chance of recovery remains 4/5 whatever the other patients do. On the basis of the past experience of the fatality rate we can calculate the probability of each possible event, as follows:

(1) The chance that each patient individually recovers is 4/5, so the chance that all three recover is $4/5 \times 4/5 \times 4/5 = 64/125 = 0.512$.

(2) The chance that A and B recover but C dies is $4/5 \times 4/5 \times 1/5$.
The chance that A and C recover but B dies is $4/5 \times 1/5 \times 4/5$.
The chance that B and C recover but A dies is $1/5 \times 4/5 \times 4/5$.
Thus the chance that two recover and one dies is
$3 \times (4/5 \times 4/5 \times 1/5) = 48/125 = 0.384$.

(3) The chance that one recovers and two die is found similarly as $3 \times (4/5 \times 1/5 \times 1/5) = 12/125 = 0.096$ the multiplier 3 arising from the fact that any of the 3 patients may be the one to recover.

(4) Finally the chance that all three die is $1/5 \times 1/5 \times 1/5 = 1/125 = 0.008$.

These values are tabulated in Table 12.1. It is seen that the probabilities add to 1.0, indicating certainty that one of these four events must happen.

In such a small sample, the only event that could favour the treatment would be the recovery of all patients for only that outcome gives a lower fatality rate than the 20 per cent expected on past experience, but the two-tailed P-value of such a result (calculated as explained in Chapter 11) is 1.0, indicating that we must always get a result at least as far as that (in probability terms) from our 20 per cent past experience. We must use a two-tailed test because, although we hope that our drug will be beneficial, we cannot be certain that it will not be harmful. Indeed, even on so small a

Table 12.1 The possible events for three patients, when the probability of dying is 0.2.

Event	*Probability of event*	*Fatality rate (per cent)*	*Two-tailed P*
Three recover	0.512	0.0	1.000
Two recover, one dies	0.384	33.3	0.488
One recovers, two die	0.096	66.7	0.104
Three die	0.008	100.0	0.008
Total	1.000		

trial as this, if all three patients died we should certainly have to take notice of $P = 0.008$ as a significant result.

Sample of thirty

Having seen, in a simple case, the method of approach, we can now return to our original problem of 3 deaths in 30 patients. The probability we need is that with which this result *or a more extreme one* might be expected to occur even if our treatment were quite ineffective. Table 12.2 shows how the probabilities are calculated, by multiplying the appropriate powers of 4/5 and 1/5, just as in Table 12.1, and the number of ways in which the event can happen.

Clearly there is only one way in which all 30 could recover, and there are 30 ways in which 29 can recover and 1 die for any one of the 30 may be the unlucky one. It is less obvious that there are 435 ways in which 28 can recover and 2 die, but this can be derived from $(30 \times 29)/2$ as given in Table 12.2, and similarly for the greater numbers of deaths—the pattern of the formula is easy to remember.

In Column (4) of Table 12.2 we have the cumulative probability working from each end. We find the tail area probability corresponding to the observed number of deaths to be 0.1227, but in using a two-tailed test we

Table 12.2 The possible events for thirty patients, when the probability of dying is 0.2.

Number of deaths	*Calculation of probability*	*Probability*	*Cumulative probability*
(1)	*(2)*	*(3)*	*(4)*
0	$(4/5)^{30}$	0.0012	0.0012
1	$(4/5)^{29} \times (1/5) \times 30$	0.0093	0.0105
2	$(4/5)^{28} \times (1/5)^2 \times (30 \times 29)/2$	0.0337	0.0442
3	$(4/5)^{27} \times (1/5)^3 \times (30 \times 29 \times 28)/(2 \times 3)$	0.0785	0.1227
4	$(4/5)^{26} \times (1/5)^4 \times (30 \times 29 \times 28 \times 27)/(2 \times 3 \times 4)$	0.1325	etc.
5	etc.	0.1723	
6		0.1794	
7		0.1538	
8		0.1106	etc.
9		0.0676	0.1287
10		0.0355	0.0611
11		0.0161	0.0256
12		0.0064	0.0095
13		0.0022	0.0031
14		0.0007	0.0009
15+		0.0002	0.0002

need to add in the appropriate probability from the other tail also. This probability is 0.0611 (because to take one more term would give 0.1287 which is greater than the observed tail 0.1227). So we have $P = 0.1227 + 0.0611 = 0.1838$. Thus if the drug were totally ineffective, and the probability of death for each patient were still 1/5, the chance of observing a result as extreme as we have found (or more extreme) is more than 18 per cent. Clearly this is not small enough to let us claim that we have demonstrated a real effect.

The general case. The binomial distribution

The distribution for which we have been calculating particular values above is called the *binomial distribution* because it is closely allied to the binomial theorem of algebra.

To give a general formula it is necessary to introduce a mathematical function called the factorial, indicated in symbols by the exclamation mark following a number or an algebraic variable. It is defined as:

$1!$ (pronounced 'factorial 1') $= 1$

and $n!$ (pronounced 'factorial n') $= n \times (n-1)!$

Thus $2! = 2 \times 1! = 2 \times 1 = 2$

$3! = 3 \times 2! = 3 \times 2 = 6$

$4! = 4 \times 3! = 4 \times 6 = 24$ etc.

Sometimes we find 0! appearing in a formula. Since, by definition,

$$n! = n \times (n-1)!$$

we have

$$1! = 1 \times 0!$$

and consequently

$$0! = 1$$

a fact that many people find surprising when they first meet it.

Care needs to be taken when making calculations involving factorials of large numbers, because they get big so quickly. For example, 20! is already greater than 1 000 000 000 000 000 000, but in most formulae where they are used much cancellation of factors is possible as will be seen below.

The formula for the binomial frequency distribution says that if the probability that each particular patient dies is p, and the probability that each particular patient recovers is q, then the probability of observing exactly r deaths in a sample of n is

$$\frac{n!}{r!\,(n-r)!} p^r q^{n-r}$$

In the example above we had $p = 0.2$, $q = 0.8$ ($p + q$ must always equal 1 of course) and $n = 30$. To find the probability of exactly 3 deaths, the formula is

$$\frac{30!}{3!\,27!} \times (0.2)^3 \times (0.8)^{27}$$

but instead of calculating the huge value of 30! we note that

$$30! = 30 \times 29 \times 28 \times 27!$$

and 27! cancels with 27! in the denominator, so we get

$$\frac{30 \times 29 \times 28}{3 \times 2} \times (0.2)^3 \times (0.8)^{27}$$

just as in Table 12.2.

The formula applies, of course, not only to death and survival of patients but to any statistical values expressed as the numbers falling into two groups—those who did have a certain characteristic and those who did not.

The normal approximation

In Fig. 12.1 the values of the probabilities from Table 12.2 are plotted as a bar chart. If a curve were drawn to connect the tops of the bars, it would be

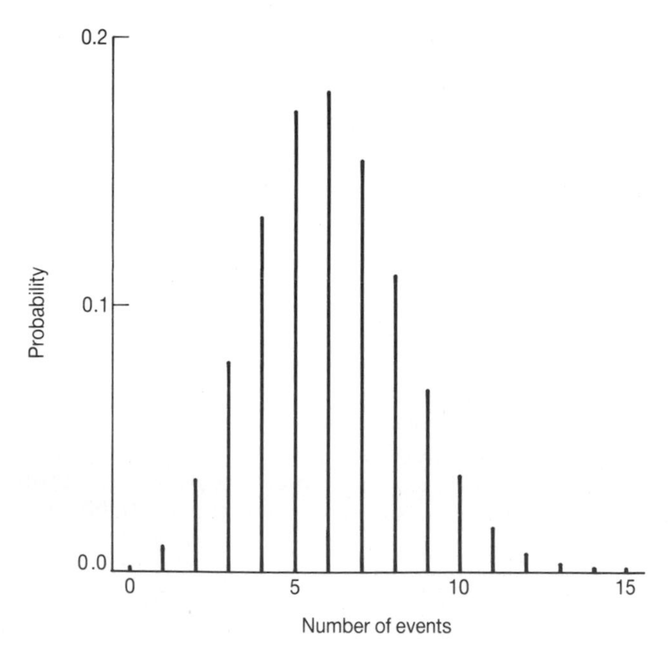

Fig. 12.1 Binomial distribution for $n = 30$, $p = 0.2$.

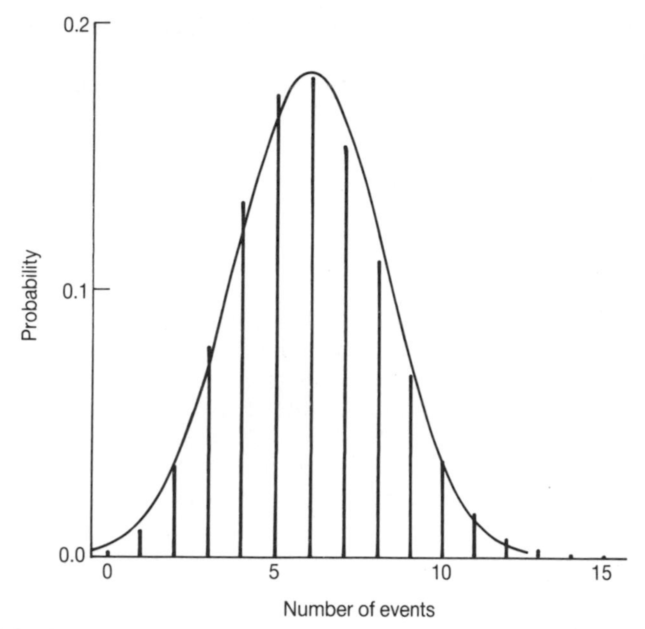

Fig. 12.2 The same binomial distribution with corresponding normal distribution.

remarkably close to a normal curve. There is indeed some skewness there, but Fig. 12.2 shows that the approximation is quite good, and this is always so provided that n is large, and neither p nor q is too close to 0. A good working rule is that the approximation should not be used if either np or nq is less than 5; nor should it in any circumstances if the computing power is available to calculate the terms of the exact binomial distribution. If, however, one has to rely on hand calculations and the conditions are met, the normal approximation to the binomial is worth while.

To use the approximation we need to know that the mean of a binomial distribution is np and its standard deviation is $\sqrt{(npq)}$, and a normal curve with these same values of mean and standard deviation is appropriate.

In the example above we get

$$np = 30 \times 0.2 = 6.0$$

$$\sqrt{(npq)} = \sqrt{(30 \times 0.2 \times 0.8)} = \sqrt{4.8} = 2.1909$$

To find the probability of 3 deaths or fewer we need a continuity correction, as explained in Chapter 9. This means that, as we are trying to distinguish between 3 or fewer, and 4 or more, we use 3.5 as the boundary point, so we calculate

$$(3.5 - 6.0)/2.1909 = -1.141$$

Tables of the normal curve show the tail area less than -1.141 standard deviations to be 0.1269, a good approximation to the correct 0.1227 found in Table 12.2. For a two-tailed P value, the normal distribution being continuous, we double the one-tailed value, which in general will give an overestimate—here $2 \times 0.1227 = 0.2454$ instead of 0.1838, but the conclusion that the observed difference cannot be regarded as significant remains.

An example of digit preference

We are now ready to look again at the figures of the heights of 1111 male factory workers. As mentioned in Chapter 9, 554 of them had heights (in centimetres) whose final digit was 0, 3, 5 or 8, the other 557 showing 1, 2, 4, 6, 7 or 9. It was suggested that there was 'digit preference' in the measurements—was that suggestion justified or could it be a chance effect? The null hypothesis is of a binomial distribution with $p = 0.4$ (because the supposedly preferred digits are 4 out of 10). We want to know the probability of 554 or more out of 1111 (doubled up to get a two-tailed test), so we use the continuity-corrected value 553.5. We also have

$$np = 1111 \times 0.4 = 444.4$$
$$\sqrt{(npq)} = \sqrt{(1111 \times 0.4 \times 0.6)} = 16.3291$$
$$\frac{553.5 - 444.4}{16.3291} = 6.68$$

The conditions for the normal curve approximation are well met, and using the probability of exceeding 6.68 standard errors (in either direction) gives

$$P < 0.000\,000\,000\,1$$

We need to be cautious in interpreting this result because we formed the hypothesis of digit preference, and selected which digits we thought were preferred, *after* seeing the data. This is always a dangerous thing to do. It would be preferable, if it were practicable, to regard these as initial data that had allowed the hypothesis to be formed, and then to collect a new set of data, derived under the same conditions, to test for it.

However, it is ridiculous to say (as some people do) that we should not even consider testing the hypothesis on the same data that led to it. Where the probability is as low as that observed here, we can view it with the greatest of caution and still decide that digit preference almost certainly existed.

Confidence limits for a proportion

If, instead of considering the *number* of 'preferred' digits, 554, we consider the *proportion* of them $554/1111 = 0.4986$, we may want to give confidence limits for the equivalent proportion in the population. Provided that we are within the circumstances where the normal approximation is valid, this is easily done. The appropriate standard error is $\sqrt{(pq/n)}$, but now we cannot use p and q from a specified hypothesis, as we are asking what the sample itself tells us. As the best estimates we have to use the observed proportions, taking $p = 0.4986$ and $q = 1 - p = 0.5014$. The normal curve two-tailed 95 per cent value is 1.96, so we get 95 per cent limits as

$$0.4986 \pm 1.96 \sqrt{(0.4986 \times 0.5014/1111)}$$

i.e. 0.4836 and 0.5136, and we can notice, as the significance test told us, that the 'no preference' value of 0.4 is excluded by a long way.

If, however, we are outside the circumstances where the normal approximation is good, the equivalent calculations will lead to inaccurate answers. For instance, returning to the example of 3 deaths from 30 patients, we have observed $3/30 = 0.1$, but if we try to get 95 per cent confidence limits around this proportion by the normal approximation we get

$$-0.007 \text{ and } 0.207$$

where the lower limit is obvious nonsense and, in fact, the upper one is not good either.

Exact methods, beyond the scope of this book, give us

$$0.021 \text{ and } 0.265$$

It is important, therefore, not to use the approximate method where it is not justified, i.e. where either np or nq is less than 5.

Summary

A statistical value which is of particular importance is the proportion, or the number, of a sample of a given size that exhibited a particular feature. For a given probability that any individual from a population will have the feature, the binomial frequency distribution can be used to calculate the probability that any given number in the sample will do so. For large samples, binomial probabilities can be approximated by means of the normal distribution.

Chapter 13

Counting improbable events

As seen in the last chapter, a binomial frequency distribution is to be expected when we are observing a *count* of favourable (or unfavourable) cases in a number of observations of a given sample size. The distribution is determined by the two values n (the sample size) and p (the probability that any particular observation is of the specified kind—favourable or unfavourable). We have to take care, in the circumstances, to distinguish between two different sample sizes that appear within the discussion. The first is the value n from which the count is made to give one observation of the binomial distribution. The second is the number of such binomial observations that we have made.

For example, if we examine 200 records of births each month, taken at random from all births recorded in the country, and find out how many of those 200 babies are recorded as having congenital abnormalities, each month's observation will be from a binomial distribution having $n = 200$. The sample size from that binomial distribution will be the number of months for which the exercise is continued.

Suppose that the chance that any individual baby has such an abnormality is improbable, perhaps 1 in 100 (i.e. $p = 0.01$). We should then, on the

Table 13.1 Probabilities in three binomial distributions, each with $np = 2$.

(1)	*(2)*	*(3)*	*(4)*
r	$n = 200$ $p = 0.01$	$n = 2000$ $p = 0.001$	$n = 20\,000$ $p = 0.0001$
0	0.1340	0.1352	0.1353
1	0.2707	0.2707	0.2707
2	0.2720	0.2708	0.2707
3	0.1814	0.1805	0.1805
4	0.0902	0.0902	0.0902
5	0.0357	0.0361	0.0361
6	0.0117	0.0120	0.0120
7	0.0033	0.0034	0.0034
8	0.0008	0.0009	0.0009
9	0.0002	0.0002	0.0002

average, find 2 in each sample of 200, but calculating the binomial probabilities we find the distribution as given in column (2) of Table 13.1, showing that 13.40 per cent of the time we should find no case in our monthly sample of 200, 27.07 per cent of the time we should find 1 case, and so on.

Suppose, however, that instead of taking samples of 200 babies and looking for something with a prevalence of 1 in 100, we took samples of 2000 and looked for something with a prevalence of 1 in 1000. Again we should find an average of 2 per sample; calculating the distribution this time we find column (3) of Table 13.1. The distributions in columns (2) and (3) are remarkably similar. They are not identical, but if we were merely presented with samples from one or the other it would be extremely difficult to detect which had been used. Is there then something about binomial distributions with a mean value of 2 that makes them all very alike?

If we try a still bigger n and smaller p, while keeping $np = 2$, we find column (4) of Table 13.1, showing virtually the identical distribution to that in column (3).

The Poisson distribution

Would the distribution remain this shape then for all higher values of n, and lower values of p, provided that np keeps the value 2? The answer is 'Yes'. For all large n and small p the binomial distribution is closely approximated by the Poisson distribution (taking its name from Poisson the French mathematician), which has the great advantage of depending only on the product np, and not on n and p individually.

The Poisson distribution for $np = 2$ is shown in Table 13.2, which may be compared with Table 13.1 to see just how close the approximation is. In spite of this very close arithmetical resemblance, the formula for a Poisson distribution looks very different from the binomial formula. If we have a

Table 13.2 Probabilities in a Poisson distribution, with $m = 2$.

r	$m = 2.0$
0	0.1353
1	0.2707
2	0.2707
3	0.1804
4	0.0902
5	0.0361
6	0.0120
7	0.0034
8	0.0009
9	0.0002

large sample, a small chance that any individual has a particular characteristic (such as an abnormality), and the average number with that characteristic in a sample is m (where $m = np$ in the preceding text) then the probability of observing exactly r with the particular characteristic in a sample is

$$\frac{m^r}{r!} e^{-m}$$

where the exclamation mark indicates the factorial function dealt with in the last chapter, and e is a number (which seems to be one of the basic building blocks of the universe) defined as

$$\mathrm{e} = 1 + \frac{1}{1!} + \frac{1}{2!} + \frac{1}{3!} + \frac{1}{4!} + \ldots = 2.718 \text{ approximately,}$$

the terms going on for ever, but becoming rapidly smaller so that to find its value to any given degree of accuracy is not difficult.

To calculate the first term of a Poisson distribution, for $r = 0$, we need e^{-m} and this is best found from a table such as Table IV or from a calculator or a suitable computer program. Any scientific pocket calculator will give it (probably under either e^x or exp as its notation). After that first term, calculation is much easier than for a binomial distribution. We find each successive value by multiplying the previous value by m/r.

Thus for our example where $m = 2$, we first enter -2 into a calculator and find e^{-2} as 0.1353. We then proceed as

r		Poisson probability
0		0.1353
1	$0.1353 \times 2/1 =$	0.2707
2	$0.2707 \times 2/2 =$	0.2707
3	$0.2707 \times 2/3 =$	0.1804
4	$0.1804 \times 2/4 =$	0.0902

and so on, remembering that the 2, 2, 2, 2 multipliers must be replaced by the value of m for the particular case, while the 1, 2, 3, 4 divisors are always the same.

Advantages of the Poisson distribution

Because it depends only on the single constant m $(=np)$ instead of on n and p individually, calculations and tables of the Poisson distribution are easier to handle than those of the binomial distribution. It also has the advantage that there are situations where we wish to ignore, or do not even know, the value of n, but we do know that it is large and we do know m, or at least an estimate of it as the average number observed.

For example, suppose we are investigating some particular congenital abnormality over the whole of a country. We may not know, or need to know, the total number of births so long as we know it to be reasonably constant from week to week. If, on the average, 2 cases of the abnormality are seen each week, then we should expect the actual number observed to follow a Poisson distribution with $m = 2$, and any apparent change from such a distribution can be investigated to see whether it tells us something useful.

In the case of a count from a radioactive source, we may believe that the number of atoms that could disintegrate is huge almost beyond imagination, while the chance that any individual atom does so is correspondingly minute, so again we should expect a Poisson distribution and the simulated observations of Table 10.1 were in fact thus derived (with $m = 3$).

Very large *p*

If, instead of observing a very rare event we are observing a very common one, i.e. instead of p less than 0.001, say, we have p greater than 0.999, then we can use a Poisson distribution by switching our attention to the opposite event. For example, instead of counting the proportion of men who are more than 150 cm tall, we can switch our attention to the proportion who are less than 150 cm tall. The Poisson distribution will then fit, provided the sample size is large enough.

When to use the Poisson distribution

If n is huge, and we do not even know its value, as in the example of a radioactive count, the Poisson distribution can be regarded as correct in its own right rather than as an approximation to the binomial distribution.

If, however, the situation is one where we have values for n and p individually, and the computing facilities are available to use the binomial distribution itself, then the binomial should be used—there is no point in approximating, even with a very good approximation, if we do not have to. If such computing facilities are not available, the Poisson approximation is very useful, provided the conditions are met. For a given value of p, the larger n is, the better; for a given value of n, the smaller p is, the better—so there is no restriction on np. In general the approximation is quite good if p is less than 0.1, and very good if p is less than 0.01, in each case provided that n/p is at least 10 000.

The normal approximation

We saw, in the last chapter, that a binomial distribution is well approximated by a normal curve with mean np and standard deviation $\sqrt{(npq)}$ (where $q = 1 - p$), provided that neither np nor nq is less than 5. It follows that if n is large, p is small, and np at least 5, a Poisson distribution can be approximated by a normal distribution, with mean np and standard deviation $\sqrt{(np)}$, since if p is tiny, $q = 1$ for all practical purposes.

This standard deviation was very closely realised in practice in the figures of Table 10.1, derived from a Poisson distribution with $np = 3$, for the observed standard deviation was 1.7389 whereas $\sqrt{3} = 1.7321$. The normal approximation would not be good, though, in that case: $np = 3$ is too small, and the distribution is too skewed for any symmetrical distribution to give a good fit.

Summary

If an event is improbable for any individual (p small) but the number of individuals is very large (n large), the number of events observed can be represented by the Poisson distribution with the appropriate mean (np), which is a simpler distribution to use than the binomial.

Chapter 14

Differences between averages

As noted in Chapter 11, we often wish to consider one sample against another, rather than just a single sample against some preconceived standard. We saw there that if the individual observations are paired, so that each value in one sample has a corresponding observation in the other sample, such as being derived from the same patient before and after treatment, then we can turn the problem of comparing two samples into a one-sample problem by forming the differences of each pair of observations and considering that sample of differences.

Paired and unpaired *t* tests

Whenever we wish to compare the means of two samples of the same size and there is a logical method of pairing the observations such that each in one sample has a particular paired observation in the other, then the above procedure is the best and is known as a *paired t test*. The method of pairing need not necessarily refer to the same patient, but can be any logical criterion, provided that it does not in any way depend upon the values of the observations to be tested.

Suppose, however, that pairing is not feasible, because the two samples are of different sizes, or because there seems to be no logical pairing between individual readings, how can we then compare the two means? We have two sample sizes, two means, and two standard deviations and wish to use them to judge whether the population means are different. We can subtract one sample mean from the other and say that if the two population means are really the same, the expected result would be zero, but we need a standard error of the difference of two means if we are to form any judgement from the observed difference.

Let us see how differences of means do, in fact, behave. Using Table 10.1 again, we may regard sample 1 and sample 2 as two samples each of size 5 and subtract one mean from the other. They have means of 2.8 and 2.4 and a difference of $2.8-2.4=0.4$. Samples 3 and 4 have means of 1.6 and 4.4, giving a difference of -2.8, and so on. Forming a frequency distribution of these 50 differences, we find that it has a mean of 0.14 and a standard deviation of 1.12, and such figures could be used to form a confidence interval for the difference of population means, or to make a significance test of whether it differs from zero. In practice, though, we do not have such a distribution of 50 differences, but just a single difference between two means; thus we have no such standard deviation of differences but need to estimate what it would be if we did have it, by means of a standard error. If the two samples are equal in size we can calculate the required standard error of the difference, by squaring the individual standard errors of each of the two means, adding, and taking the square root of the answer.

If we had had just the first two samples of 5 in Table 10.1 we should have found:

	Sample 1	*Sample 2*
n	5	5
mean	2.8	2.4
s.d.	2.168	1.140

The standard errors of the two means are $2.168/\sqrt{5}$ and $1.140/\sqrt{5}$, so the standard error of their difference is

$$\sqrt{(2.168^2/5 + 1.140^2/5)} = 1.095$$

and $t = (2.8 - 2.4)/1.095 = 0.365$

This t has 8 degrees of freedom, or in general, $2n-2$, where each sample is of size n. This value of t corresponds to $P = 0.72$, showing that the two samples could very well have come from one and the same population, which is not surprising as we know that in fact they did.

If the two samples are unequal in size, we need a rather more complicated formula for the standard error of the difference of their means namely, if the sizes are n_1 and n_2 and the corresponding standard deviations are s_1 and s_2,

$$\sqrt{\left(\frac{(n_1-1)s_1^2 + (n_2-1)s_2^2}{n_1+n_2-2}\left(\frac{1}{n_1}+\frac{1}{n_2}\right)\right)}$$

which gives more weight to the larger sample. Dividing the difference of means by this gives a t value having $n_1 + n_2 - 2$ degrees of freedom.

The need for equal variability

One advantage of the paired t test, when it can be used, is that it forms a legitimate test for equality of means even if the standard deviations in the two populations are very different. Of course, if the question is not 'are the population means the same?' but 'do the samples both come from the same population?' then, if the population standard deviations are not the same, the answer is clearly that they do not, but sometimes it is desired simply to look for a difference of means irrespective.

However, in the unpaired *t* test, it is a necessary assumption of the test that the population standard deviations are the same (or, in practice, not too different). In thc example above, the standard deviations of 2.168 and 1.140 are rather different, but on samples of only 5, this could well be no more than a chance difference.

Unpaired tests of means can be done (using other tests) in cases where standard deviations are very different, but they lead one deeply into areas of statistical controversy and are beyond the scope of this book. The most that can be done here is to warn that the circumstances need watching out for, and if they arise expert statistical help may be needed.

Comparison of a paired *t* test and an unpaired *t* test

It may be helpful to compare the two sorts of *t* test for a difference of means in similar examples. Table 14.1 repeats the figures from Table 11.1 giving the systolic blood pressure of 9 patients before and after treatment. In this case we use a paired test of the mean difference. Table 14.2 repeats the same 9 before-treatment figures, but this time compares them with readings from 11 comparable normal people. As there is no suitable pairing of observations available, we must use an unpaired test, using the difference of the means instead of the mean of the differences.

It should be noted that in Table 14.1 the two readings on each line refer to the same patient, so we have suitable pairs for subtraction, but in Table 14.2 the two figures on each line bear no special relationship to each other; they merely happen to have been put on the same line of the table. The values in either column of this table could be changed in the order in which they appear, without changing the other column to match, and the facts

Table 14.1 The systolic blood pressure (mm) of 9 patients before and after treatment with a specific drug.

Before treatment	*After treatment*	*Difference*
132	136	4
160	130	−30
145	128	−17
114	114	0
125	115	−10
128	117	−11
154	125	−29
134	136	2
123	111	−12

Mean = −11.44 mm
Standard deviation = 12.43 mm
Standard error of mean = $12.43/\sqrt{9}$ = 4.14 mm
t (8 degrees of freedom) = −11.44/4.14 = −2.76

Table 14.2 The systolic blood pressure (mm) of 9 patients with a specified illness compared with the values observed in 11 comparable normal persons.

	9 patients	*11 normals*
	132	139
	160	107
	145	98
	114	140
	125	115
	128	136
	154	123
	134	129
	123	126
		110
		105
Mean	135.00 mm	120.73 mm
Standard deviation	15.12 mm	14.63 mm
Estimate of common variance	$= (8 \times 15.12^2 + 10 \times 14.63^2)/18\ \text{mm}^2$	
	$= (1830.00 + 2140.18)/18\ \text{mm}^2$	
	$= 3970.18/18\ \text{mm}^2$	
	$= 220.57\ \text{mm}^2$	
Standard error of difference of means	$= \sqrt{(220.57/9 + 220.57/11)}$ mm	
	$= \sqrt{(24.51 + 20.05)}$ mm	
	$= 6.68$ mm	
t (18 degrees of freedom)	$= (135.00 - 120.73)/6.68$	
	$= 2.14$	

represented and the analysis would remain unchanged.

In the paired t test the standard error of the mean of differences is derived from 9 difference values, and thus has 8 degrees of freedom. In the unpaired test the standard error of the difference of means is derived from 9 values in one sample, giving 8 degrees of freedom, and 11 in the other, giving another 10 degrees of freedom, or 18 degrees of freedom in all. Reference to tables of the t distribution gives the corresponding P values as $P = 0.025$ for the paired test, or 1 chance in 40, and $P = 0.046$ for the unpaired test, or about 1 chance in 22. It is thus rather unlikely that either difference is due to chance alone.

Distribution-free (or non-parametric) tests

Relevant to this discussion there are, with small numbers of observations, some quite different and very simple tests of significance. They are applied not to the averages but to the distribution of values and are increasingly to be seen in current medical literature. They are known as *distribution-free* or non-parametric methods (because unlike the tests already discussed, they do not assume that the values in the population from which the sample has been drawn are normally distributed). In the present setting, there are two that need description—the Wilcoxon signed rank test and the Wilcoxon rank sum test.

The Wilcoxon signed rank test

In Table 14.1 we have the 9 observations of blood pressure before and after treatment. The changes that occurred are shown in Column (3). Ignoring at first the sign of the change we can list them in order from smallest to largest, and we can assign to each its appropriate rank number. In the last row we insert the original sign of the change, i.e. whether the blood pressure rose or fell (there was one patient who exhibited no change; this observation is omitted from the test as it provides no sign). Thus we have:

Change	0	2	4	10	11	12	17	29	30
Rank	—	1	2	3	4	5	6	7	8
Sign of rank		+	+	−	−	−	−	−	−

Finally we sum the positive ranks (patient's blood pressure rose) $= 1+2=3$, and the negative ranks (patient's blood pressure fell) $= 3+4+5+6+7+8=33$. We have now to consider these two figures 3 and 33. With 8 observations the total of all the ranks is $1+2+3+\ldots+8=36$, and this total can be divided into two sections in 37 different ways (from 0 and 36 to 36 and 0). If the treatment has no effect (the null hypothesis) then we might expect the upward changes in blood pressure that occur by chance to equal the downward changes. In other words, the sums of signed ranks would, in the long run, be 18 and 18. But with a small number of observations we shall see, by the way of chance, departures from that equality. Is it, then, probable that a difference between the positive and negative sums of 3 to 33 could have arisen by chance? Or is the domination of the negative ranks over the positive 'statistically significant'? The answer is provided by the table of values given in Table V, which refers to the larger of the two signed rank sums i.e. we now ignore the 3 and check 33 against the table.

Looking at the values given for 8 paired observations we see that 33 is significant at the 0.05 level, i.e. it would occur by chance not more than once in twenty times. Any less extreme division of the 36 ranks, e.g. 4 and 32, would not reach this level of significance and, accepting such a level as necessary, would not contradict the null hypothesis that the treatment had no effect.

If, to take another example, we had had 19 pairs of observations, then the sum of all ranks would be 190 and the division of this total into 46 and 144 would be statistically significant at the 0.05 level and into 32 and 158 at the 0.01 level. If the larger figure were less than 144, this would not be significant.

The Wilcoxon rank sum test

The above test was applicable to *paired* observations. In comparing two separate sets of observations the Wilcoxon rank sum test is appropriate. For example, Table 14.2 gives the blood pressure of 9 patients with a specified illness and of 11 comparable normal persons. Taking these 20 observations we list them from lowest to highest and give to each its appropriate rank. Thus, with the blood pressures of the patients in italics and those of the normal persons in bold type, we have the figures shown in Table 14.3.

In allotting the ranks we have to take into account the fact that there are two identical observations. Maybe, however, these values were rounded off in the process of tabulation and, if we refer back to the original observations, we could find that they were slightly different, e.g. 122.7 and 123.3. If this be so, then we can give them their appropriate ranks 7 and 8. But if reference back is not possible or still does not distinguish between the two (or more) values, then we must give the average of the ranks involved. Thus the two values of 123 occupy the position of ranks 7 and 8 and are each given a value of 7.5. (Such tied values slightly weaken the validity of the test of significance and would make it unreliable if there were many identical values.)

We now sum the ranks for the 11 normal persons and for the 9 patients and reach values of 93.5 and 116.5 (there is, in this test, no question of a sign). The question at issue is whether such totals as these could have easily arisen by chance. The answer is provided in Table VI in which provision is already made for the fact that the two groups were of different sizes and in which the figures relate to the *smaller* of the groups.

Thus, looking at the row for 9 and 11 observations we see that in the *smaller* of the two groups we could expect to see a value as low as 68 or as high as 121 only once in 20 trials, i.e. at the 0.05 level. As our observed value is 116.5, i.e. between 68 and 121, the difference between the two groups is not statistically significant at a level of probability of 0.05.

Two points of importance should be noted. The t test applied to these

Table 14.3

Blood pressure (mm)	*Rank*	*Blood pressure (mm)*	*Rank*
98	1	*128*	11
105	2	**129**	12
107	3	*132*	13
110	4	*134*	14
114	5	**136**	15
115	6	**139**	16
123	7.5	**140**	17
123	7.5	*145*	18
125	9	*154*	19
126	10	*160*	20

figures gave a 'statistically significant' result—it was almost precisely at the 0.05 level. The Wilcoxon test has given a 'not statistically significant' result—the total of 116.5 was below the 0.05 level of 121. Thus it should be realised that two different tests of significance may give different answers—particularly in borderline cases; and secondly, that the *t* test is in general a more sensitive test for genuine differences than is the rank test.

One disadvantage of these rank tests is that they are merely tests of significance and do not supply associated confidence limits, and all the extra information and insight that such limits provide, as the *t* test does.

Summary

The averages of two samples may be compared, using a paired *t* test if the observations have a natural pairing, or an unpaired *t* test otherwise. It is assumed in these tests that the observations come from normal distributions, but with large samples these tests are reliable without that condition being precisely met. The unpaired test also has an assumption of equal standard deviations in the two populations.

Distribution-free tests may be used as alternatives. They avoid the assumption of normally distributed values, but are less sensitive than *t* tests at detecting differences between means.

Chapter 15

Differences between proportions: the 2 × 2 table

In Chapter 12 statistical methods for examining a proportion were developed based upon a knowledge of the proportion to be expected from some past experience—e.g. if past experience shows that on the average 20 per cent of patients die, how great a discrepancy from that 20 per cent may be expected to occur by chance in samples of a given size? In practical statistical work the occasions upon which such past experience is available as a safe and sufficient guide are very rare. As a substitute for past experience the experimenter takes a control group and uses it as the standard of comparison against the experimental group. For instance, as a result of the collection of data in this way, we may have:

(1) 50 patients with a specific disease treated by the customary orthodox methods had a fatality rate of 20 per cent.
(2) 50 similar patients with that disease treated by the customary orthodox methods *plus* a new drug had a fatality rate of 10 per cent.

Is this difference more than is likely to arise merely by chance? It is clear that *both* these percentages will, by the play of chance, vary from sample to sample; if the new drug were quite useless we should, if the results of a large number of trials were available, sometimes observe lower fatality rates in groups of patients given that treatment, sometimes lower fatality rates in control groups, and sometimes no difference at all. In the long run—i.e. with very large samples—we should observe no material difference between the fatality rates of the two sets of patients; if the new drug is useless (but innocuous) the difference we expect to observe is, clearly, 0. Our problem is to determine how much variability will occur round that value of 0 in samples of given sizes, how large a difference between the two groups is likely to occur by chance.

It helps to start with smaller figures to develop the method and then return to the original example. Let us first suppose then that, instead of the two proportions being 10 out of 50 and 5 out of 50, they had been 3 out of 9 and 1 out of 10, a 33 per cent rate in one group compared with only a 10 per cent rate in the other. This is a considerable difference, but based on such small numbers, might it have arisen merely by chance?

The use of conditional probabilities

Before looking at the solution of this problem, a diversion is necessary to point out that there are many places in statistics where it is possible to divide an overall situation into a number of sub-situations, where it is possible to calculate probabilities *within* each sub-situation, we know which sub-situation has occurred, and the knowledge of which sub-situation it is, does not in itself give any answer to the problem at issue.

To give more concrete substance to the argument, let us suppose that we are going to make a train journey and we want to know the probability that the train will be late. If (though this may be hard to imagine) we know nothing about what train it is going to be, then records of all the railways in the world would be relevant; if, however, we know that it is a British train, then only records of British trains are relevant and, for this purpose, the rest of the world can be ignored. If, furthermore, we know that it is to be the 09.25 train from Euston to Carlisle, then all other British trains become irrelevant too. The records of how often *that* train is late can be used to estimate the probability for this occasion.

Thus we do not want the *overall* probability that a train is late. We want the *conditional* probability that *our* train is late, given that it is the 09.25 from Euston to Carlisle. It does not matter whether we chose that particular train for ourselves, or whether someone else chose it for us, or whether we were put on it by some random process. What matters is not why we are going on that train, but only that that is the train we are going on.

Similarly in our present problem, we have Table 15.1. The equivalent of 'which train we are on' is that we have 19 patients altogether, of whom 9 were in Group 1 and 10 in Group 2, and that overall 4 died and 15 recovered. How we got those particular marginal totals is irrelevant to our problem, which is, given that those happen to be our marginal totals, does the body of the table give us any reason to suppose that the population fatality rate differs in Group 1 and Group 2?

Table 15.1

	Died	*Recovered*	*Total*
Group 1	3	6	9
Group 2	1	9	10
Total	4	15	19

The reason that we need to specify this in such detail is that a number of people at present seem to have the idea that we can get improved methods, within this problem, by taking into account that we might have been on a different train (to pursue our analogy). We believe these arguments to be false—once we know what our marginal totals are, it is logical to argue solely within those.

Fisher's test

Within the observed marginal totals, there are only 5 possible tables that could have been observed, as shown in Table 15.2. Their probabilities on the null hypothesis (that the population fatality rate is the same in the two groups) are also given. These are derived from the formula that the probability of any particular such table is given by the ratio of the factorials of all the 9 numbers in the table, the 4 marginal totals in the numerator of the fraction, the 4 numbers in the body of the table and the grand total in the denominator.

The observed result is the fourth one, with probability of 0.2167 but, as usual, we require the probability of the observed result or a more extreme one. Thus $P = 0.2167 + 0.0325 + 0.0542 = 0.3034$, meaning that, if the null hypothesis were true, we should expect a result at least as discrepant as this nearly 1 time in 3. There is thus no reason to doubt the null hypothesis, but the numbers are really too small for a firm opinion.

The probabilities are not as fearsome to calculate as they might appear. Although, as already pointed out, factorials are generally such huge numbers that it is unwise to use them as such, a lot of cancelling can be

Table 15.2 The 5 possible tables with marginal totals 4, 15, 9 and 10 and their probabilities on the null hypothesis.

			Probability
0	9	9	
4	6	10	$\frac{9!\ 10!\ 4!\ 15!}{0!\ 9!\ 4!\ 6!\ 19!} = 0.0542$
4	15	19	
1	8	9	
3	7	10	$\frac{9!\ 10!\ 4!\ 15!}{1!\ 8!\ 3!\ 7!\ 19!} = 0.2786$
4	15	19	
2	7	9	
2	8	10	$\frac{9!\ 10!\ 4!\ 15!}{2!\ 7!\ 2!\ 8!\ 19!} = 0.4180$
4	15	19	
3	6	9	
1	9	10	$\frac{9!\ 10!\ 4!\ 15!}{3!\ 6!\ 1!\ 9!\ 19!} = 0.2167$
4	15	19	
4	5	9	
0	10	10	$\frac{9!\ 10!\ 4!\ 15!}{4!\ 5!\ 0!\ 10!\ 19!} = 0.0325$
4	15	19	

done, and each probability can be derived simply from the previous one; see Appendix D for details.

Larger numbers

Returning to the original problem, we had a fatality of 5 out of 50 in one group, and 10 out of 50 in the other. Table 15.3 shows the relevant part of the probability distribution, taking the observed table and the 5 more extreme ones in the same tail, and calculating the probabilities in the simplified way. We also need, for a two-tailed test, probabilities from the other tail, but whenever the two row totals are equal or the two column totals are equal (here 50 and 50 in the two groups) the distribution is symmetrical, so we need only find one tail and double it. Thus

$$P = 2 \times (0.0000 + 0.0002 + 0.0017 + 0.0094 + 0.0340 + 0.0859) = 0.2624$$

Table 15.3 The first 6 tables with marginal totals 15, 85, 50 and 50 and their probabilities on the null hypothesis.

				Probability
0	50	50		
15	35	50	$\dfrac{50!\ \ 85!}{35!\ \ 100!}$	= 0.0000
15	85	100		
1	49	50		
14	36	50	$0.000008885 \times \dfrac{50 \times 15}{1 \times 36}$	= 0.0002
15	85	100		
2	48	50		
13	37	50	$0.0001851 \times \dfrac{49 \times 14}{2 \times 37}$	= 0.0017
15	85	100		
3	47	50		
12	38	50	$0.001716 \times \dfrac{48 \times 13}{3 \times 38}$	= 0.0094
15	85	100		
4	46	50		
11	39	50	$0.009392 \times \dfrac{47 \times 12}{4 \times 39}$	= 0.0340
15	85	100		
5	45	50		
10	40	50	$0.03396 \times \dfrac{46 \times 11}{5 \times 40}$	= 0.0859
15	85	100		

About once in four times we should expect at least as discrepant a result if the null hypothesis were true. There is no reason, from these figures, to doubt the null hypothesis.

It should be noted that, although the probability of each table is quoted to four decimal places after the point, greater accuracy than this must be kept in the calculations when the numbers are very small. Thus the first probability is quoted as 0.0000, which is all we need to know for the purpose of adding it to other probabilities to get the P value, but to try to derive other probabilities by multiplying and dividing 0.0000 by other numbers would be a profitless exercise. For that purpose we need to know it more accurately as 0.000 008 885. Similarly the second probability is shown as 0.0002, but as 0.000 185 1 for further calculations.

The χ^2 approximation

If the computing power is available it is best to use the above exact method for all 2×2 tables, but for tables containing large numbers, if the computing power is not available, there is an approximate method that is easier to use and gives very accurate answers provided that certain conditions are met.

This is called the χ^2 method. (χ is the Greek letter 'chi', pronounced 'ki'.) It is worth knowing about anyway, because it is the usual method for tables having more than two rows or columns, as detailed in the next chapter. It is derived by saying that, if the null hypothesis is true, we should expect the number of deaths in each group (or whatever we are observing in the particular case) to be in proportion to the size of the group. In our example, we had two groups of 50, so we should have expected the 15 deaths to be divided equally as 7.5 in each group. This gives an expected table

7.5	42.5	50
7.5	42.5	50
15	85	100

to compare with the observed

5	45	50
10	40	50
15	85	100

It should be noted that, just as with the exact method, we are arguing *conditionally* on the observed marginal totals.

The general formula is then

$$\chi^2 = \text{Sum}\,\frac{(\text{Observed} - \text{Expected})^2}{\text{Expected}}$$

In the particular case of the 2×2 table, however, it is preferable to make a continuity correction (as previously explained in Chapter 9). This is known here as Yates' correction and consists of moving each observed value $\frac{1}{2}$ unit nearer to the corresponding expected value, i.e. treating thc observed table as if it were

5.5	44.5	50
9.5	40.5	50
15	85	100

This is not 'cooking the data'. The data remain exactly as before; we are merely making the adjustment within the calculation of χ^2 so that it will give a better approximation to the truth. People sometimes worry about what to do if the observed values are already within $\frac{1}{2}$ a unit of the expected ones, but in that case we can say at once that $P = 1.0$, and it is unnecessary to calculate χ^2 at all.

With this correction we have:

$$\chi^2 = \frac{(5.5-7.5)^2}{7.5} + \frac{(9.5-7.5)^2}{7.5} + \frac{(44.5-42.5)^2}{42.5} + \frac{(40.5-42.5)^2}{42.5} = 1.2549$$

This χ^2 has 1 degree of freedom because, for given marginal totals, as soon as the value is known for 1 cell in the body of the table, the other 3 values can be determined from it.

Referring to a table of χ^2 probabilities (Table III) we find that, for 1 degree of freedom, $\chi^2 = 0.45$ would give $P = 0.50$, whereas $\chi^2 = 1.64$ would give $P = 0.20$. $\chi^2 = 1.2549$ lies between these so $0.50 > P > 0.20$ (i.e. 0.50 is greater than P, and P is greater than 0.20). That is as close a result as we can give from that particular χ^2 table, but precise calculations show, in fact, that (on 1 degree of freedom) $\chi^2 = 1.2549$ gives $P = 0.2628$, astonishingly close to the 0.2624 that was found by the exact method. It should not be expected always to be as accurate as this.

An alternative way of calculating χ^2 for a 2×2 table, which does not explicitly find the expected values, may sometimes be found convenient. If the table is

a	b	$a+b$
c	d	$c+d$
$a+b$	$b+d$	n

where n is $a+b+c+d$, the formula is

$$\chi^2 = \frac{(|ad-bc| - \frac{1}{2}n)^2\, n}{(a+b)\,(c+d)\,(a+c)\,(b+d)}$$

where $|ad-bc|$ is the mathematical notation meaning 'use either $ad-bc$ or $bc-ad$, whichever is positive'. Thus, in the

5	45	50
10	40	50
15	85	100

example we have

$$\chi^2 = \frac{(45 \times 10 - 40 \times 5 - \frac{1}{2} \times 100)^2 \times 100}{50 \times 50 \times 15 \times 85} = 1.2549$$

exactly as by the other method.

A condition for use of the χ^2 method is that none of the four expected values should be less than 5. However, if that condition is not met, these are just the circumstances where the exact method is not too difficult.

Matched pairs

So far we have analysed our results on the assumption that each of the two groups contains 50 individuals but that no individual in one group has any resemblance to a particular individual of the other group. Sometimes this is not the case; the patients may have been matched case by case so that one of each matched pair went into each group. Just as with *t* tests, if there is such matching it should be taken into account in the analysis.

If our two groups of 50, in which 5 died in Group 1 and 10 died in Group 2, consisted of matched pairs, then the original table (Table 15.4) would be an incorrect way to look at the evidence. Instead we use a table such as Table 15.5. Notice that the 5, 45, 10, 40 figures that previously formed the body of the table now appear as totals, while the body now shows how things occurred within pairs.

This table is different in kind from the other and it would be quite wrong to perform either Fisher's test or its χ^2 approximation on it. Instead we note that the 4 pairs where each of the individuals died, and the 39 pairs

Table 15.4

	Died	*Recovered*	*Total*
Group 1	5	45	50
Group 2	10	40	50
Total	15	85	100

Table 15.5

			Group 1	
		Died	*Recovered*	*Total*
Group 2	Died	4	6	10
	Recovered	1	39	40
	Total	5	45	50

where each of the individuals recovered, tell us little about differences between the groups but only whether or not we did our pairing well. The real evidence for any group differences comes from the 7 pairs where the two outcomes differed. In the absence of any real group difference these should be as likely to be either way round, and we can use a binomial test (with $n = 7$, $p = 0.5$) to see whether the preponderance of 6 against 1 indicates a significant difference. In this context such a binomial test is often called McNemar's test.

Using the binomial formula (Chapter 12) we find

$$P = 2 \times (0.5^7 + 7 \times 0.5^7) = 0.125$$

the multiplier 2 giving a two-tailed test as a binomial distribution is always symmetrical if $p = 0.5$. If the number of discrepant pairs is 10 or more, the normal approximation to the binomial will usually be adequate, if the exact binomial result cannot easily be calculated.

Summary

In practical statistical work a problem that frequently arises is the significance of a difference between two proportions. The difference between two such values in a pair of samples will fluctuate from one pair of samples to another, and though the samples may be drawn from one and the same population we shall not necessarily observe no difference between them. It is convenient to display the results in the form of a 2×2 table. For unpaired observations Fisher's test (or the χ^2 approximation to it) should be used; for paired observations, McNemar's test on the table of pairs.

Chapter 16

Contingency tables

In the last chapter attention was devoted to the comparison of proportions in two groups. Occasions frequently arise, however, when we need to compare the characteristics of more than two groups. Such comparisons may be made using a more general form of the χ^2 test used in that chapter. Tables for which this approach is most suitable are those with two or more rows and two or more columns, where each individual in a sample can be classified to fall into one of the rows, and into one of the columns, but either the rows or the columns (or both) have no logical order.

To take a concrete example, Table 16.1 shows 258 male factory workers divided into 4 columns by which of 4 factories employed them, and into 3 rows by whether they were non-smokers, cigarette smokers or pipe smokers. Each of these categories could be arranged in any order—had we interchanged factories 1 and 2, for example, the meaning of the table would be identical. On the other hand, if we had classified by *amount* of smoking (perhaps 1–5, 6–20, 21+ cigarettes per day), then there would have been a logical order to the rows. If both the row classification and the columns classification have logical orders, a trend test (as described later) would be preferable, but here neither of them do.

We might wish to ask ourselves whether this sample of workers suggests differences in smoking habits from factory to factory or whether the observed differences might have arisen by chance. It is clear that if another sample of 258 were taken at random from the same population we would be unlikely to find the same numbers in the body of the table. Each of them is certain to vary from one sample to another, and the smaller the sample the more, as has been previously shown, they are likely to vary. The question at issue then becomes this: Is it likely that the magnitude of the

Table 16.1 258 factory workers classified by factory and smoking habits.

	Factory 1	*Factory 2*	*Factory 3*	*Factory 4*	*Total*
Pipe smokers	8	7	3	8	26
Cigarette smokers	22	32	56	36	146
Non-smokers	26	22	27	11	86
Total	56	61	86	55	258

observed differences could arise merely by chance in taking samples of the given size? Is it likely, in other words, that, if we had observed the whole population from which the sample was taken, the proportions would be the same in each factory?

To answer that question the assumption is made that the proportion ought to be identical in each factory. We then seek to determine whether that assumption (the null hypothesis) is a reasonable one by measuring whether the differences actually observed from the uniform figure might frequently or only infrequently arise by chance in taking samples of the recorded size. If we find that the departure from uniformity is one that might frequently arise by chance, then we must conclude that the differences might vanish if we took another sample. We must, therefore, be cautious in drawing deductions from them. If, on the other hand, we find that the differences from our assumed uniformity are such as would only arise by chance very infrequently, then we may abandon our original null hypothesis that each of the groups ought to show the same pattern. For if that hypothesis were true an unlikely event would have occurred, and it is reasonable to reject the unlikely event and say that we think the differences observed are real, in the sense that they would not be likely to disappear (though they might be modified) if we took another sample of equal size.

The technique involved

As with the 2×2 table, we *condition* on the observed marginal totals, that is to say we work out the expected values, on the assumption that the null hypothesis is true, to give the same marginal totals as those observed. The technique involves:

(1) calculating how many of each smoking category would have been observed in each factory, if the proportion were the same in each;
(2) calculating the differences between these numbers expected on the hypothesis of no relationship and the numbers actually observed;
(3) calculating whether these differences are of a magnitude that might reasonably be observed by chance alone.

The first step is to find the expected value in each cell. Consider as an example pipe smokers in factory 1. Since, overall, there were 26 pipe smokers out of 258, if the proportion were constant that proportion would have to be 26/258 in each factory. As 56 men were observed from factory 1, the number of pipe smokers would have to be $56 \times 26/258 = 5.64$. By similar reasoning we can derive the expected number for each cell as the *column total × row table/grand total*. These expected numbers are shown in Table 16.2.

We are now ready to use the χ^2 formula as given in the last chapter. That is we calculate the sum of all the values of

$$\frac{(\text{observed number} - \text{expected number})^2}{\text{expected number}}$$

Table 16.2 Expected numbers corresponding to the observed numbers in Table 16.1, on the null hypothesis of independence of factory and smoking habits.

	Factory 1	*Factory 2*	*Factory 3*	*Factory 4*	*Total*
Pipe smokers	5.64	6.15	8.67	5.54	26
Cigarette smokers	31.69	34.52	48.67	31.12	146
Non-smokers	18.67	20.33	28.67	18.33	86
Total	56	61	86	55	258

In the present example this gives

$$\chi^2 = \frac{(8-5.64)^2}{5.64} + \frac{(22-31.69)^2}{31.69} + \frac{(26-18.67)^2}{18.67} + \ldots = 16.96$$

A χ^2 value is meaningless, however, without also knowing its degrees of freedom. How many have we here? Since we are keeping the marginal totals constant, it is clear that as soon as we know any three values in a row the fourth one is determined. Similarly as soon as we know any two values in a column the third one is determined. Thus if we fill in 6 values the totals tell us what all the other values must be. So we have 6 degrees of freedom, the general formula being

$$(\text{number of rows} - 1) \times (\text{number of columns} - 1)$$

Reference to Table III shows that any χ^2 value, on 6 degrees of freedom, exceeding 16.81 gives $P<0.01$, and a suitable computer program will tell us more precisely that $P = 0.0094$, showing a significant difference from the null hypothesis.

Having found that the differences are unlikely to be due to chance, it is clear that the χ^2 test, in itself, has told us nothing about what they are due to. In seeking to know that, a first step would be to examine the contribution to χ^2, i.e. (observed – expected)^2/expected, for each cell of the table individually (Table 16.3). It is seen that the main contributions come from the relatively small numbers of pipe smokers in factory 3, of non-smokers in factory 4, and of cigarette smokers in factory 1, together with the relatively large number of non-smokers in factory 1.

It may be, in this instance, that we do not need to know the causes of these differences; we merely need to note that if we are examining

Table 16.3 Contributions to χ^2 from the observed and expected values in Tables 16.1 and 16.2.

	Factory 1	*Factory 2*	*Factory 3*	*Factory 4*	*Total*
Pipe smokers	0.98	0.12	3.71	1.09	5.90
Cigarette smokers	2.96	0.18	1.11	0.76	5.02
Non-smokers	2.88	0.14	0.10	2.93	6.05
Total	6.83	0.44	4.91	4.79	16.96

differences in the illness rates between the factories it would be unwise to do so without taking smoking differences into account.

Important conditions for validity of χ^2

It needs emphasising that a χ^2 test is valid only for comparing *frequencies* in the cells of a table, i.e. an actual count of the number of observations in each cell. It must *never* be performed on percentage figures, because the evidence from a table depends upon the size of the numbers involved. If two frequencies are expected to be identical, an observed 2 to 1 ratio is in no way surprising if the actual numbers are 2 of one and 1 of the other, but very surprising indeed if they are 2000 of one and 1000 of the other, yet the percentage comparison would be the same each time.

Furthermore such a test must *never* be performed on the values of measurements on a scale. There was a newspaper correspondence some years ago on the question of whether there was more rain on Thursdays than on other days of the week, in which the originator of the hypothesis (that there was) put forward a χ^2 value calculated from the number of inches of rain he had observed on each day; but had he used millimetres instead of inches he would have got a totally different value. Indeed he could have got any value he wished merely by changing his units of measurement. Such a test is obviously complete nonsense. If he had merely classified each day as wet or dry (with a careful definition of those terms) and then counted how many of each he observed for each day of the week, then he would have been dealing with frequencies and a χ^2 test could have been validly used.

An additional requirement for a χ^2 test to be valid is that the expected numbers should not be too small: a good practical rule is to say that no expected value may be less than 1.0, and no more than 10 per cent of the expected values may be less than 5.0. If this requirement is not met, the help of a statistician is advisable.

The hourly distribution of births

In Table 16.4 distributions are given of a series of live and still births according to the time of day at which they took place. With live births the figures show that a high proportion of the total took place during the night and a smaller proportion during the early afternoon and evening. With still births rather the reverse was the case; the proportion during the night was somewhat low, while in the morning and afternoon the number was rather high, though the differences are not very uniform.

Are these differences between the two distributions such as might well have occurred by chance? The table shows the expected numbers on the assumption of independence, derived from *row total* × *column total/grand total*. From these we can calculate $\chi^2 = 32.84$ on 7 degrees of freedom, and $P<0.0001$. Looking at the contributions to χ^2 in Table 16.5, a word of

Table 16.4 Distribution of live and still births over the day. Comparison of live birth and still birth distributions. (Adapted from J.V. Deporte, *Maternal Mortality and Stillbirths in New York State*.)

	Observed		*Expected*	
Time interval	*Live births*	*Still births*	*Live births*	*Still births*
Midnight–	4 064	126	4 030.9	159.1
0300–	4 627	157	4 602.3	181.7
0600–	4 488	142	4 454.2	175.8
0900–	4 351	195	4 373.4	172.6
1200–	3 262	150	3 282.4	129.6
1500–	3 630	178	3 663.4	144.6
1800–	3 577	144	3 579.7	141.3
2100–midnight	4 225	180	4 237.7	167.3
Total	32 224	1 272	32 224.0	1 272.0

Table 16.5 Contributions to χ^2 from the observed and expected values in Table 16.4.

Time interval	*Live births*	*Still births*
Midnight–	0.272	6.886
0300–	0.133	3.358
0600–	0.256	6.499
0900–	0.115	2.907
1200–	0.127	3.211
1500–	0.305	7.715
1800–	0.002	0.052
2100–midnight	0.038	0.964
Total $\chi^2 = 32.84$		

warning is necessary. The largest values are all in the still births column, but this is only because there are so many more live births than still births in total. In any given row of the table, the contributions to χ^2 must show 25.33 (i.e. 32 224/1272) times as great a contribution from still births as from live births.

We can also look at these same data in another way, however. Taking live births and still births separately, we can ask for each whether they were evenly spread across the day, subject only to chance variation, or whether they lacked such uniformity.

Table 16.6 shows the calculation for the live births only. The expected numbers are quite different from those in Table 16.4. Expected numbers are always expected on the basis of some hypothesis. Here the hypothesis is different, so the expected numbers are different too. It may be noted that, as the expected value is the same throughout, the quickest way of calculating χ^2 is to sum the squares of the (observed minus expected) values and divide this total by the expected number, instead of making separate divisions by the expected number in each instance.

Table 16.6 Distribution of live births over the day. Comparison of live birth distribution with uniformity. (Adapted from J.V. Deporte, *Maternal Mortality and Stillbirths in New York State*.)

Time interval	*Live births*		*Observed – expected*
	Observed	*Expected*	
Midnight–	4 064	4 028	+36
0300–	4 627	4 028	+599
0600–	4 488	4 028	+460
0900–	4 351	4 028	+323
1200–	3 262	4 028	−766
1500–	3 630	4 028	−398
1800–	3 577	4 028	−451
2100–midnight	4 225	4 028	+197
Total	32 224	32 224	

We have 7 degrees of freedom here, because there are 8 groups, but 1 degree of freedom is lost in taking the expected total as 32 224 to match the observed total. We find $\chi^2 = 413.0$ which, on 7 degrees of freedom, is enormously significant. Similar calculations can be made for the still births, producing $\chi^2 = 23.8$, also on 7 degrees of freedom, giving $P = 0.0012$, so as well as being differently distributed from each other, neither distribution can be regarded as uniform across the day. The differences from the uniformity that we presumed ought to be present are therefore more than would be ascribed to chance, and we conclude that neither live nor still births are distributed evenly over the twenty-four hours in these records. The differences of the live births from uniformity are more striking than those of the still births, for they show a systematic excess during the hours between 9 p.m. and 12 noon and a deficiency between 12 noon and 9 p.m.; with the still births there is some change of sign from one period to another which makes the differences from uniformity less clearly marked—perhaps due to the relatively small number of observations. Inspection of these differences themselves adds considerably to the information provided by χ^2. The latter value tells us that the differences are not likely to be due to chance; the differences themselves show in what way departure from uniformity is taking place, and may suggest interpretations of that departure.

It may be noticed that the recording of these times of day apparently breaks one of our basic rules, namely that no observation should be ambiguous as to which group it appears in. Here 'midnight' apparently appears in two time intervals. There is no real ambiguity, however, because every birth has a date attached to it: if the date of a midnight birth is given as that of the preceding day it clearly goes in the last group; if of the succeeding day, in the first group.

Interpretation of the associations found

It must be fully realised that χ^2 gives no evidence of the *meaning* of any associations found. For instance the value of χ^2 found above was such that we concluded that live birth or still birth was not independent of time of day. The interpretation of that association is quite another matter. Even if there is a direct cause-and-effect relationship, we do not know which of these two things is the cause and which the effect, and we always need to consider whether the cases examined were effectively equivalent in all other relevant respects.

The value of the χ^2 test is that it prevents us from unnecessarily seeking for an explanation of, or relying upon, an 'association' which may quite easily have arisen by chance. But if the association is not likely to have arisen by chance, we are not, as with all tests of 'significance', thereby exonerated from considering different hypotheses to account for it. If we use some form of treatment on mild cases of a disease and compare the fatality experienced by those cases with that shown by severe cases not given the treatment, the χ^2 value will certainly show that there is an association between treatment and fatality. But clearly that association between treatment and fatality is only an indirect one. We should have reached just the same result if our treatment were quite valueless, for we are not comparing like with like and have merely shown that mild cases die less frequently than severe cases. Having applied the significance test, we must always consider with care the possible causes to which the association may be due.

It must be observed also that the value of χ^2 does not measure the *strength* of the association between two factors but only whether they are associated at all in the observations under study. Given sufficiently large numbers of observations, the test may show that two factors are associated even though the degree of relationship may be very small and of no practical importance whatever.

The additive characteristic of χ^2

One further characteristic of χ^2 is useful in practice. Suppose we had three tables showing the incidence of attacks upon different groups of inoculated and uninoculated persons, observed, say, in different places, and each table suggests an advantage to the inoculated, but in no case by more than could fairly easily have arisen by chance—e.g. the χ^2 values are 2.0, 2.5 and 3.0, and, with 1 degree of freedom in each case, the *P*s are 0.157, 0.114 and 0.083. The systematic advantage of the inoculated suggests that some protection is conferred by inoculation. We can test this uniformity of result, whether taken together these tables show a significant difference between the inoculated and uninoculated, by taking the sum of the χ^2 values and consulting the χ^2 table again with this sum and the sum of the degrees of freedom—namely $\chi^2 = 2.0 + 2.5 + 3.0 = 7.5$ and degrees of freedom 3. $P = 0.058$ so that we must still conclude that the three sets of differences, though very suggestive, are not quite beyond what might fairly frequently arise by chance in samples of the size observed.

The trend of values

It may finally be noted that the χ^2 test in the form so far discussed gives an answer as to whether the magnitude of the differences in a contingency table could easily have arisen by chance, but it clearly takes no account whatever of the order of those differences. We should have reached precisely the same value for χ^2 if the rows or the columns had been transposed in any way. Where the rows and columns have logical orders there may sometimes appear to be an orderly progression that needs to be taken into account.

If a plain χ^2 test shows that the differences, whatever may be their order, are *unlikely* to have arisen by chance, no serious difficulty arises in its use. One has then to interpret the figures, which are unlikely to be due to chance, and the only difference between the two situations is that often it will be easier to reach a plausible explanation with an orderly progression than with a disorderly one. In short, one has significant departures of one kind or another from expectation and they have to be interpreted.

If, on the other hand, the χ^2 test shows that the results do *not* differ significantly from the uniformity assumed, i.e. that the observed differences between the groups may well be due to chance, then it is sometimes argued, very properly, that, since the orderly progression 'makes sense' and the disorderly does not, the order should in some way be brought into the picture.

To take order into account, a test of trend, as given below, must be applied. Without such a test in the two situations it is, however, reasonable to argue thus:

(1) *In cases of disorder*. The χ^2 test has shown that these differences might easily be due to chance; further, their order is not systematic, so even if they were real differences no sensible explanation of their meaning would be easy to develop; it is therefore preferable to accept them as due to chance.

(2) *In cases of order*. The χ^2 test has shown that these differences might easily be due to chance; their order is, however, systematic and a rational explanation of the trend is simple; it is therefore reasonable to argue somewhat cautiously from them, and well worth while pursuing further data.

A test of order in proportions

A commonly occurring problem arises from the order of a series of proportions (such as those of Table 16.7) and the application of a test, which is not difficult to calculate and use, is therefore important and worth illustrating.

The figures, based upon real data, compare a teacher's assessment of intelligence of a number of children (in 4 categories) with a doctor's

Table 16.7 869 children classified by intelligence and nutrition.

	Assessment of intelligence				
	Very poor	*Poor*	*Good*	*Very good*	*Total*
(i) Number of children with satisfactory nutrition	245	228	177	219	869
(ii) Number of children with unsatisfactory nutrition	31	27	13	10	81
(iii) Total number of children observed	276	255	190	229	950
(iv) Percentage in each intelligence group that had unsatisfactory nutrition (and the difference of the group from the mean value for all groups)	11.2319 (+2.7056)	10.5882 (+2.0619)	6.8421 (−1.6842)	4.3668 (−4.1595)	8.5263
(v) Intelligence 'score' assigned to each group	−1	0	+1	+2	
(vi) Difference of intelligence score of the group from mean intelligence score	(−1.3916)	(−0.3916)	(+0.6084)	(+1.6084)	

independent assessment of whether their nutrition was satisfactory or not. The logical progression is seen in that the percentage with unsatisfactory nutrition falls monotonically with increase of intelligence. Even if unlikely to be due to chance, there is no suggestion here of which is the cause and which the effect, nor whether there is a common cause of which both are effects, but before we even start thinking about causes and effects, we need to know whether the apparent relationship may be only apparent or whether there really is anything to explain.

To calculate a χ^2 value which takes into account the *progression* of the percentages, the qualitative assessment of intelligence must first be placed upon a quantitative scale, as shown in line (v) of Table 16.7. Such scores are, clearly, to some extent subjective and arbitrary and, as Cochran has pointed out, 'Some scientists may feel that the assignment of scores is slightly unscrupulous, or at least they are uncomfortable about it. Actually', he adds, 'any set of scores gives a *valid* test, provided that they are constructed without consulting the results of the experiment. If the set of scores is poor, in that it badly distorts a numerical scale that really does underlie the ordered classification, the test will not be sensitive. The scores should therefore embody the best insight available about the way in which the classification was constructed and used'. (*Biometrics*, 1954, **10**, 417)

Having adopted this quantitative scale we can calculate in its terms the mean intelligence of all the children observed, namely $(276\times(-1)+255\times 0+190\times 1+229\times 2)/950=0.3916$. The position of each group can then be determined in relation to this mean, as in line (vi). Similarly the proportion with poor nutrition in each group can be calculated as an excess, or defect, in relation to the figure of 8.5263 per cent for the observed population as a whole (see line (iv)). Note that it is advisable to keep the figures to considerable accuracy during the calculations. Rounding to the nearest whole number, for example, could lead to much inaccuracy in the result.

From the data in this form we can calculate (1) a measure of the degree of relationship of intelligence to nutrition and (2) the χ^2 value associated with that degree of relationship. The measure of the degree of relationship of intelligence to nutrition is calculated as follows. For each intelligence group separately the difference between its own intelligence score and the mean intelligence score for all children is multiplied by the corresponding difference between its own percentage with unsatisfactory nutrition and the mean percentage in the whole population with unsatisfactory nutrition; the resulting product must be 'weighted', or multiplied, by the number of children in that particular group, i.e. to whom this product applies. These sums for the four intelligence groups are then added together to form the numerator of the required measure of relationship.

Its denominator requires for each intelligence group the *square* of the difference between its own intelligence score and the mean intelligence of all children, again weighted by the number of children in that particular group, and then the sum of these values for all the four groups.

Thus we have as the measure of relationship:

$$\frac{\textit{Sum}\ [(\text{score} - \text{mean score})\ (\% - \text{mean}\ \%)\ (\text{number in group})]}{\textit{Sum}\ [(\text{score} - \text{mean score})^2\ (\text{number in group})]}$$

Tabulating gives the figures in Table 16.8, which show that the fraction giving the measure of relationship is −2971.8/1236.3. Calling it in general terms numerator/denominator, the χ^2 value to test its significance is (numerator)^2/(denominator × pq), where p is the percentage of marked persons in the total observations (i.e. with unsatisfactory nutrition) and q is the percentage of unmarked persons (i.e. with satisfactory nutrition).

χ^2 for trend therefore equals

$$(2971.8)^2/[(1236.3)\ (8.5623)(91.4737)] = 9.16$$

and it has *one degree of freedom*. Consulting Table III gives P considerably less than 0.01, or highly significant. Actually $P = 0.0024$.

Table 16.8 Calculations for χ^2 for trend.

Intelligence group	*No. in group(a)*	*Score − mean score (b)*	*% −mean% (c)*	*(a) × (b) × (c)*	*(a) × (b)^2*
Very poor	276	−1.3916	+2.7056	−1039.2	534.5
Poor	255	−0.3916	+2.0619	−205.9	39.1
Good	190	+0.6084	−1.6842	−194.7	70.3
Very good	229	+1.6084	−4.1595	−1532.0	592.4
Total	950			−2971.8	1236.3

Summary

The χ^2 test is particularly useful for testing the presence, or absence, of association between characteristics which cannot be quantitatively expressed. It is not a measure of the strength of an association, though inspection of the departure of the observed values from those expected on the no-association hypothesis will often given some indication, though not a precise numerical measure, of that degree. As with all tests of significance, the conclusion that a difference has occurred which is not likely to be due to chance does not exonerate the worker from considering closely the various ways in which such a difference may have arisen. The calculation of χ^2 must always be based upon the absolute numbers. In this discussion the mathematical development of the test and the foundation of the table by means of which the value of χ^2 is interpreted in terms of a probability have been ignored. The test can be applied intelligently without that knowledge, provided the rules for calculation of the values of χ^2 and its degrees of freedom are followed, and the usual precautions are taken in interpreting a difference. It must also be borne in mind that with small expectations the probability derived from the test may be inaccurate and must therefore be cautiously used (or some other method of analysis applied).

Chapter 17

Means of more than two groups

This chapter, and some others of this book, is not designed actually to teach the reader the best ways of performing the operations involved, but to give an insight into what those operations are doing, so that the reader will be able to follow the ideas when they are met, for example, in published papers. The ideas of this chapter are simple cases of the technique called *analysis of variance* in which the variability observed in a collection of observations is split up into components to show how much of it appears to be due to each of a number of causes.

Generalisation of *t* tests

In comparing the means of two samples, we have seen in Chapter 14 that an unpaired, or a paired, *t* test is useful according to the nature of the observations. Now we wish to say, suppose there are more than two groups, what then?

The equivalent of an unpaired test is known as a *one-way* analysis of variance, and of a paired test as a *two-way* analysis of variance.

One-way analysis

Table 14.2 contained figures of two samples of systolic blood pressure, where pairing was not possible. Table 17.1 contains similar figures with

Table 17.1 Systolic blood pressure (mm) in 10 patients, divided into 3 groups, (hypothetical figures).

	Group 1	*Group 2*	*Group 3*	
	132	139	126	
	160	107	110	
	128	98	106	
		144		
Mean	140	122	114	*Grand mean* 125

three groups, who have been differently treated. The numbers of observations are deliberately kept small for ease of demonstration; we should not often wish to reach conclusions on such small samples in reality. The three groups have different means, but are those differences greater than would be expected merely by chance given the amount of variability observed within each group? To answer that question, we need to consider the variability *between groups* in comparison with the variability *within groups*.

For this purpose, it turns out to be helpful to consider not the standard deviations themselves but the sums of squares of deviations about the mean that lead to them. That is to say, taking the formula for standard deviation.

$$s = \sqrt{(\text{Sum } (x - \bar{x})^2/(n-1))}$$

we ignore, for the present purpose, the division by $n-1$ and the taking of the square root, and concentrate just on

$$\text{Sum } (x - \bar{x})^2$$

For brevity's sake, we shall refer to this merely as the *sum of squares* and take 'of deviations about the mean' to be implied.

Calculating the sum of squares of all 10 observations, we find

$$(132 - 125)^2 + (160 - 125)^2 + \text{etc.} = 3480$$

Table 17.2 shows the same observations again, but separated into parts, each being formed of the grand mean 125, followed by the difference of the group mean from the grand mean, which is +15 in group 1, −3 in group 2, and −11 in group 3, followed finally by the necessary adjustment to get back to the original observation.

We can now say that if our observations had had the same variability as that actually observed between the groups, but *no* variability within the groups, we should have got Table 17.3, whereas if there had been the same variability as that actually observed within the groups, but *no* variability between the groups, we should have got Table 17.4

From Table 17.3 (between the groups) the sum of squares is

$$(+15)^2 + (+15)^2 + (+15)^2 + (-3)^2 + \cdots + (-11)^2 = 1074$$

These numbers are already deviations from the mean, so they may be squared directly as shown. Similarly from Table 17.4 (within the groups)

Table 17.2 The figures from Table 17.1, shown as a sum of grand mean, group effect and residual.

	Group 1	*Group 2*	*Group 3*	
	125 + 15 − 8	125 − 3 + 17	125 − 11 + 12	
	125 + 15 + 20	125 − 3 − 15	125 − 11 − 4	
	125 + 15 − 12	125 − 3 − 24	125 − 11 − 8	
		125 − 3 + 22		
Mean	140	122	114	*Grand mean* 125

Table 17.3 The figures from Table 17.1, shown as a sum of grand mean and group effect, if there were no residual variability.

	Group 1	*Group 2*	*Group 3*	
	125 + 15	125 − 3	125 − 11	
	125 + 15	125 − 3	125 − 11	
	125 + 15	125 − 3	125 − 11	
		125 − 3		
Mean	140	122	114	*Grand mean* 125

Table 17.4 The figures from Table 17.1, shown as a sum of grand mean and residual variability, if there were no group effect.

	Group 1	*Group 2*	*Group 3*	
	125 − 8	125 + 17	125 + 12	
	125 + 20	125 − 15	125 − 4	
	125 − 12	125 − 24	125 − 8	
		125 + 22		
Mean	125	125	125	*Grand mean* 125

the sum of squares is

$$(-8)^2 + (+20)^2 + (-12)^2 + \cdots + (-8)^2 = 2406$$

and we observe the, at first sight remarkable, fact that $1074 + 2406 = 3480$, the original sum of squares of all the observations.

Furthermore we find the corresponding degrees of freedom to be additive too, for with 3 groups there are 2 degrees of freedom between the groups, whereas with group sample sizes of 3, 4 and 3 we have $2 + 3 + 2 = 7$ degrees of freedom within the groups, giving $2 + 7 = 9$ overall, corresponding to the original 10 observations. Putting these numbers into tabular form we have Table 17.5. The figure called *mean square* is calculated by dividing the relevant sum of squares by its corresponding degrees of freedom.

It can be shown mathematically that if, in the population sampled, there were really no difference between groups, but merely an overall variance within the population, then the two mean squares derived are independent estimates of that variance, so we should expect the ratio of the two to be 1.0 or thereabouts. We calculate this ratio as

$$F = 537.0/343.7 = 1.562$$

Table 17.5

	Sum of squares	*Degrees of freedom*	*Mean square*
Between groups	1074	2	537.0
Within groups	2406	7	343.7
Total	3480	9	

In the two-group case we referred the value of t to the t distribution to get a P value; now we have to refer F to the appropriate F distribution. Tables of F are much more long-winded than those of t because they need to take into account two separate degrees of freedom figures, 2 and 7 in the current example. They are therefore not given in this book, but it can be found that $F = 1.562$, on 2 and 7 degrees of freedom corresponds to $P = 0.27$. This is certainly not low enough to make us doubt the null hypothesis.

Two-way analysis

In Table 17.1, as in unpaired data for a t test, the columns were meaningful, giving the three groups into which the data were divided, but the rows had no meaning but were merely a convenient way of setting out the results. Thus, for example, if the third column had been ordered 110, 106, 126 instead of 126, 110, 106 there would have been no difference in the meaning or in the analysis.

Often, however, as in a paired t test, rows as well as columns have a meaning. Thus in Table 17.6, we have further figures of systolic blood pressure, with one observation for each of three treatments for each of four racial groups. We may wish to judge whether the figures demonstrate any differences in treatment, while avoiding any disturbance that might result from racial differences. Table 17.7 shows how each observation may be expressed in terms of a grand mean 134.5, followed by a treatment effect +3.25, +0.5 or −3.75 for treatments 1, 2 and 3 respectively, followed by a

Table 17.6 Systolic blood pressure (mm) in 12 patients, divided into 3 treatment groups and 4 racial groups (hypothetical figures).

	Treatment 1	*Treatment 2*	*Treatment 3*	*Mean*
Race 1	132	125	139	132
Race 2	160	128	105	131
Race 3	145	154	139	146
Race 4	114	133	140	129
Mean	137.75	135	130.75	134.5

Table 17.7 The figures from Table 17.5, shown as a sum of grand mean, treatment effect, race effect and residual.

	Treatment 1	*Treatment 2*	*Treatment 3*	*Mean*
Race 1	134.5 + 3.25 − 2.5 − 3.25	134.5 + 0.5 − 2.5 − 7.5	134.5 − 3.75 − 2.5 + 10.75	132
Race 2	134.5 + 3.25 − 3.5 + 25.75	134.5 + 0.5 − 3.5 − 3.5	134.5 − 3.75 − 3.5 − 22.25	131
Race 3	134.5 + 3.25 + 11.5 − 4.25	134.5 + 0.5 + 11.5 + 7.5	134.5 − 3.75 + 11.5 + 7.5	146
Race 4	134.5 + 3.25 − 5.5 − 18.25	134.5 + 0.5 − 5.5 + 3.5	134.5 − 3.75 − 5.5 + 14.75	129
Mean	137.75	135	130.75	134.5

race effect −2.5, −3.5, +11.5 or −5.5 for races 1, 2, 3 and 4 respectively. Finally we have the *residual* component necessary to reach the observed value of each observation.

Tables 17.8–10 show what the observations would have been if: (1) only the treatment differences were present; (2) only the racial differences were present; (3) only the residual variability were present. The sum of squares for the original table and for each of these reduced tables may be calculated, and again will be found to be additive, giving Table 17.11.

The degrees of freedom are 2 for 3 treatments and 3 for 4 races. The 6 degrees of freedom for the residual variability come from the same formula.

$$(\textit{number of rows} - 1) \times (\textit{number of columns} - 1)$$

that we used for degrees of freedom in contingency tables.

If we are mainly interested in treatment differences, we can calculate $F = 49.75/333.42$; if we also want to know whether our sample values

Table 17.8 The figures from Table 17.6, shown as a sum of grand mean and treatment effect, if there were no race effect, or residual variability.

	Treatment 1	*Treatment 2*	*Treatment 3*	*Mean*
Race 1	134.5 + 3.25	134.5 + 0.5	134.5 − 3.75	134.5
Race 2	134.5 + 3.25	134.5 + 0.5	134.5 − 3.75	134.5
Race 3	134.5 + 3.25	134.5 + 0.5	134.5 − 3.75	134.5
Race 4	134.5 + 3.25	134.5 + 0.5	134.5 − 3.75	134.5
Mean	137.75	135	130.75	134.5

Table 17.9 The figures from Table 17.6, shown as a sum of grand mean and race effect, if there were no treatment effect, or residual variability.

	Treatment 1	*Treatment 2*	*Treatment 3*	*Mean*
Race 1	134.5 − 2.5	134.5 − 2.5	134.5 − 2.5	132
Race 2	134.5 − 3.5	134.5 − 3.5	134.5 − 3.5	131
Race 3	134.5 + 11.5	134.5 + 11.5	134.5 + 11.5	146
Race 4	134.5 − 5.5	134.5 − 5.5	134.5 − 5.5	129
Mean	134.5	134.5	134.5	134.5

Table 17.10 The figures from Table 17.6, shown as a sum of grand mean and residual variability, if there were no treatment effect, or race effect.

	Treatment 1	*Treatment 2*	*Treatment 3*	*Mean*
Race 1	134.5 − 3.25	134.5 − 7.5	134.5 + 10.75	134.5
Race 2	134.5 + 25.75	134.5 − 3.5	134.5 − 22.25	134.5
Race 3	134.5 − 4.25	134.5 + 7.5	134.5 + 7.5	134.5
Race 4	134.5 − 18.25	134.5 + 3.5	134.5 + 14.75	134.5
Mean	134.5	134.5	134.5	134.5

Table 17.11

	Sum of squares	*Degrees of freedom*	*Mean square*
Between treatments	99.5	2	49.75
Between races	543.0	3	181.00
Residual	2000.5	6	333.42
Total	2643.0	11	

indicate racial differences, we can calculate $F = 181.00/333.42$. Since the expected value of F on the null hypothesis is 1.0, and these values are both less than 1.0, there is no need to find any P values. We can say at once that there is no indication of either row or column variability in excess of what would be expected by chance.

Analysis of variance in practice

There are many different patterns in which data may be found, not merely the simple examples here of one-way classification by groups, or two-way classification by rows and columns with one observation per cell. More complicated patterns each have their corresponding pattern of analysis of variance, but the objective is always the same of splitting the observations into additive components representing the various variables that may be affecting them (plus a residual component after all the others) and then saying for each component 'Suppose only that component existed, what would be the observed sum of squares?'

It should be emphasised that the methods employed in this chapter have been designed as giving insight into the underlying ideas. In practice such methods would not be used, because mathematical formulae have been devised that enable the appropriate sums of squares to be found without actually splitting each observation into such component parts.

Summary

Where means of more than two groups are to be compared, equivalents of unpaired and paired t tests are to be found in one-way and two-way analysis of variance. This splits the sums of squares of deviations about the mean into components each relating to the variability associated with a particular variable that may be affecting the observations.

The methods employed in this chapter have been used as being enlightening about the process, and should not be taken as the way to go about it in practice.

Chapter 18

Correlation and regression

A problem with which the scientific worker is frequently faced is the relationship between two or more characteristics of a population. For instance, in Fig. 18.1 we have a *scatter diagram* showing the result of a single measurement of FEV (forced expiratory volume in 1 second), and a single measurement of FVC (forced vital capacity), for each of 232 workers at a particular factory. Each dot in the diagram represents these two values for one worker. As always, the basic information to be used in any further work lies in the original figures, not in the diagram derived from them, but the figures are too bulky to reproduce here and the diagram gives a good impression of them. Such a diagram reveals to the eye (1) the presence (or absence) of any relationship between two characteristics and (2) broadly

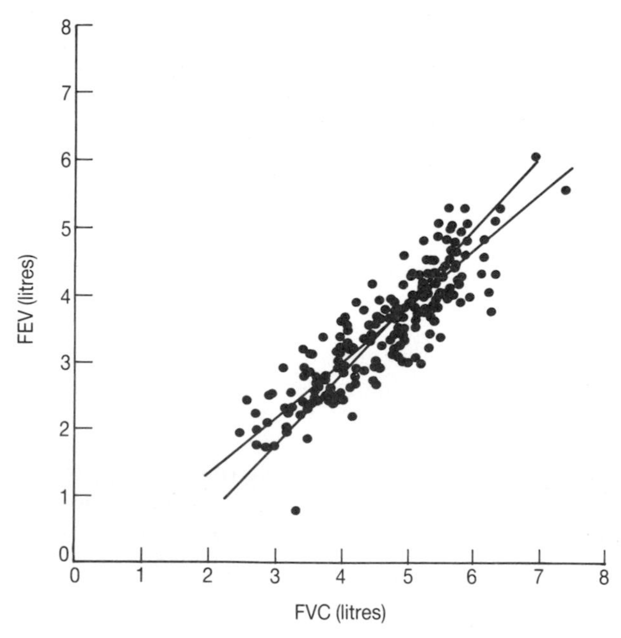

Fig. 18.1 A scatter diagram showing lung function measurements on male factory workers.

the nature of that relationship—whether it appears to be well described by a straight line (linear relationship) or requires something more elaborate such as some form of curve (curvilinear relationship). It is regrettable that it is difficult to look at things in a similarly clear way when three variables are involved instead of just two, and impossible to do so when four or more are involved.

In the present instance it appears from the distribution of the points that the relationship between the FEV and the FVC could reasonably be described by a straight line, and that the amount of scatter about the line would be reasonably constant throughout its length.

The two regression lines

Suppose, first, that the FEV were much more difficult or expensive to measure than the FVC. We might then, for future observations, wish to measure the FVC only and to estimate the FEV from it. We should have to take into account the scatter in the observations in saying how imprecise our estimate was, but the best estimate we could make would evidently be the mean value of FEV observed for the given value of FVC. Dividing the observations into groups by FVC, we can find the mean FEV for each group, as shown in Table 18.1

Suppose, however, that we had the contrary situation and wished to estimate the FVC for a given value of FEV. Thc same procedure leads to the means in Table 18.2

Each set of means of one variable, for given values of the other, is reasonably represented by a straight line—but the two lines are quite clearly not the same line, and Fig. 18.1 shows the two lines in addition to

Table 18.1 Mean FEV when grouped by FVC. Note that the grouping is not into equal intervals, but to give approximate equality of number of observations in each group.

FVC group (litres)	*Number of observations*	*Mean FEV (litres)*
0.82–	19	3.20
2.41–	19	3.53
2.59–	19	3.84
2.89–	19	4.20
3.10–	20	4.36
3.37–	21	4.61
3.64–	20	5.03
3.81–	20	5.02
4.00–	19	5.40
4.21–	19	5.36
4.38–	19	5.56
4.82–6.08	18	5.95

Table 18.2 Mean FVC when grouped by FEV. Note that the grouping is not into equal intervals, but to give approximate equality of number of observations in each group.

FEV group (litres)	*Number of observations*	*Mean FVC (litres)*
2.50–	19	2.15
3.38–	19	2.65
3.60–	18	2.73
3.94–	20	3.09
4.14–	20	3.23
4.52–	21	3.52
4.86–	20	3.77
5.11–	20	3.91
5.20–	19	4.07
5.36–	19	4.12
5.60–	19	4.57
5.81–7.39	18	4.78

the points. When people first meet this, they usually feel that something must have gone wrong, and that there ought to be one line that best fits the observations really. If they have done experiments in physics, where exact relationships do exist between variables and the only variability about a fitted line is due to errors of measurement, such a feeling may be reinforced by that experience. Here, we are not in that situation. Errors of measurement do exist, of course, but the main variability comes from the genuine biological differences between people. As soon as we have such substantial scatter, there are always two different lines, one fitting mean y for given x, the other fitting mean x for given y. They are known as *regression lines*.

The word 'regression' is a somewhat odd usage. A dictionary definition is 'act of returning' which is not what it means here. Historically it comes from early statistical research, in the nineteenth century, when it was noticed that tall fathers, on average, had sons shorter than themselves, whereas short fathers, on average, had sons taller than themselves. This was referred to as 'regression towards mediocrity' and it seemed to imply that, before long, all men would be of equal height. It was odd that in the history of the world it had not yet happened. They had found one of the two regression lines, but at first failed to notice the other one, which shows that tall sons, on average, have fathers shorter than themselves, whereas short sons, on average, have fathers taller than themselves. This makes it look as though things are becoming more extreme in successive generations. In fact the two phenomena are compatible and perfectly consistent with no change of the overall distribution between generations—the word 'regression', however, has remained with us.

The fact of the two lines becomes easier to understand by looking at two variables that are hardly connected with each other at all. In Fig. 18.2 we

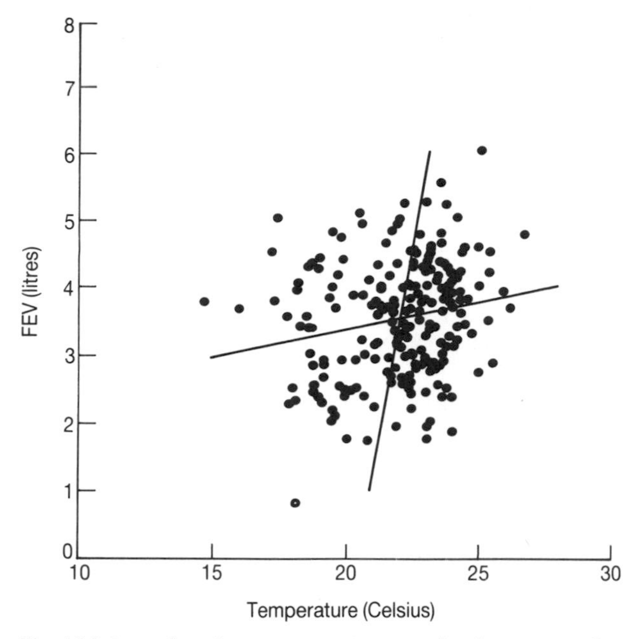

Fig. 18.2 Lung function versus temperature for the same workers.

compare the same FEV readings with the surrounding temperature at the time that they were taken. Since the measuring instrument for the FEVs contained an adjustment for temperature, we should expect very little relationship if this adjustment was correctly made, and that is what we see. Here the two regression lines are nearly vertical and nearly horizontal, showing the mean temperature to be about 22°C irrespective of what the FEV may be, whereas the mean FEV is about 3.5 litres irrespective of what the temperature may be. In such a case the two distinct lines seem intuitively sensible. When there is some relationship the two lines approach each other, until with a perfect relationship (no variability about the line) they become identical.

Calculation of regression lines

To find a regression line we need to calculate the *covariance*. This is very similar to a variance but uses a sum of products instead of a sum of squares. Recall that a variance is

$$\text{Sum } (x-\bar{x})^2/(n-1)$$

and that for calculations by hand it is easier to use the equivalent formula

$$(\text{Sum } (x^2) - (\text{Sum } x)^2/n)/(n-1)$$

but that this latter formula can be dangerously inaccurate if used in adverse circumstances, so it should not be used thoughtlessly in computer programs.

For the covariance of x and y we have

$$\text{Sum } (x-\bar{x})(y-\bar{y})/(n-1)$$

and, an easier formula for calculations by hand,

$$(\text{Sum } xy - (\text{Sum } x)(\text{Sum } y)/n)/(n-1)$$

But again this latter formula can be unsafe. The similarity of this to calculating a variance is obvious. It is worth noting that a variance must always be positive (or zero), but a covariance can be negative, and will be if the regression line slopes downwards instead of upwards.

The *regression coefficient* of y on x (denoted by $b_{y.x}$) means the average change in y for a unit change in x and is given by

$$b_{y.x} = \frac{\text{covariance of } x \text{ and } y}{\text{variance of } x}$$

The equation of the regression line is then

$$y-\bar{y} = b_{y.x}(x-\bar{x})$$

This line is valid *only* for estimating y from a given value of x. If we wish to estimate x from a given value of y, the other line must be found by interchanging x and y in the formulae.

Using the data from which Fig. 18.1 was derived, we find

$$\begin{aligned} \text{mean FEV} &= 3.5464 \text{ litres} \\ \text{mean FVC} &= 4.6676 \text{ litres} \\ \text{variance of FEV} &= 0.7443 \text{ litres}^2 \\ \text{variance of FVC} &= 0.8504 \text{ litres}^2 \\ \text{covariance of FEV and FVC} &= 0.7022 \text{ litres}^2 \end{aligned}$$

Hence the regression lines are

$$\text{FEV} - 3.5464 = \frac{0.7022}{0.8504}(\text{FVC} - 4.6676)$$

$$\text{FVC} - 4.6676 = \frac{0.7022}{0.7443}(\text{FEV} - 3.5464)$$

That is to say

$$\begin{aligned} \text{FEV} &= 0.8257\text{FVC} - 0.3078 \\ \text{FVC} &= 0.9434\text{FEV} + 0.8572 \end{aligned}$$

where FEV and FVC are measured in litres each time.

Observed and controlled variables

In examples such as those already considered, each of the variables concerned represents observations from a frequency distribution. Both regression lines then make sense. Sometimes, however, one of the variables is not a frequency distribution at all but merely a sequence of settings controlled by the investigator.

For example in a chemical laboratory a scientist might investigate the speed of a chemical reaction by choosing to take 10 observations at each of 10°C, 15°C and 20°C. The reaction times at each temperature would then be three frequency distributions, but the temperatures themselves would be a matter of choice and not an observed distribution at all. The regression equation of reaction time on temperature then makes sense, and is calculated in the same way as in the case where both variables are observed distributions, but the other regression line (of temperature on reaction time) makes no sense, and should not be calculated or used in any way.

Confidence limits and significance

Having found the regression coefficient of y on x, its standard error is

$$\sqrt{\left(\left(\frac{\text{variance of } y}{\text{variance of } x} - (\text{regression coefficient})^2\right) \Big/ (n-2)\right)}$$

and may be taken as following a t distribution with $n-2$ degrees of freedom.

Thus for the regression of FEV and FVC above we had $n = 232$

$$\begin{aligned} \text{variance of FEV} &= 0.7443 \text{ litres}^2 \\ \text{variance of FVC} &= 0.8504 \text{ litres}^2 \\ \text{covariance} &= 0.7022 \text{ litres}^2 \\ \text{regression coefficient} &= 0.7022/0.8504 \\ &= 0.8257 \text{ litres/litre} \end{aligned}$$

To find the standard error of this regression coefficient we calculate

$$\frac{0.7443}{0.8504} - (0.8257)^2 = 0.1934$$

$$\sqrt{(0.1934/230)} = 0.0290 \text{ litres/litre}$$

For 95 per cent confidence limits for the regression coefficient we get 1.97 as the appropriate value of t for 230 degrees of freedom (Table II) and calculate

$$0.8257 \pm 1.97 \times 0.0290$$

i.e. 0.7686 to 0.8828 litres/litre.

It is so evident in this case that there is a real slope to the line that it is really pointless to make a formal significance test of the null hypothesis that the population regression is zero. In less obvious cases, however, we might wish to do so. The regression coefficient divided by its standard error is

$$0.8257/0.0290 = 28.47$$

which, of course, is wildly significant as would be expected with such data.

It is worth recalling that the $n-1$ divisor, used in calculating a standard deviation, comes from the fact that we are considering variability about a mean and one observation gives us no information about that, but the second observation gives us the first bit of information about the variability. Here we are fitting a sloping line and are interested in the scatter about the line (as well as in the line itself). But in those circumstances we need two points before we can draw a line at all, so it is the third point that gives us the first bit of information about such variability. This explains the $n-2$ degrees of freedom used above.

The correlation coefficient

Sometimes it is useful, in addition to the two regression coefficients, to have a measure of the *strength* of association between two variables. This is given by the *correlation coefficient* (r) defined as

$$r = \frac{\text{covariance of } x \text{ and } y}{\sqrt{(\text{variance of } x \times \text{variance of } y)}}$$

This gives a value of 1 if the points all lie on a straight line with positive slope, -1 if they all lie on a straight line with negative slope, irrespective of whether those slopes are steep or shallow, 0 if there is no relationship at all between the variables, and intermediate values where there is a relationship but it is less than perfect. Positive slope means that the two variables tend to be big together or small together, whereas negative slope means that when one is big the other is small.

Using this formula on the data of Figs 18.1 and 18.2, we find for the first

$$r = 0.8826$$

and for the second

$$r = 0.1926$$

demonstrating a strong, but not perfect, positive relationship between the FEV and FVC measurements, but much less relationship between the FEV and temperature measurements, as would be expected from the relevant pictures.

It should be noted that the correlation coefficient makes sense only if both regression coefficients do, that is if both the variables being compared are observed frequency distributions. If one of the variables is controlled by an experimenter rather than observed, the correlation coefficient should not be used.

We may wish to test whether there is any connection at all between two variables, or whether the null hypothesis of no correlation in the population is tenable. For this purpose we note that if there is zero correlation in the population, it must mean that the population covariance is zero, and thus the population regression coefficients will also be zero, and testing one of these (it does not matter which), using the test given above, will serve the purpose.

Precautions in use and interpretation

In using and interpreting the correlation coefficient certain points must be observed.

(1) *The relationship must be representable by a straight line.* In calculating this coefficient we are, as has been shown, presuming that the relationship between the two factors with which we are dealing is one which a straight line adequately describes. If that is not approximately true then this measure of association is not an efficient one. For instance, we may suppose the absence of a vitamin affects some measurable characteristic of the body. As administration of the vitamin increases, a favourable effect on this body measurement is observed, but this favourable effect may continue only up to some optimum point. Further administration leads, let us suppose, to an unfavourable effect. We should then have a distinct *curve* of relationship between vitamin administration and the measurable characteristic of the body, the latter first rising and then falling. The graph of the points would be shaped roughly like an inverted U and no straight line could possibly describe it. Efficient methods of analysing that type of curvilinear relationship have been devised and the correlation coefficient should not be used. Plotting the observations, as in Fig. 18.1 relating to FEV and FVC, is a rough but reasonably satisfactory way of determining whether a straight line will adequately describe the observations. If the number of observations is large it would be a very heavy task to plot the individual records, but one may then plot the *means of columns* in place of the individual observations—e.g. the mean height of children aged 6–7 years was so many centimetres, of children aged 7–8 years so many centimetres—and see whether those means lie approximately on a straight line. Nowadays, of course, the time-consuming work of plotting many points can often be avoided by viewing them on a computer screen instead.

(2) *The line must not be unduly extended.* If the straight line is drawn and the regression equation found, it is dangerous to extend that line beyond the range of the actual observations upon which it is based. For example, in schoolchildren height increases with age in such a way that a straight line describes the relationship reasonably well. But to use that line to predict the height of adults would be ridiculous. If, for instance, at school ages height increases each year by 4 centimetres, that increase must cease as adult age is reached. The regression equation gives a measure of the relationship between certain observations; to presume that the same relationship holds beyond the range of those observations would need justification on other grounds.

(3) *Association is not necessarily causation.* The correlation coefficient is

a measure of *association*, and, in interpreting its meaning, one must not confuse association with causation. Proof that *A* and *B* are associated is not proof that a change in *A* is directly responsible for a change in *B* or vice versa. There may be some common factor *C* which is responsible for their associated movements. For instance, in the towns of a country in which tuberculosis is still common, it might be shown that the incidence of respiratory tuberculosis and overcrowding were correlated with one another, the former being high where the latter was high and vice versa. This is not necessarily evidence that the incidence of tuberculosis is directly due to overcrowding. Possibly, and probably, towns with a high degree of overcrowding are also those with a low standard of living and therefore of nutrition. This third factor may be the one which is responsible for the level of the tuberculosis rate, and overcrowding is only indirectly associated with it or, as discussed below, both factors may independently contribute to it.

It follows that the meaning of correlation coefficients must always be considered with care, whether the relationship is a simple direct one or due to the interplay of other common factors. In statistics we are invariably trying to disentangle a chain of causation and several factors are likely to be involved. Time correlations are particularly difficult to interpret but are particularly frequent in use as supposed evidence of causal relationships—e.g. the recorded increase in the death rate from cancer of the lung is attributed to an increase in some environmental factor such as air pollution from diesel engines. Clearly such concomitant movement might result from quite unrelated causes and the two characteristics might actually have no relationship whatever with one another except in time. Merely to presume that the relationship is one of cause and effect is fatally easy; to secure satisfactory proof or disproof is often a task of very great complexity.

(4) *Sampling variability*. As with all statistical values, the correlation coefficient must be regarded from the point of view of sampling errors. In taking a sample of individuals from a population it was shown that the mean and other statistical characteristics would vary from one sample to another. Similarly if we have *two* measures for each individual the correlation between those measures will differ from one sample to another. For instance, if the correlation between the age and weight in all schoolchildren were 0.85, we should not always observe that value in samples of a few hundred children; the observed values will fluctuate around it. Similarly if two characteristics are not correlated at all so that the coefficient would, if we could measure all the individuals in the population, be 0, we shall not necessarily reach a coefficient of exactly 0 in relatively small samples of those individuals. The coefficient observed in such a sample may have some positive or negative value. In practice, we have, then, to answer this question: could the value of the coefficient we have reached have arisen quite easily by chance in taking a sample, of the size observed, from a population in which there is no correlation at all between the two characteristics? For example, in a sample of 145 individuals we find the correlation between two characteristics to be +0.32. Is it likely that these two characteristics are not really correlated at all; that if we had taken a very much larger sample of observations the coefficient

would be 0 or approximately 0? As already stated, a significance test of the regression coefficient is the simplest answer to the question.

Confidence limits for the correlation coefficient

Confidence limits can be found using a device called the z transformation, which is quite easy to use if you have a calculator that includes 'natural logarithms' among its functions. The appropriate key is likely to be marked either '$\log_e$' or 'ln'. If you do not have such a key but you do have ordinary 'common logarithms' (key marked 'log') then all you need to do is to find the common logarithm and multiply it by 2.3026. The reason that two sorts of logarithms exist is that common logarithms are much more convenient in arithmetic work, but natural logarithms are much more convenient in algebraic work.

We then need the formula

$$z = \tfrac{1}{2} \text{ natural logarithm of } \frac{1+r}{1-r}$$

For our FEV and FVC data we had $r = 0.8826$, giving

$$z = \tfrac{1}{2} \text{ natural logarithm of } \frac{1.8826}{0.1174} = 1.3874$$

z is, to a good approximation, normally distributed with standard error $1/\sqrt{(n-3)} = 1/\sqrt{(232-3)} = 0.06608$. So 95 per cent confidence limits for z are

$$1.3874 \pm 1.96 \times 0.06608$$

i.e. 1.2579 to 1.5169.

To get back from z to r, we need the reverse formula

$$r = \frac{e^{2z} - 1}{e^{2z} + 1}$$

where the number e is the same one used in the Poisson distribution. If $z = 1.2579$, $e^{2z} = 12.3763$, and if $z = 1.5169$, $e^{2z} = 20.7773$.

So the 95 per cent limits for r are

$$\frac{11.3763}{13.3763} = 0.8505 \text{ and } \frac{19.7773}{21.7773} = 0.9082$$

Rank correlation

As with the t tests, where we had the alternative Wilcoxon tests, which used rank numbers instead of the actual values of the observations and were valid without any need to assume that the population values were normally distributed, so here. The distribution of the ordinary correlation coefficient depends upon the underlying distributions of the observations, usually taken to be normal, but that assumption can be avoided by using rank values.

There are two different versions that may be used, known as the Spearman rank correlation and the Kendall rank correlation. We shall concentrate here on the Kendall version as being the worthier of the two. We shall illustrate the idea on a small sample, using a method that makes the process clear; it would be rather difficult to do by hand by this method on a large sample, but that need not deter us from gaining the insight of what the method involves.

Taking a random sample of 6 people from the distribution of FEV and FVC readings, the figures in Table 18.3 were obtained.

Table 18.3

Person	*FEV*	*FVC*
A	4.02	5.13
B	2.69	4.54
C	4.34	5.07
D	3.55	4.50
E	2.56	2.96
F	3.66	5.11

Arranging each column in order of size we get Table 18.4, where the two readings from the same person are each time joined by a straight line. We find the number of places where lines cross (c, for crossings) and calculate the maximum number of such crossings there could be (m, for maximum), which is given by the formula

$$m = \tfrac{1}{2}n(n-1)$$

Here we have $n = 6$

$$m = \tfrac{1}{2} \times 6 \times 5 = 15$$

and $c = 3$

Denoting Kendall's coefficient r_K (r being the usual symbol for a correla-

Table 18.4

FEV	*FVC*
4.34	5.13
4.02	5.11
3.66	5.07
3.55	4.54
2.69	4.50
2.56	2.96

tion and the subscript K indicating the Kendall rank version), the formula is then

$$r_K = \frac{m - 2c}{m}$$

It is evident that, like an ordinary correlation coefficient, perfect agreement will produce a value of 1 (because $c = 0$), whereas perfect negative agreement will produce a value of -1 (because $c = m$). Here we have

$$r_K = \frac{15 - 2 \times 3}{15} = 0.6$$

Table VII then tells us that for $n = 6$, we need r_K to be greater than 0.75 for 5 per cent significance, greater than 0.8 for 2 per cent significance, or greater than 0.9 for 1 per cent significance. These are two-tailed figures so values of less than -0.75, -0.8, or -0.9, would also give 5, 2 or 1 per cent significance respectively.

Similarly, taking another random sample of 6 people, but using the FEV and temperature readings, gave Table 18.5.

As before, arranging each column in order of size, and drawing lines to connect the two readings for the same person, we get Table 18.6.

Note that the 'straight' lines should be bent slightly, as necessary to prevent more than two from crossing at any one point. This does not change the number of crossings, and makes that number easier to count.

Table 18.5

Person	*FEV*	*Temperature*
A	3.12	22.0
B	3.89	24.0
C	3.06	19.2
D	2.46	22.5
E	5.13	20.5
F	4.37	18.7

Table 18.6

FEV	*Temperature*
5.13	24.0
4.37	22.5
3.89	22.0
3.12	20.5
3.06	19.2
2.46	18.7

We have $n = 6$

$$m = \tfrac{1}{2} \times 6 \times 5 = 15$$

$$c = 9$$

$$r_K = \frac{15 - 2 \times 9}{15} = -0.2$$

A simple test for correlation

If the aim is merely to demonstrate that two variables are correlated, without (at least at first) bothering about the values of the correlation and regression coefficients, a quick simple test is available by splitting the observations into four compartments using the two medians as the borders, and then testing as a 2×2 table.

For example, in Figs. 18.3 and 18.4 random samples of 20 of the previous observations of FEV and FVC, and of FEV and temperature, are shown with the median lines drawn in, and the number of points counted in each of the four compartments. (Any point falling *exactly* on a median line, including the double-median point itself if the number of observations is odd, is not counted).

From the first distribution we get

1	9	10
9	1	10
10	10	20

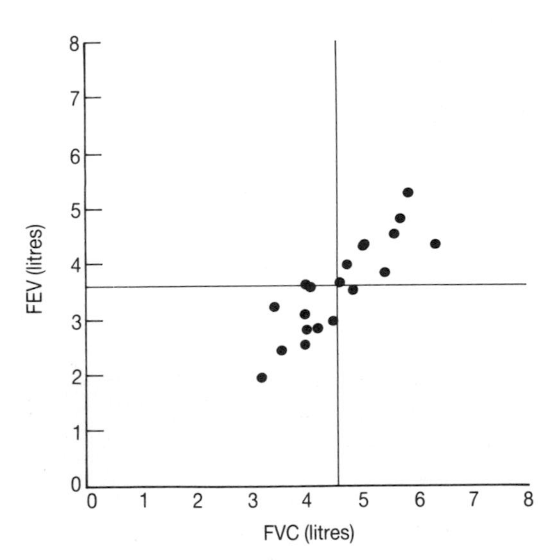

Fig. 18.3 A random sample of 20 from Figure 18.1 with the two median lines.

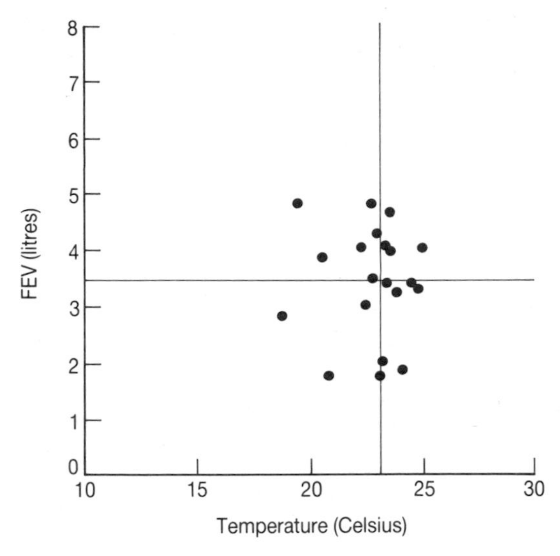

Fig. 18.4 A random sample of 20 from Figure 18.2 with the two median lines.

giving $P = 0.0011$, showing the clear correlation there. From the second we get

6	4	10
4	6	10
10	10	20

giving $P = 0.66$, showing no significant correlation.

Multiple regression

In medicine, associations between the prevalence of disease or mortality and environmental factors are not uncommon, e.g. the prevalence of typhoid fever and the insanitary nature of a water supply. However, in many, if not most, real-life situations a simple one-to-one 'cause and effect' mechanism is unlikely to prevail. The effect is more likely brought about by a number of causes, some inter-related, some independent of each other.

For instance it might be found that deaths due to bronchitis and pneumonia each week were associated with, and possibly influenced by, the mean air temperature of that week. Even for so simple (or apparently simple) a situation, very great care would be needed in interpretation, for there is a hidden time variable that could be important. Bronchitis and pneumonia could be more prevalent in the winter than in the summer for reasons quite unconnected with temperature, and the correlation with temperature would then be spurious.

There are many variables here, all correlated with each other. Maybe the mortality rate is affected by relative humidity, by the prevailing degree of air pollution, by the number of hours of daylight, etc.

If we have the records of all such features for each week, then methods are available in almost every statistical computer package to construct a *multiple regression equation*. This will take the form:

$$
\begin{aligned}
&(\text{weekly deaths} - \text{mean weekly deaths})\\
&\quad = a\,(\text{temperature} - \text{mean temperature})\\
&\quad + b\,(\text{relative humidity} - \text{mean relative humidity})\\
&\quad + c\,(\text{index of air pollution} - \text{mean air pollution})\\
&\quad + d\,(\text{hours of daylight} - \text{mean hours of daylight})
\end{aligned}
$$

where the regression coefficients a, b, c and d may each, of course, be positive or negative, and each has its associated standard error. Each shows the amount by which the weekly deaths will (on the average) change when the relevant variable changes by one unit while the other features remain unchanged. However, as with simple regression, seriously wrong results can be derived if the relation of deaths to each explanatory variable is not, at least approximately, a straight line.

By such methods of analysis we can hope to (1) disentangle the complex causative mechanism at work, (2) estimate the relative importance of each contributing cause, and (3) when more than one cause is operating, predict more accurately the weekly level of mortality when changes take place in these variables. With modern computing methods, multiple regression has become a very widely used method. Unfortunately the computer output is often merely taken at face value, with little attempt to take the necessary care in interpretation or to contemplate possible fallacies. In our particular example, for instance, the hidden time variable is still present and may be important for its correlation with other variables that have not been considered.

Summary

The correlation coefficient is a useful measure of the degree of association between two characteristics, but only when their relationship is adequately described by a straight line. The equation to this line, the regression equation, allows the value of one characteristic to be estimated when the value of the other characteristic is known. The error of this estimation may be large even when the correlation is very high. Evidence of association is not necessarily evidence of causation, and the possible influence of other common factors must be remembered in interpreting correlation coefficients.

Chapter 19

Bayesian methods

Over recent years 'Bayesian' methods have become more popular among medical statisticians; the aim of this chapter is to give some indication of what these methods are and the arguments that arise involving them. If fuller details are wanted they must be sought elsewhere, but it is hoped that this brief review will enable the reader to gain at least some understanding of what is going on.

Likelihood

In everyday speech the words 'probability' and 'likelihood' mean virtually the same thing. The same is true in statistics, but 'likelihood' has been given a precise special meaning, and it is unwise to use the word in a statistical context unless that special meaning is intended.

Suppose we believe that a patient suffering from a given disease has 1 chance in 4 (a probability of 0.25) of dying from it. We have observed 6 such patients and 4 of them died—a surprisingly high number, which we wish to examine more closely. If the true chance actually is 0.25, as assumed, the appropriate probabilities of observing anything from 0 to 6 deaths are given in the second row of Table 19.1, as derived from the relevant binomial distribution.

However, our belief that 0.25 is the true chance might be wrong, so we might want to consider other possible values as shown in the other rows of the table. These other four rows are, of course, only a selection from an

Table 19.1 Probabilities from binomial distributions.

Probability of an individual patient dying	*Number of deaths observed among 6 patients*						
	0	1	2	3	4	5	6
0.10	0.5314	0.3543	0.0984	0.0146	0.0012	0.0001	—
0.25	0.1780	0.3560	0.2966	0.1318	0.0330	0.0044	0.0002
0.50	0.0156	0.0937	0.2344	0.3125	0.2344	0.0937	0.0156
0.75	0.0002	0.0044	0.0330	0.1318	0.2966	0.3560	0.1780
0.90	—	0.0001	0.0012	0.0146	0.0984	0.3543	0.5314

infinite number of possible rows that could be presented.

If, instead of considering the values in different columns of the table for a given row, we switch our attention to considering the values in different rows for a given column (namely the column headed 4—the number of deaths actually observed) it is then that we call the values *likelihoods* instead of *probabilities*. This change of wording to indicate the different viewpoint is useful in keeping our ideas clear.

In the medical field one place where likelihoods are relevant is in problems of diagnosis. If we form a similar table to Table 19.1, giving a list of possible diagnoses down the side, and a list of signs and symptoms across the top, can we derive the information to fill in the values across each row—that is, for a given diagnosis, the probabilities of each set of signs and symptoms? If we can, then we can read the values off vertically, for a patient with given signs and symptoms, to get the likelihood associated with each possible diagnosis. The mistake must not be made, however, of thinking that these are the probabilities that a diagnosis is correct. To get those we need further information—of the *prior probabilities*. That is to say we need to know the probability, before we see the patient at all, that any given patient will have a particular diagnosis. Such prior probabilities may, perhaps, be estimated from the prevalence of various diseases at the particular time and place concerned.

If we have that information, then *Bayes' theorem* tells us how to combine prior probabilities and likelihoods to produce *posterior probabilities*. These are then the probabilities of each particular diagnosis being the true one.

Bayes' theorem

Bayes' theorem (named after the Rev. Thomas Bayes, 1702–61) says that to get posterior probabilities, you multiply prior probabilities by likelihoods and then adjust the results (keeping them in proportion) so that they add to 1. To see how this works we shall first of all look at a simple, though non-medical and unrealistic, example. In spite of its lack of reality, it is a good example in making the process clear; having done that we shall return to the medical implications.

The example is of a coin that has been found and we wish to know whether it is double-headed, double-tailed or an ordinary one having one head and one tail, but the only information we are allowed is that the coin should be tossed and we are told whether it landed to show heads or tails.

To get started we need to know the prior probability that it is a coin of each of the three possible types. Let us assume that there is one chance in ten thousand that it is double-headed, one in ten thousand that it is double-tailed, and 9998 in ten thousand that it is ordinary. Where did these probabilities come from? We simply made them up—a point to which we shall return. Accept them for the moment. Now for the first toss: 'heads' is reported, and Table 19.2 shows the resulting calculation. The likelihoods come from the fact that if the coin is double-headed it is certain to show 'heads', if it is ordinary there is a 50 per cent chance, if it is double-tailed it

Table 19.2 Use of Bayes' theorem: *Toss 1.*

Hypothesis	*Prior probability* (1)	*'Heads' likelihood* (2)	*Product (1) × (2)* (3)	*Posterior probability (3)/Total of (3)* (4)
Double-headed	0.0001	1.0	0.0001	0.0002
Ordinary	0.9998	0.5	0.4999	0.9998
Double-tailed	0.0001	0.0	0.0000	0.0000
Total	1.0000		0.5000	1.0000

is impossible. The result shows little change from the original, except that double-tailed has been ruled out, its probability having been transferred to double-headed.

We can notice in Table 19.2 an example of the possible mistake mentioned above of thinking that likelihoods are probabilities. Thus the 1.0 figure at the head of column (2) does not mean that if the coin shows heads it is certainly double-headed—that would clearly be nonsense—but only that if it is double-headed it is certain to show heads.

If, on the next toss, we again get 'heads', Table 19.3 shows what happens. The posterior probabilities from the first toss are now the prior probabilities for the second, and so on. If we continue to get 'heads', over and over again, Table 19.4 shows that after the 14th toss, double-headed

Table 19.3 Use of Bayes' theorem: *Toss 2.*

Hypothesis	*Prior probability* (1)	*'Heads' likelihood* (2)	*Product (1) × (2)* (3)	*Posterior probability (3)/Total of (3)* (4)
Double-headed	0.0002	1.0	0.0002	0.0004
Ordinary	0.9998	0.5	0.4999	0.9996
Double-tailed	0.0000	0.0	0.0000	0.0000
Total	1.0000		0.5001	1.0000

Table 19.4 Use of Bayes' theorem: *Toss 14.*

Hypothesis	*Prior probability* (1)	*'Heads' likelihood* (2)	*Product (1) × (2)* (3)	*Posterior probability (3)/Total of (3)* (4)
Double-headed	0.4504	1.0	0.4504	0.6210
Ordinary	0.5496	0.5	0.2748	0.3790
Double-tailed	0.0000	0.0	0.0000	0.0000
Total	1.0000		0.7252	1.0000

Table 19.5 Use of Bayes' theorem: *Toss 15*.

Hypothesis	*Prior probability* (1)	*'Tails' likelihood* (2)	*Product (1) × (2)* (3)	*Posterior probability (3)/Total of (3)* (4)
Double-headed	0.6210	0.0	0.0000	0.0000
Ordinary	0.3790	0.5	0.1895	1.0000
Double-tailed	0.0000	1.0	0.0000	0.0000
Total	1.0000		0.1895	1.0000

has actually become more probable than an ordinary coin. However, if the 15th toss shows 'tails', the whole thing collapses into certainty that the coin must be ordinary after all (Table 19.5). It would be unusual, in less artificial circumstances, for a calculation to end like this with one hypothesis certain and all the others impossible. In real-life examples, the most that can normally be expected is that one hypothesis should become much more probable than the others.

The medical relevance

Basically the same procedure can be used to find the probability of a given diagnosis being correct, provided that: (1) it is possible to get estimates of the prior probabilities, that a patient has any given diagnosis; (2) we know the probabilities of the various signs and symptoms for the various diagnoses. The process will be complicated, however, by the fact that the different signs and symptoms will be correlated with each other, and these correlations must be allowed for in the calculations.

Where are the prior probabilities to come from? In the unrealistic example above, we merely invented them. In real life it should be possible to do better than that, by using figures of prevalence, for example, but not forgetting to update them when necessary (as during an epidemic).

Suppose we are applying a screening test for cancer, where the prevalence of cancer among those being tested is 2 per cent and the test being used has a sensitivity of 90 per cent and a specificity of 95 per cent (i.e. the

Table 19.6 Bayes' theorem applied to a hypothetical test for cancer. Analysis if result of test is positive.

Actual condition	*Prior probability* (1)	*Test result likelihood* (2)	*Product (1) × (2)* (3)	*Posterior probability (3)/Total of (3)* (4)
Cancer	0.02	0.90	0.018	0.269
Clear of cancer	0.98	0.05	0.049	0.731
Total	1.00		0.067	1.000

Table 19.7 Bayes' theorem applied to a hypothetical test for cancer. Analysis if result of test is negative.

Actual condition	*Prior probability*	*Test result likelihood*	*Product (1) × (2)*	*Posterior probability (3)/Total of (3)*
	(1)	*(2)*	*(3)*	*(4)*
Cancer	0.02	0.10	0.002	0.002
Clear of cancer	0.98	0.95	0.931	0.998
Total	1.00		0.933	1.000

test gives a positive result 90 per cent of the time if cancer is present—10 per cent false negatives—and a negative result 95 per cent of the time if cancer is not present—5 per cent false positives). Table 19.6 then shows the analysis, following a positive test result, and Table 19.7 following a negative test result.

Again these results make clear that likelihoods should not be mistaken for probabilities. If cancer is present, there is a 90 per cent chance of a positive test, but (because of the low prevalence) a positive test gives only a 26.9 per cent chance that cancer is present.

Pros and cons

Many statisticians feel that the whole concept is too vague, and that it should be possible to derive the meaning of data without resort to prior probabilities. They say that if results are to depend upon such a concept, then they do not trust those results, whereas if results do not depend on it, why introduce it?

The opposing school reply that nobody in his senses would try to interpret results without taking previous information into account and it is better to do so formally rather than informally. Furthermore, by trying several different sets of prior probabilities on the same data, if the results are widely different it shows that the data are not themselves supplying a firm answer, whereas if the results are much the same each time, then those results are trustworthy. For the present the matter remains controversial.

Summary

The name 'likelihood' is used to denote probabilities when regarded in terms of fixed data, and various hypotheses that might have led to those data. Bayes' theorem is concerned with the combination of prior probabilities and likelihoods to derive posterior probabilities. The aim of this chapter has not been to describe such methods completely, but to give the doctor some indication of what the methods are trying to do so that, if a statistician wishes to use them, there will be some knowledge of the aims.

Chapter 20

Standardised death rates and indices

In using death rates in comparison with one another, or as a measure of the success of some treatment or procedure, it must be remembered that such rates are usually affected considerably by the age and sex constitution of the population concerned. The fact that the death rate of an attractive seaside resort on the south coast of England was, in a given year, 13.3 per 1000, while the rate in the same year in an unattractive and by no means well-to-do area in east London was only 10.3, is no evidence of the salubrity of the latter. The greater proportion of old people living in the south coast resort *must* lead to a higher death rate there, since old people, however well-housed and fed, die at a faster rate than young people. The census, in fact, shows that there were at the time 2½ times as many people in the south coast resort as in the specified London area at ages over 75 years, 70 per cent more at ages 50–74, and 10 per cent less at ages 10–40. Any population containing many people round about the ages 5–20, where the death rate is at its minimum, must have a lower *total* death rate than that of a population containing many old people, at which point of life the death rate is relatively high, even though comparisons at *every* age show an advantage to the latter. The fictitious figures of Table 20.1 may be taken as an example.

Comparison of the two districts shows that *B* has in *every age-group* a lower death rate than *A*. Yet its death rate at all ages, the *crude* death rate, is more than double the rate of *A*. The fallacy of the crude rates lies in the fact that like is not being compared with like: 72 per cent of *B*'s population is over age 45 and only 26 per cent of *A*'s population; in spite of *B*'s relatively low death rates at these ages over 45, the *number* of deaths registered must be higher than in *A*'s smaller population and therefore its total death rate must be high.

The same problem arises frequently when we need to compare the death rates prevailing in some area at *different times*. For example, the crude death rate, i.e. at all ages, of females in England and Wales was 11.3 per 1000 in the year 1969 and virtually the same, 11.14 per 1000, in 1987. Yet in these 18 years the mortality at *every* stage of life had declined, and usually considerably, as shown in Table 20.2.

Table 20.1 A comparison of death rates.

	District A				District B			
Age-group (years)	*Population*		*Deaths*	*Death rate per 1000*	*Population*		*Deaths*	*Death rate per 1000*
0–	500	74%	2	4	400	28%	1	2.5
15–	2000		8	4	300		1	3.33
30	2000		12	6	1000		5	5
45–	1000	26%	10	10	2000	72%	18	9
60–	500		20	40	2000		70	35
75+	100		15	150	400		50	125
All ages	6100		67	11.0	6100		145	23.8

Table 20.2 Death rates per 1000 of females in England and Wales in 1969 and 1987.

	Age in years								
Year	0–	5–	15–	25–	35–	45–	55–	65–	75+
1969	3.56	0.24	0.41	0.63	1.74	4.38	10.58	28.59	100.41
1987	1.99	0.16	0.30	0.48	1.13	3.23	9.13	22.75	82.92
1987 as % of 1969	56	67	73	76	65	74	86	80	83

The explanation lies, of course, in the changes in the age distribution of the population between these dates. Thus in 1969 women of 55 years or more formed 27.7 per cent of the total female population. In 1987 the proportion had risen to 29 per cent. The increased numbers at the stages of life at which the death rates are substantial are having a dominating effect upon the crude rate, giving the misleading effect of no improvement.

It follows that in many countries where there is an increasing proportion of persons living at later ages (owing to a fall in the birth rates and death rates) it is certain that the crude death rate from all causes must rise in spite of the fact that mortality at each age continues to decline. The only really satisfactory answer to this problem is to compare the death rates at different ages—as shown in Tables 20.1 and 20.2. *No single figure, however derived, can ever fully replace them or succinctly summarise numerous contrasts.*

The standardised rate

At the same time a legitimate desire is often felt for a single mortality rate, summing up the rates at ages and enabling satisfactory comparisons to be made between one rate and another. For this purpose the standardised death rate is generally used. For its calculation (by what is known as the *direct* method) the mortality rates at ages in the different districts (or at different times) are applied to some common standard population, to discover what would be the total death rate in that standard population if it were exposed first to *A*'s rates and then to *B*'s rates at each age. These total rates are clearly fictitious, for they show what *would* be the mortality in *A* and *B* if they had populations which were equivalent in their age distributions instead of their actual differing populations. But these fictitious rates are more comparable with one another, and show whether *B*'s rates at ages would lead to a better, or worse, total rate than *A*'s rates if they had populations of the same age type.

For example, if the standard population taken for the districts *A* and *B* of Table 20.1 were as shown in Table 20.3, then population *A*'s death rates would lead to a total of 235 deaths and *B*'s rates to 201 deaths, giving standardised rates at all ages of 19.6 and 16.8 per 1000. Taking a population of the same age distribution thus shows the more favourable mortality experience of *B*, and the fallacy of the crude rate is avoided.

This example is, of course, a very exaggerated one and such gross differences in population are unlikely to occur in practice. On the other hand, the differences that do occur in practice are quite large enough to make the use of crude rates seriously misleading—as shown in Table 20.2.

To take another example, the crude death rate, in England and Wales, of women from cancer (of all forms) was 188 per 100 000 in the years 1951–55 and had risen to 231 twenty five years later; thus a rise in mortality of 23 per cent was recorded during that short span of years. The corresponding standardised rates were 185 and 194, a rise of only 5 per cent. In other words the large number of women living in the older age groups (where cancer is more frequent) in the later period compared with the earlier was largely responsible for the increase in the crude rates, and no more subtle environmental factor need be sought.

Comparisons of death rates may also be affected by the sex proportions of the populations considered, for at most ages and from many causes females tend to suffer a lower mortality rate than males. Standardisation, therefore, is sometimes made for both sex and age. Some methods in general use will now be illustrated in detail. They are equally applicable to all causes of death or to specific causes.

Table 20.3 An imagined 'standard population'.

Age in years	*Number of persons*
0–	500
15–	2500
30–	3000
45–	3000
60–	2500
75+	500

The direct method

In Columns (2) and (3) of Table 20.4 are given for males and females separately the populations for a specific year for (a) Large Towns and (b) Rural Districts of England and Wales. The deaths that these populations experienced in that year are given in Columns (4) and (5). Combining these two items of information leads to the death rates of Columns (6) and (7). In Fig. 20.1 the death rates of the rural districts are plotted as percentages of the rates in the large towns; it is clear that the former group

Table 20.4 Mortality in large towns and rural districts in a specific year.

			Large towns			
	Estimated population (to nearest hundred)		*Number of deaths from all causes*		*Death rates per 1000*	
Age-group in years	*Males*	*Females*	*Males*	*Females*	*Males*	*Females*
(1)	*(2)*	*(3)*	*(4)*	*(5)*	*(6)*	*(7)*
0–	461 000	447 000	1 835	1 320	3.98	2.95
5–	998 400	968 900	359	257	0.36	0.26
15–	1 011 600	1 071 100	931	653	0.92	0.61
25–	1 030 700	1 104 800	1 165	773	1.13	0.70
35–	845 500	980 300	2 215	1 647	2.62	1.68
45–	727 900	864 300	6 631	4 425	9.11	5.12
55–	590 900	680 800	15 103	11 648	25.56	17.11
65–	413 100	496 200	29 161	23 256	70.59	46.87
75+	108 700	172 800	16 637	21 925	153.05	126.88
All ages	6 187 800	6 806 200	74 037	65 904	11.96	9.68
All ages, both sexes	12 994 000		139 941		10.77	

			Rural districts			
	Estimated population (to nearest hundred)		*Number of deaths from all causes*		*Death rates per 1000*	
Age-group in years	*Males*	*Females*	*Males*	*Females*	*Males*	*Females*
(1)	*(2)*	*(3)*	*(4)*	*(5)*	*(6)*	*(7)*
0–	255 900	246 000	814	566	3.18	2.30
5–	598 100	575 600	179	121	0.30	0.21
15–	595 000	517 200	428	253	0.72	0.49
25–	587 400	566 400	599	312	1.02	0.55
35–	481 500	529 400	896	656	1.86	1.24
45–	422 300	470 100	2 770	1 782	6.56	3.79
55–	366 800	398 200	7 314	5 041	19.94	12.66
65–	243 300	259 900	13 912	9 621	57.18	37.02
75+	94 500	123 700	12 727	14 284	134.68	115.47
All ages	3 644 800	3 686 500	39 639	32 636	10.88	8.85
All ages, both sexes	7 331 300		72 275		9.86	

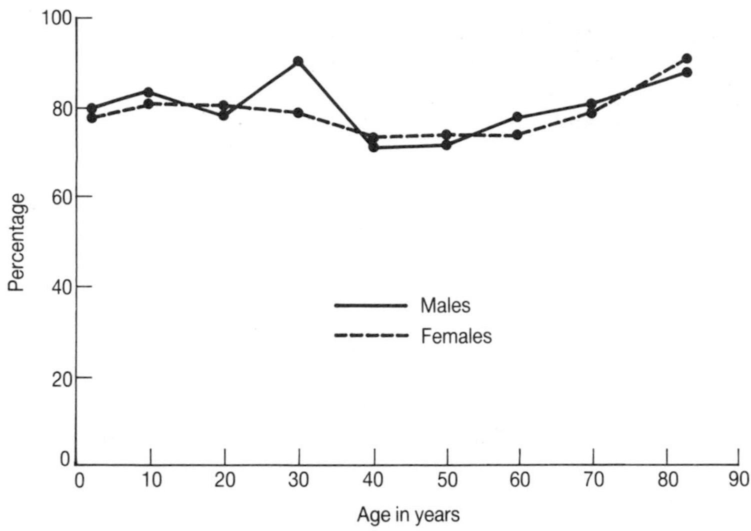

Fig. 20.1 The death rates in the rural districts, expressed as percentages of the rates in the large towns.

had a substantial advantage at all ages and for both sexes.

Returning to Table 20.4 two things will be observed. First, in both groups of areas and in every one of the age groups the death rate of females is less than the death rate of males. Clearly the population that contains the greater number of females will benefit in its total death rate. In fact, 52 per cent of the population are females in the large towns against 50 per cent in the rural districts. Secondly, the death rate is relatively high at ages 0–5 and rises steeply after age 55. The population that contains the smaller number of persons at these ages will thereby benefit in its total death rate. In fact, both have 7 per cent of their total population at ages 0–5 but in the large towns 19 per cent of the population is of age 55 and over compared with 20.3 per cent in the rural districts. Correspondingly, more of the population of the large towns belongs to the age group 5–35, when the death rate is relatively low, than is the case in the rural districts.

Although these age and sex differences are not huge, their result is that the crude death rate of the rural districts is 9.86 per 1000; that of the large towns 10.77 per 1000. The relatively large advantages in the age groups have been much reduced in the total.

The direct method of standardisation endeavours to meet this situation by seeing what total death rates the two sets of rates at ages would lead to if they existed *not* in two populations of different age and sex constitution, but in *identical* populations. The first step, then, is the choice of the identical, or standard, population to be used. In the upper half of Table 20.5 the population of England and Wales in 1961 has been taken for that purpose. By simple proportion it is an easy matter to determine the number of deaths that would occur in this population, first at the death rates of the large towns and then at the rates of the rural districts (as given in Table 20.4). For instance, in the large towns 1835 males died at ages 0–4

Table 20.5 Standardisation of the mortality in large towns and rural districts. Direct method, using as standard the population of England and Wales in (1) 1961, (2) 1987.

(1) *Standard population 1961*			*Deaths that would occur in standard population at mortality rates of large towns*		*Deaths that would occur in standard population at mortality rates of rural districts*	
Age-group in years (1)	*Males (2)*	*Females (3)*	*Males (4)*	*Females (5)*	*Males (6)*	*Females (7)*
0–	1 882 000	1 783 000	7 491	5 265	5 987	4 102
5–	3 544 000	3 377 000	1 274	878	1 061	710
15–	3 072 000	3 015 000	2 827	1 838	2 210	1 475
25–	2 932 000	2 878 000	3 314	2 014	2 990	1 585
35–	3 115 000	3 175 000	8 161	5 334	5 797	3 934
45–	3 165 000	3 280 000	28 832	16 793	20 760	12 433
55–	2 516 000	2 911 000	64 307	49 805	50 169	36 852
65–	1 431 000	2 105 000	101 015	98 658	81 825	77 923
75+	689 000	1 296 000	105 454	164 437	92 793	149 653
All ages	22 346 000	23 820 000	322 675	345 022	263 592	288 667
All ages, both sexes	46 166 000		667 697		552 259	
Standardised death rate per 1000			14.46		11.96	

(2) *Standard population 1987*			*Deaths that would occur in standard population at mortality rates of large towns*		*Deaths that would occur in standard population at mortality rates of rural districts*	
Age-group in years (1)	*Males (2)*	*Females (3)*	*Males (4)*	*Females (5)*	*Males (6)*	*Females (7)*
0–	1 653 500	1 573 100	6 582	4 645	5 260	3 619
5–	3 163 800	2 995 000	1 138	778	947	630
15–	4 095 000	3 932 200	3 769	2 397	2 946	1 924
25–	3 641 900	3 589 000	4 116	2 511	3 714	1 977
35–	3 484 200	3 463 900	9 128	5 820	6 484	4 292
45–	2 731 700	2 720 300	24 885	13 927	17 918	10 312
55–	2 589 700	2 736 900	66 191	46 826	51 639	34 648
65–	1 989 600	2 499 000	140 447	117 124	113 766	92 508
75+	1 143 100	2 240 700	174 956	284 302	153 950	258 740
All ages	24 492 500	25 750 100	431 212	478 330	356 624	408 650
All ages, both sexes	50 242 600		909 542		765 274	
Standardised death rate per 1000			18.10		15.23	

out of a population of 461 000, whereas in the standard population the total in that age group is 1 882 000. Therefore the number that would occur if the observed death rate were applied to the standard population is (1835/461 000) × 1 882 000 = 7491 to the nearest whole number, which is quite accurate enough for the purpose. 7491 is shown as the first value in Column (4) of Table 20.5, and all the other figures in Columns (4)–(7) of that table have been calculated similarly. It is found that at the rates of the large towns the standard population would have experienced a total of 667 697 deaths, and at the rates of the rural districts a total of 552 259 deaths. As the standard population amounts in all to 46 166 000 persons (Columns (2) and (3)), the rates of the large towns would, therefore, lead to a rate of all ages in this standard population of (667 697/46 166 000) × 1000 = 14.46 per 1000. Similarly the rates of the rural districts would lead to a rate at all ages of (552 259/46 166 000) × 1000 = 11.96 per 1000. These are the two standardised death rates, using the 1961 population as standard.

The choice of the standard population

It will be noted, however, that different standardised rates must result if a different population is chosen as the standard. For instance, in the lower half of Table 20.5 the population of England and Wales in 1987 has been used and the two standardised rates on that basis are 18.10 and 15.23 per 1000 in place of the previous 14.46 and 11.96. In consequence the question that immediately arises is what standard population ought to be adopted. The standardised death rate is, it will be realised, a fiction. It is not the total death rate that actually exists in an area but the rate that the area would have if, while retaining its own rates at ages, it had instead of its real population one of some particular chosen type. The fiction is useful because, as already seen, it enables summary comparisons to be made between places or between epochs, and these comparisons are free from distortion which arises from age and sex differences in the existing populations. The object throughout is, therefore, *comparison*; a standardised death rate alone has little meaning. Accordingly it does not really matter if by the use of a different standard population the standardised death rates of the areas under study are changed, *so long as their relative position is not changed materially*. In the example of Tables 20.4 and 20.5 it is seen that the crude rate of the rural districts is 92 per cent of the crude rate of the large towns, a very poor expression of the differences at ages revealed in Fig. 20.1. Using the 1961 census population of England and Wales, the standardised rates are 14.46 in the large towns and 11.96 in the rural districts. The latter is 83 per cent of the former, a very much better expression of the differences at ages. Changing to the 1987 population of England and Wales changes the standardised rates to 18.10 in the large towns and 15.23 in the rural districts. The latter is 84 per cent of the former, so that the *relative* position is virtually unchanged by the alteration in the standard.

In practice, the comparative position between the standardised rates of two areas, or between two points of time, will not, when the standard population is changed, invariably be as close as is given in this example, and sometimes, indeed, serious differences may result, though probably, experience indicates, not very frequently.

The indirect method

The direct method of standardisation illustrated above requires a knowledge for each age group of (a) the numbers of persons, and (b) the number of deaths in the population for which a standardised death rate is needed. Sometimes all that information is not available, or sometimes the populations at the different ages are so small that the death rates are subject to large fluctuations through the presence or absence of merely a few deaths. In such instances the indirect method can be applied. The first step in this method is the selection of a series of *standard death rates* at ages. In Table 20.6 the rates in England and Wales in 1987 have been chosen (Columns (4) and (5)). These rates are then applied (by simple proportion) to the population at ages in the area for which the standardised rate is sought, to determine what deaths would have occurred in that area if it had had these standard rates. For example, Column (2) shows that in the large towns there were 461 000 males at ages 0–5. The death rate of males of these ages in England and Wales in 1987 was 2.55 per 1000. Therefore, at this rate, there would have been $(461\,000 \times 2.55)/1000 = 1176$ deaths in the large towns. This procedure is carried out for each age group for each sex and the deaths expected at the standard rates are totalled. It is thus found that if these standard rates had been operative in both groups of areas there would have been 83 742 deaths in the large towns and 53 630 in the rural districts, or rates of mortality at all ages of 6.44 and 7.32 per 1000 inhabitants (dividing the 'expected' deaths by the total populations of the two sets of areas). These rates are called the *index death rates*, for their level is an index of the type of population from which they have been derived. For instance, if the population contains a large proportion of old people its index death rate will be greater than that of a population composed more of young people. In the particular example taken, the index rates show, in fact, that the rural population is of a type to produce, under standard conditions of mortality, a higher death rate at all ages than the urban population. In other words, their crude rates cannot be safely compared, and some adjustment of them is necessary to allow for the difference in population type revealed by their index rates.

The adjustment consists of increasing or diminishing the recorded crude mortality rate to compensate for the advantage or disadvantage disclosed by the index rate; the next step, therefore, is to determine how much compensation must be made in each case. The standard rates of 1987 are, of course, derived from the deaths and the mean population in England and Wales in those years, and with that type of population they lead to a total death rate of 11.25 per 1000. If the populations of the large towns and rural districts had also been of exactly that type then their index death rates

Table 20.6 Standardisation of the mortality in the large towns and rural districts. Indirect method using as standard the death rates of England and Wales in 1987.

	Large towns					
	Estimated population (to nearest hundred)		*Standard death rates per 1000*		*Number of deaths that would occur at standard rates*	
Age-group in years	*Males*	*Females*	*Males*	*Females*	*Males*	*Females*
(1)	*(2)*	*(3)*	*(4)*	*(5)*	*(6)*	*(7)*
0–	461 000	447 000	2.55	1.99	1 176	890
5–	998 400	988 900	0.12	0.16	120	158
15–	1 011 600	1 071 100	0.40	0.30	405	321
25–	1 030 700	1 104 800	0.88	0.48	907	530
35–	845 500	980 300	1.67	1.13	1 412	1 108
45–	727 900	864 300	5.01	3.23	3 647	2 792
55–	590 900	680 800	15.97	9.13	9 437	6 216
65–	431 100	496 200	41.22	22.75	17 028	11 289
75+	108 700	172 800	110.18	82.92	11 977	14 329
All ages	6 187 800	6 806 200	—	—	46 109	37 633
All ages, both sexes	12 994 000		11.25		83 742	

	Rural districts					
	Estimated population (to nearest hundred)		*Standard death rates per 1000*		*Number of deaths that would occur at standard rates*	
Age-group in years	*Males*	*Females*	*Males*	*Females*	*Males*	*Females*
(1)	*(2)*	*(3)*	*(4)*	*(5)*	*(6)*	*(7)*
0–	255 900	246 000	2.55	1.99	653	490
5–	598 100	575 600	0.12	0.16	72	92
15–	595 000	517 200	0.40	0.30	238	155
25–	587 400	566 400	0.88	0.48	517	272
35–	481 500	529 400	1.67	1.13	804	598
45–	422 300	470 100	5.01	3.23	2 116	1 518
55–	366 800	398 200	15.97	9.13	5 858	3 636
65–	243 300	259 900	41.22	22.75	10 029	5 913
75+	94 500	123 700	110.18	82.92	10 412	10 257
All ages	3 644 800	3 686 500	—	—	30 699	22 931
All ages, both sexes	7 331 300		11.25		53 630	

would naturally have come out as 11.25 as well. In that case no correction to their crude rates would be necessary, for under standard conditions of mortality their populations give the same total death rates as that which exists in the standard population. But, in fact, their index death rates have come out to be 6.44 and 7.32, and therefore under the selected standard conditions of mortality both areas have populations which lead to too low a death rate. To allow for this the crude rates of each must be lowered. The precise degree to which they must be lowered is measured by the ratio of the death rate at all ages in the standard to the index rates in the areas, namely 11.25/6.44 and 11.25/7.32.

In short, we find by means of the index rates what kind of population exists in different areas—favourable or unfavourable to mortality—and then apply a 'handicap' to each crude rate to compensate for the advantage or disadvantage disclosed. These handicaps are known as the *standardising factors*. The factor in the large towns is, accordingly, 11.25/6.44 = 1.747, and in the rural districts 11.25/7.32 = 1.537. The crude death rate of the large towns is 10.77 and of the rural districts 9.86 and therefore, applying to them their respective factors, the standardised rates are $10.77 \times 1.747 = 18.82$ and $9.86 \times 1.537 = 15.15$. According to this result the rural mortality is 81 per cent of the urban.

The choice of the standard rates

As with the direct method, the standardised rates will depend in part upon the standard rates that are used, and the question arises whether their *relative* positions will be materially changed by the change in the standard. Again the purpose is comparison, and again it will be found that the precise choice of standardised rates makes little difference to such a comparison provided that nothing outrageous is chosen, but standard rates that arc broadly characteristic of thc times and places involved.

Other aspects of the indirect method

One advantage of the indirect method is that if we have calculated the standardising factor for a particular population as registered in, say, the census year, then we can continue to apply that factor to its annual crude death rate so long as we have reason to suppose that the age and sex distribution of its population has not much changed. The method then involves merely the multiplication of the crude rate by the factor already determined, and is particularly rapid. The continued use of the factor is, of course, unjustified if the population type is changing, e.g. by migration inwards and outwards.

It will be observed that if different causes of death are being studied, then a standardising factor must be calculated specifically for each cause. For example, if an area has a population which favours a relatively high death rate from cancer then its population is of an old-age type, for it is at these old ages that cancer mainly prevails. Its factor for cancer will be less than 1 to allow for its *unfavourable* population in this respect. On the other hand, this type of population will favour a relatively low death rate from the infectious diseases that prevail in childhood. Its factor for such causes of death must therefore be greater than 1 to allow for its *favourable* population in this respect.

The standardised mortality ratio

With this indirect method it will in some circumstances be possible, and indeed advantageous, to dispense entirely with the fictitious standardised rates and to proceed immediately to some comparative index. Thus using as standard rates the mortality by sex and age in England and Wales in 1990 we might calculate the number of deaths that would have occurred at these rates in the population of the country as it was constituted 50 years previously. The expected total number of deaths we can compare directly with the number that actually occurred to give an index of the change in mortality that has taken place, while allowing for the concurrent changes that have occurred in the sex and age distribution of the population. Thus we have a *standardised mortality ratio* (or SMR) to express the rise or fall in mortality. Similarly the rates of all England and Wales in 1990 could be applied to the population of a particular area of the country, e.g. London. The number of deaths expected at these rates can be immediately compared with the number that actually occurred in London, in 1990, to give, again, an easily understood standardised mortality ratio of observation to expectation.

This form of comparison has been used extensively by the Registrar-General of England and Wales with the mortality in the years 1950–52 as a standard. Thus the standardised mortality ratio for any year shows the number of deaths registered in that year as a percentage of those which would have occurred had the sex/age mortality of the years 1950–52 operated on the sex/age population of the year in question.

This particular index has also been used extensively in the study of occupational mortality. Thus we may take the mortality rates of all males as the standard rates. Applying these to the population of a particular occupational group, we find the number of deaths that would have occurred in that group if it had had the standard rates. These expected deaths we can compare with the observed deaths, expressing the latter as a percentage of the former. For instance, according to the *Occupational Mortality Supplement*, 1970–72, of the Registrar-General of England and Wales, the number of deaths from all causes of farmers, foresters and fishermen was 10 041 at ages 20–64. If they had experienced the death rates of all males in those three years they would have had 11 034 deaths. This group, therefore, experienced 91 per cent of their expected mortality at the standard rates, differences in age distribution having been allowed for. On the other hand, the observed deaths amongst miners and quarrymen were 144 per cent of their expected mortality at the standard rates. The procedure for the calculation of this type of standardised mortality ratio is demonstrated in detail in Table 20.7 in which it is applied to men included in Social Class I of the population (which comprises the more affluent and professional members). Allowing for their age distribution the SMR is 77 per cent. In this table, Column (4) is calculated as $3 \times$ Column (2) $\times$ Column (3) because Column (2) contains annual death rates but we wish to compare with observed deaths over a 3-year period.

The underlying reasons for observed differences between occupations is,

Table 20.7 Calculation of the standardised mortality ratio.

Age-group in years	*Annual death rates of all males per 1000 in 1970–72*	*Census population of men in Social Class I in 1971*	*Expected deaths of men in Social Class I at rates of all males*	*Observed deaths of men in Social Class I in 1970–72*
(1)	*(2)*	*(3)*	*(4)*	*(5)*
15–	0.92	95 190	263	193
25–	1.00	214 680	644	431
35–	2.31	171 060	1 185	854
45–	7.20	137 080	2 961	2 079
55–64	20.54	100 000	6 162	5 029
Total	—	—	11 215	8 586

Standardised mortality ratio 8586/11 215 = 77 per cent

of course, another matter, outside the scope of this chapter. They may be related to such features as specific occupational hazards or absence of hazards, the general living conditions consequent upon an occupation (through, for example, low or high rates of pay), and—of considerable importance—selective factors influencing the entry of the physically fit or unfit to a particular occupation and leading also to the withdrawal from it of the fit or unfit to take up some other occupation.

A general application

This form of standardised comparison can also be usefully applied in an experimental situation, e.g. a clinical trial, if the groups to be compared differ in some important feature. Thus the total number of deaths (or other event) observed in the treated group can be compared with the number that would have occurred if the age-specific fatality rates in the control group had been operating.

An example is shown in Table 20.8, derived from a study made by Inman and Vessey of the deaths of women from thrombosis and embolism in relation to the use of oral contraceptives (*Br. Med. J.*, **ii**, 193, 1968).

In their records there were 26 deaths from pulmonary embolism in which there were no known predisposing conditions in the women. As many as 16 of these women, however, had been taking oral contraceptives at the time of the onset of their fatal illness. How, then, does this frequency of use, 16 in 26, compare with the frequency that might be expected from the habits of the general population of women of childbearing age? As a measure of these habits, information was derived from nearly 1000 control subjects and it will be seen that the use of oral contraceptives was related to two factors—age and parity. Looking along each row, the frequency of use fell steadily with increasing age and, looking down each column, rose steeply with increasing parity.

Table 20.8 The frequency of use of oral contraceptives in control subjects (percentages) and, in parentheses, the number of women dying from pulmonary embolism without known predisposing cause.

	Age in years		
Parity	*20–24*	*25–34*	*35–44*
0	8.9 (2)	5.8 (1)	0.0 (3)
1	15.2 (–)	8.8 (–)	10.2 (–)
2	34.4 (–)	25.0 (1)	9.9 (5)
3+	43.8 (1)	29.9 (4)	17.4 (9)

It follows that in any comparison of the frequency of use of oral contraception in the controls and in the observed deaths, these two characteristics must be taken into full account. We can do so by applying the specific age/parity rates of use in the controls to the corresponding numbers of women in the age/parity groups who had died from a pulmonary embolism. Thus, converting the rate from per 100 to per unit, the expected numbers using oral contraception are:

$$(2\times 0.089) + (1\times 0.058) + (3\times 0.000) + (1\times 0.250) + (5\times 0.099) + (1\times 0.438) + (4\times 0.299) + (9\times 0.174) = 4.2$$

There was, therefore, a considerable excess of women using oral contraceptives in the group dying from pulmonary embolism over the number that would have been expected from the characteristics of the control group (16 observed to 4.2 expected). In the process of reaching this comparison age and parity have been allowed for.

The statistical significance of such a comparison can be calculated by using a Poisson distribution with mean 4.2, and finding the probability corresponding to 16 or more observed. This gives $P = 0.00009$. This must be regarded as an approximation, because the conditions for a Poisson distribution are not strictly observed (the probability is not constant for all individuals), but it is unlikely to be misleading.

Summary

Rates of mortality and morbidity are usually affected considerably by the age and sex constitution of the population concerned. The comparison of crude rates at all ages is therefore likely to be misleading as a measure of the mortality, or morbidity, experiences of an area in relation to such features as the sanitary environment. In view of population changes they may be equally misleading in comparing the crude death rate of one year with that of another. Some form of average of the death rates at ages is required which allows for the fact that populations differ in their structure. This average is customarily reached by a process of standardisation, a process which leads to a weighted average and thereby to more comparable indices. But it should be remembered that a standardised rate, or index, *is* only an *average*. The basic death rates at ages which contribute to it are likely to be very much more informative than any single summary figure that can be derived from them. In particular, at this present time the death rates at ages below 60 years are so low in the UK that any standardised rate must be dominated by the mortality levels in the old.

Chapter 21

Life tables and survival after treatment

In the assessment of the degree of success attending a particular treatment given to patients over a series of years the life-table method is sometimes an effective procedure. Before illustrating its application to such data, consideration of the national life table and its use in public health work will be of value. A life table, it must be realised, is only a particular way of expressing the death rates experienced by some particular population during a chosen period of time. For instance, a life table for England and Wales based upon the mortality experience of men in the year 1987 contains six columns as shown in Table 21.1. The essence of the table is this: suppose we observe 100 000 infants all born on the same day and dying as they passed through each year of life at the same rate as was

Table 21.1 Extract of a life table based upon the death rates of men in England and Wales in 1987.

Age x	l_x	d_x	p_x	q_x	$\mathring{e}_x$
0	100 000	1 271	0.98729	0.01271	71.0
1	98 729	84	0.99915	0.00085	71.0
2	98 645	51	0.99948	0.00052	70.0
3	98 594	37	0.99962	0.00038	69.1
4	98 557	.	.	.	.
.	.	.	.	.	.
.	.	.	.	.	.
.	.	.	.	.	.
50	92 758	.	.	.	24.3
.	.	.	.	.	.
.	.	.	.	.	.
.	.	.	.	.	.
85	14 153	.	.	.	4.3
.	.	.	.	.	.
.	.	.	.	.	.
.	.	.	.	.	.
100	165	.	.	.	1.9
.	.	.	.	.	.

experienced at each of these ages by the male population of England and Wales in 1987, in what gradation would that population disappear? How many would be still left alive at age 25, at age 56, etc.? How many would die between age 20 and age 30? What would be the chance of an individual surviving from age 40 to age 45? What would be the average length of life experienced by the 100 000 infants? Such information can be obtained from these different columns. The basis of the table is the value known as q_x, which is the probability, or chance, of dying between age x and age $x+1$, where x can have any value between 0 and the longest observed duration of life. For instance, q_{25} is the chance that a person who has reached a 25th birthday will die before reaching the 26th birthday. These probabilities, one for each year of age, are calculated from the mortality rates experienced by the population in 1987. This probability of dying is the ratio of those who fail to survive a particular year of life to those who started that year of life; (to take an analogy, if 20 horses start in a steeplechase and 5 fail to survive the first round of the course the probability of 'dying' on that round is 5/20; 15 horses are left to start on the second round and if 3 fail to survive, the probability of 'dying' on the second round is 3/15).

The probability of dying

As pointed out above, the basic element of the life table is the probability of dying between one age and the next. Once those values are known throughout life the construction of the remainder of the table is a simple, though arithmetically laborious, process. To calculate these probabilities requires a knowledge, for the population concerned, of the numbers living and dying in each single year of life. Let us suppose that such detailed data are available and that in a particular city, there were 1500 persons enumerated (or estimated) at the middle of the year whose age was 22 years last birthday, i.e. their age was between 22 and 23. During that calendar year there were, say, 6 deaths between ages 22 and 23. Then the *death rate* at ages 22–23, as customarily calculated, is the ratio of the deaths observed to the mid-year population, i.e. 6 in 1500, which equals 4 per 1000, or 0.004 per person. In symbols, $m_x = D/P$, which gives the death rate per person. This mid-year population does not indicate precisely how many people *started* the year of life 22–23, as it is an enumeration of those who were *still alive* at the middle of the calendar year. On the average they were at that point of time $22\frac{1}{2}$ years old, since some would have just passed their 22nd birthday, some would be just on the point of having their 23rd birthday, and all intermediate values would be represented. If we may reasonably presume that the deaths occurring between ages 22 and 23 are evenly spread over that year of life, then we may conclude that half of them would have occurred before the mid-year enumeration (or estimation) and half would have occurred after it. In other words, the population that *started* out from age 22 is the 1500 survivors plus a half of the recorded deaths, or $1500+3$, this half of the deaths being presumed to have taken place before the mid-year point. The *probability of dying* is, by definition, the ratio of the deaths observed in a year of life to the number who set out on that year of life, i.e. 6 in 1503. In symbols, therefore, $q_x = D/(P+\frac{1}{2}D)$.

If, then, we know the mid-year population at each age and the deaths taking place between each age and the next, the probabilities of dying can be readily calculated from this formula. (It is, however, not very accurate in the first 2 or 3 years of life and particularly in the first year. In the first year in countries with a low infant mortality rate the deaths occur more frequently in the first 6 months of life than in the second 6 months and a more appropriate fraction would be $D/(P+\frac{4}{5}D)$.)

It is clear that there must be a simple relationship between m_x, the death rate, and q_x, the probability of dying. It may be demonstrated as follows:

$m_x = D/P$. So that $Pm_x = D$.
$q_x = D/(P+\frac{1}{2}D)$. Substituting Pm_x for D gives
$q_x = Pm_x/(P+\frac{1}{2}Pm_x)$

But P occurs in both numerator and denominator, so it may be removed to give $q_x = m_x/(1+\frac{1}{2}m_x)$; and, finally, multiplying top and bottom by 2 to get rid of the half, gives $q_x = 2m_x/(2+m_x)$.

In other words, the probability of dying may be calculated from the formula (twice the death rate)/(the death rate plus 2), where the death rate is calculated not as usual per 1000 persons but as per person; or from the formula (deaths)/(population plus half the deaths).

The construction of the life table

Having thus calculated, by one or other formula, these values of q_x for each year of life, the life table is started with an arbitrary number at age 0, e.g. 1000, 100 000, or 1 000 000. By relating the probability of a newborn infant dying before its first birthday (q_0) to this starting number, we find the number who will die in the first year of life. By subtracting these deaths from the starters we have the number of survivors that there will be at age 1. But for these survivors at age 1 we similarly know the probability of dying between age 1 and 2; by relating this probability to the survivors we can calculate how many deaths there will be between age 1 and 2. By simple subtraction of these deaths we must reach the survivors at age 2. And so on throughout the table till all are dead. Thus Table 21.1 shows that for males, the probability of dying in the first year of life is 0.01271, or in other words, according to the infant mortality rate of 1987, 1.271 per cent of our 100 000 infants will die before they reach their first birthday. The actual number of deaths between age 0 and age 1 will therefore be 1.271 per cent of 100 000, or 1271. Those who *survive* to age 1 must be 100 000 less 1271 = 98 729. According to this table, the probability of dying between age 1 and age 2 is 0.00085, or in other words 0.085 per cent of the 98 729 children aged 1 year old will die before reaching their second birthday. The actual number of deaths between age 1 and age 2 will therefore be 0.085 per cent of 98 729 = 84; those who survive to age 2 must therefore be 98 729 less 84 = 98 645. From these q_x values the l_x and d_x

columns can thus be easily constructed, l_x showing the number of individuals out of the original 100 000 who are still alive at each age, and d_x giving the number of deaths that take place between each two adjacent ages. p_x is the probability of living from one age to the next. $p_x + q_x$ must equal 1, since the individuals must either live or die in that year of life. To return to our analogy, if 5 out of 20 horses do not complete the round, clearly 15 out of 20 do survive the round. p_x, therefore, equals $1 - q_x$.

Finally, we have the column headed $\mathring{e}_x$, which is the 'expectation of life' at each age. This is an 'expectation' in the statistical sense, for it is the *average* duration of life lived beyond each age. For example, if we added up all the ages at death of the 100 000 male infants and took the average of these durations of life we should reach the figure 71.0 years. If, alternatively, we took the 98 729 infants who had lived to be 1 year old, calculated the *further* duration of life that they enjoyed, and then found the average of those durations, we should again reach the figure 71.0 years. At age 85 there are 14 153 persons still surviving, and the average duration of life that they will enjoy after that age is only 4.3 years. The expectation of life is thus the average length of life experienced after each age.

We thus have the complete life table.

Calculation of the expectation of life

The expectation of life, or average length of life given by a life table, can be calculated from the column of deaths (d_x) or from the column of survivors (l_x). Using the former, we have an ordinary frequency distribution. Thus of the males in this 1987 table there were 1271 who died between 0 and 1. We may presume they lived on the average half a year (some exaggeration, in fact). Their contribution to the total years lived by the original 100 000 is therefore ($1271 \times \frac{1}{2}$). Between ages 1 and 2 there were 84 deaths, and we may presume for these that the average length of life was $1\frac{1}{2}$ years. Their contribution to the total years lived by the original 100 000 is therefore ($84 \times 1\frac{1}{2}$). And so on until at the far end of the table when there are no more survivors. Proceeding to the final point, we then have the total years of life lived by the whole population of 100 000:

$$(1271 \times \tfrac{1}{2}) + (84 \times 1\tfrac{1}{2}) + (51 \times 2\tfrac{1}{2}) + \cdots$$

The expectation of life is the mean number of years lived, and so this sum is divided by the 100 000 persons starting at age 0 to whom it relates. In short, $\mathring{e}_x$ at birth equals the sum of all values of $d_x \times (x + \frac{1}{2})$ divided by 100 000, and in this table for males is, as previously stated, 71.0 years.

Using the survivor column (l_x), we may proceed as follows. The 98 729 survivors at age 1 have all lived a whole year of life, between age 0 and age 1. They give a contribution therefore of 98 729 whole years of life to the total years lived. But the survivors at age 2, who were 98 645 in number, have all lived another whole year of life—from age 1 to age 2. They therefore give a further 98 645 whole years of life to the total years lived. Similarly the survivors at age 3, who are 98 594, are contributing that extra number of whole years of life lived. Thus by summing the l_x column from age 1 to the final entry we have the number of whole years of life lived by the 100 000 (clearly the 100 000 must not be included in the sum, for they are the starters at 0 and at that point have lived no duration).

There is, however, a small error involved if we stop at that sum. In it we have made no allowance for the period each person lives in the year of his death. We have considered only the survivors at ages 1, 2, 3, etc. But we may presume that the 1271 who died between 0 and 1 had half a year's life before their death, that the 84 who died between 1 and 2 also had half a year's life in the year 1–2 in which they died, that the 51 who died between 2 and 3 also had half a year's life in the year 2–3 in which they died (it will be noted that their whole years of life up to age 2 have already been counted in the survivorship column and it is only the half-year in the year in which they died that has been omitted). We must therefore add to the sum of whole years lived, derived from the l_x column, these half-years lived by those dying in the precise year in which they died. In other words, we have to add in $(1271\times\frac{1}{2})+(84\times\frac{1}{2})+(51\times\frac{1}{2})+\cdots$ to the end of the table. But this implies, as everyone is dead by the end of the table, adding in $(100\,000\times\frac{1}{2})$. The final sum of years required is, therefore, given by the sum of the l_x column from 1 to the end of the table plus 100 000 times a half. The average, or expectation of life, is then this sum divided by the 100 000 at the start, which may be expressed as

$$\frac{\text{sum of } l_x \text{ column (excluding } l_0)}{100\,000}+\frac{1}{2}$$

As stated previously, the expectation of life at a later age than 0, say age 25, is the average length of life lived beyond that age by the survivors at age 25. It can be calculated by either of the above methods (the use of the l_x column being the simpler), the sum of the years lived relating only to the entries beyond age 25, and the denominator, to give the average, being, of course, the survivors at age 25.

Practical aspects of the life table

In summary, the life table provides some clear and useful indices of the current mortality rates free from any peculiarities in the age structure of the population concerned. In using it as a method of comparison of the mortality experience between place and place, or epoch and epoch, various values may be chosen. For example, we may take:

(1) The probability of dying between any two selected ages (the ratio of the total deaths between two ages to the number alive at the first age).
(2) The number of survivors at any given age out of those starting at age 0 (the l_x column).
(3) The probability of surviving from one age to another (the ratio of the survivors at a given age to the survivors at a previous age).
(4) The expectation of life (which suffers from the limitation inherent in

any average that it takes no cognisance of the variability around it).

In practice, however, it is not often possible to construct a life table by the methods described above since they require, it was seen, a knowledge of the population and deaths in single years of life. More often than not the numbers available relate to 5- or 10-yearly age groups, and some device has then to be adopted for estimating from these grouped data the appropriate numbers at single years of life. Alternatively there are available for public-health work excellent short methods of making a life table from the actual death rates observed in different age groups. To describe these methods fully is outside the scope of this chapter, the object of which is not to show how best to construct a life table in practice, but to clarify the general principles underlying its construction so that the values given by it may be understood. Taking the example given above, it shows, to reiterate, how a population would die out if it experienced as it passed through life the same death rates as were prevailing among males in England and Wales in 1987. It does not follow, therefore, that of 100 000 male children born in that year in reality only 98 594 would be alive at age 3; if the death rate were, in fact, declining then more than that number would survive; if it were rising, less than that number would survive. As the Government Actuary pointed out in relation to English Life Table No. 12, 'the l_x columns could only be interpreted as showing the survivors of 100 000 children born in the period 1960–62 if the improbable assumption were made that the 1960–62 rates of mortality will remain unaltered through their lifetimes, that is until at least the year 2070. The same applies to the expectations of life; if, in line with past experience, rates of mortality decline in future, babies born in 1961 have an expectation of life greater than $\mathring{e}_x$ as shown in English Life Table No. 12.' In other words the life table can show only what would happen under *current conditions of mortality*, but it puts those current conditions in a useful form for various comparative purposes and for estimating such things as life insurance risks (inherent in the questions that were propounded above).

That there is much misunderstanding of this point is clear when suggestions are seen, for example, that an increase in cigarette smoking among young women does not seem to be harming them as their expectation of life is still increasing. The expectation was, of course, calculated on the experience of women of all ages a few years ago. What will happen to those now in their 20s when they reach their 50s, say, cannot be known until they get there.

A cohort life table

A *cohort* life table, on the other hand, can sometimes be constructed to show the actual dying out of a defined group of persons, or cohort, all born at about the same time. Thus all people born in the year 1920 were subject to the infant mortality rate of 1920–21 and then, as the years passed, to the already recorded childhood, adolescent and adult mortality of those later years. And so in 1990 we can summarise in life-table form the true mortality of the cohort at different ages for the first 70 years of its life. We cannot, of course, go beyond that point until further years have elapsed. The cohort born in, say, 1880 could be traced to extinction.

The measurement of survival rates after treatment

We turn now to the application of the life-table method to groups of patients, treated over a period of calendar years, whose subsequent after-history is known. Let us suppose that treatment was started in 1978, that patients were treated in each subsequent calendar year and were followed up to the end of 1983 on each yearly anniversary after their treatment had been started (none being lost sight of). Of those treated in 1978 we shall know how many died during the first year of treatment, how many died during the second year of treatment, and so on to the fifth year after treatment. Of those treated in 1979 we shall know the subsequent history up to only the fourth year after treatment, in 1980 up to only the third year after treatment, and so on. Our tabulated results will be, let us suppose, as in Table 21.2

We shall demonstrate here the calculations on a year-by-year basis, suitable for working by hand. If a computer is used, and the dates of death of the patients are known, greater accuracy will be achieved by working on a day-to-day basis.

Separate calculation of the survival rates of patients treated in each calendar year becomes somewhat laborious if the number of years is extensive and has also to be based upon rather small numbers. If the constitution of the samples treated yearly and their fatality rates are not changing with the passage of time there is no reason why the data should not be amalgamated in life-table form. Indeed the great advantage of the life-table method is that we can utilise *all* the information to hand at some moment of time. In computing, say, a 5-year survival rate, we make *all* the patients contribute to the picture and do not restrict ourselves to only those who have been observed for the full five years. For clarity we can write

Table 21.2 Results of treatment (hypothetical figures).

Year of treatment	*Number of patients treated*	*Number alive on anniversary of treatment in*				
		1979	*1980*	*1981*	*1982*	*1983*
1978	62	58	51	46	45	42
1979	39	—	36	33	31	28
1980	47	—	—	45	41	38
1981	58	—	—	—	53	48
1982	42	—	—	—	—	40

Table 21.3 Figures of Table 21.2 rearranged.

Year of treatment	*Number of patients treated*	*Number alive on each anniversary (none lost sight of)*				
		1st	*2nd*	*3rd*	*4th*	*5th*
1978	62	58	51	46	45	42
1979	39	36	33	31	28	—
1980	47	45	41	38	—	—
1981	58	53	48	—	—	—
1982	42	40	—	—	—	—

Table 21.2 in the form given in Table 21.3.

All the patients have been observed for at least one year and their number is $62+39+47+58+42=248$. Of these there were alive at the end of that first year of observation $58+36+45+53+40=232$. The probability of surviving the first year after treatment is, therefore, $232/248=0.935$, or, in other words, 93.5 per cent of these patients survived the first year after treatment. Of the 40 patients who were treated during 1982 and were still surviving a year later, no further history is yet known. (If the year's history happens to be known for some of them these data cannot be used, for the history would tend to be complete more often for the dead than for the living, and thus give a bias to the results.) As the number exposed to risk of dying during the second year we therefore have 192—the 232 survivors at the end of the first year minus the 40 of whom we know no more. Of these there were alive at the end of the second year of observation $51+33+41+48=173$. The probability of surviving throughout the second year is therefore $173/192=0.901$. Of the 48 patients who were treated in 1981 and were still surviving in 1983 no later history is yet known. As the number exposed to risk of dying in the third year we therefore have 125—the 173 survivors at the end of the second year minus the 48 of whom we know no more. Of these there were alive at the end of the third year of observation $46+31+38=115$. The probability of surviving the third year is therefore $115/125=0.920$. Similarly, the number exposed to risk in the fourth year becomes $46+31=77$, and of these $45+28=73$ are alive at the end of it. The probability of surviving the fourth year is therefore $73/77=0.948$. Finally the probability of surviving the fifth year is $42/45=0.933$.

Construction of the life, or survivorship, table: anniversary data

Tabulating these probabilities of surviving each successive year, we have the values denoted by p_x in Column (2) of Table 21.4. The probability of not surviving in each year after treatment, q_x, is immediately obtained by subtracting p_x from 1. The number of patients with which we start the l_x column is immaterial but 100 or 1000, or some such number, is convenient. Starting with 1000, our observed fatality rate shows that 93.5 per cent would survive the first year and 6.5 per cent would die during that year. The number alive, l_x, at the end of the first year must therefore be 935 and the number of deaths, d_x, during that year must be 65. For these 935 alive on the first anniversary the probability of living another year is 0.901, or in other words there will be 90.1 per cent alive at the end of the second year, i.e. 842, while 9.9 per cent will die during the second year, i.e. 93. Subsequent entries are derived in the same way. (The order given in Table 21.4 is the most logical while the table is being constructed, because p_x is the value first calculated and the others are built up from it. In the final form the order given in Table 21.1 is more usual.)

By these means we have combined all the material we possess for calculating the fatality in each year of observation after treatment, and have found that according to those fatality rates approximately 69 per cent of treated patients would be alive at the end of 5 years. Having found from the available material the probability of surviving each of the separate years 1 to 5 we are, in effect, finding the probability of surviving the whole 5 years by multiplying together these probabilities: $p_0 \times p_1 \times p_2 \times p_3 \times p_4$.

As with all statistical calculations the results are, of course, subject to sampling error, but they do represent a useful summary of what was actually observed in the particular group of patients.

If we want an expectation of life for such patients we must keep on observing until all the patients are dead. At present we have no information on the expectation beyond 5 years. For this reason there is much to be said for characterising such groups by the median, rather than the mean, length of life; for that is known precisely as soon as half are dead.

Table 21.4 Figures of Table 21.2 in life-table form.

Year after treatment x	*Probability of surviving each year*	*Probability of dying in each year*	*Number alive on each anniversary out of 1000 patients*	*Number dying during each year*
x	p_x	q_x	l_x	d_x
(1)	*(2)*	*(3)*	*(4)*	*(5)*
0	0.935	0.065	1000	65
1	0.901	0.099	935	93
2	0.920	0.080	842	67
3	0.948	0.052	775	40
4	0.933	0.067	735	49
5	—	—	686	—

The percentage alive at different points of time makes a useful form of comparison. For instance, studying historical data, we find for patients with cancer of the cervix treated between 1925 and 1934 the number of survivors out of 100 in each stage of disease at the start of treatment (Table 21.5).

Table 21.5

	Stage 1	2	3	4
At end of 5 years	86	70	33	11
At end of 9 years	78	57	27	0

Exclusion of patients

If some of the patients have been lost sight of, or have in a few instances died from causes which we do not wish to include in the calculation (accidents, for example), these must be taken out of the number exposed to risk at the appropriate time—e.g. an individual lost sight of in the fourth year is included in the exposed to risk for the first three years but cannot be included for the complete fourth year. If he were taken out of the observations from the very beginning, the fatality in the first three years would be overstated, for we would have ignored an individual who was exposed to risk in those years and did not, in fact, die. If patients are being lost sight of at different times during each year or dying of excluded causes during that year, it is usual to count each of them as a half in the number exposed to risk for that year. In other words, they were, on the average, exposed to risk of dying of the treated disease for half a year in that particular year of observation.

Construction of the life, or survivorship, table: data at a specified date

Sometimes in putting data into the life-table form we have patients observed not on anniversaries, as above, but to a specified date. Let us suppose, for instance, that patients have been treated in the years 1978 to 1982 and have been followed up to 31 December 1983 (and thus not, as in the previous example, to the yearly anniversaries of their treatment). The data may be put in life-table form according to the technique set out in Table 21.6

The total number of patients treated in the 5 years and to be followed up was 194. During the first year of the follow-up 4 were lost sight of and 2

Table 21.6 Life table of patients undergoing a hypothetical treatment.

Year after treatment	*Number alive at beginning of the year*	*Number lost sight of during the year*	*Number dying of violence during the year*	*Number alive observed for only part of the year*	*Number exposed to risk of dying during the year*	*Number dying during the year*	*Proportion dying during the year*	*Proportion surviving the year*	*Proportion surviving from start of treatment to end of each year*
							q_x	p_x	
(1)	*(2)*	*(3)*	*(4)*	*(5)*	*(6)**	*(7)*	*(8)*	*(9)*	*(10)†*
0	194	4	2	0	191	24	0.126	0.874	0.874
1	164	3	0	35	145	12	0.083	0.917	0.801
2	114	0	1	42	92.5	6	0.065	0.935	0.749
3	65	1	0	23	53	3	0.057	0.943	0.707
4	38	2	1	21	26	2	0.077	0.923	0.652
etc.									

* Column (6) = column (2) minus half columns (3), (4) and (5)
† Column (10) = the product of the values of column (9): $p_0 \times p_1 \times p_2 \times \cdots$

died of violence and these deaths it is proposed to exclude as irrelevant. Since the last patient was treated in 1982 and the follow-up was to 31 December 1983, all the patients had been observed for at least one full year. The exposure to risk of dying during the first year will, therefore, be computed as 194 less half the number lost sight of and less half the number of deaths from violence, i.e. 191. In other words, we give only half a year's exposure to those who passed out of observation for these reasons during the year. During these 191 person-years of exposure there were 24 deaths, giving a probability of dying of 24/191 = 0.126. The probability of surviving the first year, is therefore, 0.874.

The number of patients entering the second year of follow-up is the original 194 less all those who have died or who have been lost sight of during the first year, i.e. 194 − 24 − 2 − 4 = 164. Of these 164 exposed during the second year there were 3 lost sight of during the year who must be allowed only a half-year's exposure, and there were also 35 *who were still alive at 31 December 1983* but who had not been exposed to risk over the *whole* of that second year. These are the patients who were treated in 1982 and who, therefore, by the end of 1983, had been exposed for one year and some fraction of a year. As usual we shall take the fraction to be, on the average, one-half. The number exposed to risk of dying during the full second year is, accordingly, 164 − $\frac{1}{2}$ of 3 lost sight of − $\frac{1}{2}$ of 35 alive at 31 December 1983, and not exposed for the whole of the second year = 145. During that second year the number of patients dying was 12 so that the probability of dying was 0.083 and the probability of surviving 0.917.

The number entering the third year of exposure is 164 minus the 3 lost sight of, the 12 who died and the 35 who passed out of observation alive at 31 December 1983, which equals 114. Of these 114 patients 1 died of violence during the third year and 42 were seen for only part of that year (i.e. those who were treated in 1981 and by 31 December 1983, had been observed for 2 full years and some fraction of a year). The number exposed to risk of dying is, accordingly, 114 less $\frac{1}{2}$ of 1 and $\frac{1}{2}$ of 42 = 92.5. With 6 deaths during the year the probability of dying is 0.065 and the probability of surviving is 0.935. And so on.

Taking the probabilities of dying and applying them to the customary hypothetical 1000 patients at start of treatment we could calculate the number alive at the end of each year as in Table 21.4. Alternatively if we require only the percentage of the total who will be surviving at the end of each year (i.e. the l_x column), the answer can be obtained by multiplying together the p_x values in Column 9 of Table 21.6. Thus the probability of surviving one year is, according to these data, 0.874; the probability of surviving two years is 0.874 × 0.917 = 0.801; of surviving three years 0.874 × 0.917 × 0.935 = 0.749; of surviving four years 0.874 × 0.917 × 0.935 × 0.943 = 0.707; and of surviving five years 0.874 × 0.917 × 0.935 × 0.943 × 0.923 = 0.652, a 5-year survival rate of 65 per cent.

As a general rule the exclusion of deaths regarded as irrelevant is undesirable, e.g. from violence in a follow-up of patients operated upon for some form of cancer. If the number of such deaths is few their inclusion (or omission) can have little effect upon the results. If the number is large it may be difficult to interpret the results when they are omitted. It is probably better to compute the survival rate with such deaths included and to compare this rate with the figure normally to be expected amongst people at those ages.

Patients lost to sight

Finally, if a relatively large number of patients is lost sight of we may be making a serious error in calculating the fatality rates from the remainder, since those lost sight of may be more, or less, likely to be dead than those who continue under observation.

For instance, if 1000 patients were observed, 300 are dead at the end of 5 years, 690 are alive, and 10 have been lost sight of, this lack of knowledge cannot appreciably affect the survival rate. At the best, presuming the 10 are all alive, 70 per cent survive; at the worst, presuming the 10 are all dead, 69 per cent survive. But if 300 are dead, 550 are alive, and 150 have been lost sight of, the corresponding upper and lower limits are 70 per cent surviving and 55 per cent surviving, an appreciable difference. To measure the survival rate on those patients whose history is known, or, what comes to the same thing, to divide the 150 into alive and dead according to the proportions of alive and dead in the 850 followed up, is certainly dangerous. The characteristic 'lost sight of' *may* be correlated with the characteristics 'alive or dead'; a patient who cannot be traced may be more likely to be dead than a patient who can be traced (or vice versa), in which case the ratio of alive to dead in the untraced cases cannot be the same as the ratio in the traced cases. Calculation of the possible upper rate shows at least the margin of error.

The life table in general

The use of the life table has been illustrated above in relation to death and survival, which is, indeed, its customary use. However, the method can be more generally applied to any defined feature in the follow-up of persons or patients, i.e. so long as such a feature occurs at one point in the course of time and can at that time be clearly defined as present or not present.

For instance, we might be concerned with a treatment newly introduced for patients suffering from multiple sclerosis. The question at issue is how long will the patient remain free from all, or certain defined, symptoms (in comparison, say, with a similar group of patients not so treated). The decisive end-point in the follow-up, then, is not death but the appearance of symptoms as previously defined and which, of course, need to be clear-cut.

In the same way one might follow up patients after an operation for a form of cancer and note the time at which recurrence occurred. In these examples one would have, in place of death, rates of appearance or recurrence in given intervals of time, and thus the probability of their occurring within so many months or years.

Comparing survival of two or more groups: the logrank method

If we have two or more groups of patients, perhaps with the same disease but randomised to different treatments, we may wish to compare their survival information and test an observed difference for significance. The logrank method is a simple way of doing this.

The logic of this method lies in the simple fact that if we have, say, 100 people consisting of 80 men and 20 women, and 10 of them die, then if the probability of death is not related to sex we should expect the deaths to be 8 men and 2 women. That is, the sex ratio should be the same among the deaths as among the group as a whole, subject to sampling variability. If it is very different, then there is evidence that sex is a relevant factor. The same logic may be used when following the survival of a population that is divided into two groups, but considering the deaths not 10 at a time but each death individually. Thus in the above example, with 1 death we should expect it to be 0.8 of a man and 0.2 of a woman. As always with statistical expectations, the fact that no such event could actually be observed is irrelevant.

Consider then the 40 patients (hypothetical data) in Table 21.7, 10 of whom, selected at random, have been given a special treatment. If the treatment has no effect, there is no reason for the death rate to be any greater in one group than in the other. So long as there are no deaths there is no evidence coming in on which to make any calculations, but we calculate the relevant expected values each time a death occurs. The first to die is in the group of 30, and the corresponding expected values are $30/40 = 0.7500$ and $10/40 = 0.2500$. Instead of 30 and 10 in the groups, we now have 29 and 10, so when the next death occurs, in the same group, the expected values are $29/39 = 0.7436$ and $10/39 = 0.2564$.

The next death, again among those in the large group, is caused by a road accident and it is decided that this is irrelevant to the trial, so the observation is omitted except that the numbers are reduced from 28 and 10 to 27 and 10. It would be wrong to go back and ignore that patient from the start, for if the patient had died at any time before the accident that death would have been counted. It is most important that the decision on any death that is to be considered as irrelevant should be taken by someone who is kept 'blind' as to which group the individual was in, but aware of all other relevant facts. Otherwise bias may creep in. In general, the rule should be to keep all deaths in the analysis, unless their irrelevance is very clear indeed.

In the example, when there are 27 and 9 patients remaining, 2 deaths occur and it is not possible to determine which occurred first. Had they both been in the same group it would be best to calculate as 1 followed by

Table 21.7 Comparing survival in two groups by the logrank method (hypothetical data).

Number of patients currently observed		*Relevant deaths*		*Expected deaths*	
Usual treatment	*Special treatment*	*Usual treatment*	*Special treatment*	*Usual treatment*	*Special treatment*
30	10				
		1	0	0.7500	0.2500
29	10				
		1	0	0.7436	0.2564
28	10				
	(1 in *usual treatment* group killed in road accident)				
27	10				
		0	1	0.7297	0.2703
27	9				
		1	1	1.5000	0.5000
26	8				
	(1 in *special treatment* group murdered)				
26	7				
		0	1	0.7879	0.2121
26	6				
		1	0	0.8125	0.1875
25	6				
		1	0	0.8065	0.1935
24	6				
		1	0	0.8000	0.2000
23	6				
	Totals	6	3	6.9302	2.0698

another 1, and it would not matter which of them was first, but with one in each group they are recorded together and the expected values are calculated as

$$2 \times 27/36 \text{ and } 2 \times 9/36$$

At the time when the analysis is to be made, there are still 23 and 6 patients remaining. At that time the observed deaths (ignoring the irrelevant two) and the expected deaths are added for each group and a χ^2 value (on 1 degree of freedom) may be calculated by the usual formula:

$$\text{Sum}\left(\left(\text{observed} - \text{expected}\right)^2/\text{expected}\right)$$

Here we have

$$\chi^2 = \frac{(6-6.9302)^2}{6.9302} + \frac{(3-2.0698)^2}{2.0698} = 0.54$$

$$P = 0.46$$

giving us no reason to claim that the special treatment had any effect.

In such a trial, if *all* the patients in one group have died, it should be terminated and analysed at once. There is no point (for purposes of the trial) in continuing to observe the deaths in the remaining group.

A similar procedure may be adopted for more than two groups. The resulting χ^2 will have (number of groups − 1) degrees of freedom. Two papers by Peto *et al.* (*Br. J. Cancer* **34**, 585 (1976) and **35**, 1 (1977)) are well worth consulting for further details.

If it is wished to analyse survival data in terms of continuous variables instead of merely by division into groups, a technique called 'Cox regression' (or 'proportional hazards regression') is required. It will be found available in certain computer packages, but is beyond the scope of this book.

Summary

Life tables are convenient methods of comparing the mortality rates experienced at different times and places. The same methods may be usefully applied to the statistics of patients treated and followed up over a number of years. The logrank method is a useful way of comparing the survival patterns of two or more groups, and testing whether they are significantly different.

Chapter 22

Measures of morbidity

In many countries with highly developed vital statistical systems, the death rate, from all causes or from specific causes, has been for more than a century the principal measure of health progress—or lack of progress—and one of the main instigators of epidemiological research. In Great Britain it has, for instance, revealed the disappearance of cholera and the great decline in the typhoid fevers; it has shown the dramatic reduction of diphtheria, and, in the last fifty years, the spectacular fall in the mortality rates of childhood. While these, and other striking gains, have at some ages been offset by new and serious problems, e.g. the rise in cancer of the lung in both sexes and the incidence of diseases of the heart in men of middle life, it has, nevertheless, become increasingly apparent that such low rates of mortality need to be supplemented by measures of morbidity. How much *sickness* is there in the population and in its many component groups, sickness which does not necessarily end fatally but which nonetheless calls for investigation and preventive measures (e.g. rheumatism in its various aspects)? It is this situation which is leading in a number of countries to the development of a system of morbidity statistics and, in general, to much discussion of the very difficult problems that they raise.* On the one hand such statistics are required for the purposes of medical administration, to indicate the population's requirements for medical care at home and in hospitals, etc.; on the other hand they should serve the needs of research into the factors that influence the incidence of illness of different kinds and into the steps that may contribute to prevention and cure.

Special problems of morbidity statistics

Compared with the statistics of death, statistics of sickness present some very substantial problems. Thus (1) death is a unique event whereas illness may occur repeatedly in the same person from the same or different

* *Morbidity Statistics*, World Health Organisation Technical Report Series No. 389, 1968 and Report by the Permanent Commission and International Association on Occupational Health, Sub-Committee on Absenteeism. *J. of the Soc. of Occ. Med.* **23**, 132 (1973).

causes; (2) death occurs at a point of time whereas illness exists over a duration of time; (3) in spite of modern problems, death is precisely defined whereas illness varies very greatly in its severity, ranging from negligible effects to a condition which is completely disabling. In the measurement of morbidity it is clear that all these aspects have to be taken into account.

Under (1) we shall have to consider whether in calculating the amount of morbidity that occurred in a given period we should add the *number of persons* ill or the *number of illnesses* or both. For some persons will be ill more than once.

Under (2) we shall have to consider whether we wish to know the number of *new illnesses that arise* in a given period or the number that were *extant* in that period, whether they first arose in it or extended into it from a previous period. Suppose person *A* has an illness lasting for 3 months, whereas person *B* is ill for the first and the third of those months, but well during the second month. If we merely count the number of illnesses, *A* had 1, *B* had 2. So *B* appears to have done twice as badly *because* he was well for a month. Interpreting such paradoxes is far from easy.

Under (3) we shall have closely to consider what *we intend to count* as morbidity in any given circumstances. Thus we may, perhaps, categorise sickness broadly as (a) congenital or acquired defects, injuries or impairments (such as a residual paralysis from a past attack of poliomyelitis, visual and auditory defects); (b) latent or incipient diseases usually not recognised by the person affected but revealed by laboratory or other tests, e.g. diabetes or tuberculosis; the question arises as to the point at which they are to be regarded as clinically manifest diseases, and variations in the answer to that question must influence the incidence of that disease (thus the number of cases of respiratory tuberculosis can increase considerably, though, in a sense, artificially, with the introduction of mass radiography); (c) manifest disease recognised by the patient or by his medical attendant. Which of these various components we intend to count will obviously depend upon the circumstances and upon our needs but, if false comparisons are not to be made, the inclusions and exclusions on each occasion must be made perfectly clear in publishing our results.

We may also need to know about carriers of a disease, who do not themselves suffer the illness, but are infected and can pass it to others.

Sources of morbidity statistics

The usual sources of morbidity statistics, and the principal special problems arising in each, are as follows:

(1) *The survey of sickness* in which a representative sample of a population maintains a diary or is interviewed and each member is asked details of the sickness experienced over some defined preceding period of time. (The point prevalence rate, described below, will relate to the sickness actually existing on the day of the survey.) The definition of

sickness is usually based on subjective rather than objective criteria and, with the interview approach, the accuracy of the records depends upon the memories of the participants for perhaps relatively minor events in their lives (one to two months is usually the limit) as well as upon their knowledge of the name or nature of their illness. On the other hand, no other method can bring to light *all* the sickness, major and minor, experienced by a population.

(2) *General practitioner statistics* of patients attending surgery or visited at home. If fully maintained these records are more reliable in amount and diagnosis than those obtained by survey, but they are necessarily limited to those sick persons who choose to consult their doctors. Many minor illnesses will thus go unobserved and the amount unobserved will undoubtedly vary with such characteristics as sex, age, social class, etc.

(3) *Hospital in-patient statistics* are usually complete within their own field and likely to provide a firm diagnosis. But usually morbidity implies a *rate* and it may be very difficult to define the population from which are drawn the admissions to a particular hospital. Who are the exposed to risk who would go to such-and-such a hospital in the event of a major sickness? The statistics can also be influenced materially from time to time by changes in medical practice, e.g. cases of illness today may be treated at home, whereas before the introduction of antibiotics they could have found their way to hospital. Similarly they may differ in space, i.e. between hospitals, according to the views of general practitioners on the need to refer to hospital patients with certain conditions and on the views of clinicians on the need to admit and treat them. Needless to say, hospitals with special facilities and skills will attract patients with specific complaints. Length of stay and the fatality rate from particular causes of illness that a hospital may reveal will be influenced by its policy—whether all or only severe cases of an illness are admitted, whether convalescent patients, with a low risk of dying, are retained or whether there is a policy of early discharge, which will exclude thereby any late deaths. A very early death may be excluded because the patient was never formally admitted. In general, therefore, the diseases to be seen in hospitals are highly selective—by the nature of the illness, its severity and its required treatment.

(4) *Sickness absence records* reveal, as their title shows, who was absent from work and, usually, why in terms of a doctor's diagnosis of the cause of the illness, but the introduction of self-certification for short illnesses does not help in this respect. However, if the certificate is an open one, the diagnosis will not invariably reflect the doctor's real opinion. It is important to note too that the incidence of sickness absence must vary according to the nature of the sick person's job. Thus a fractured radius or a mild gastroenteritis may not keep a clerk from work but make it impossible for a bus driver to report for duty. The statistics are, of course, limited to those employed in industry, etc. and therefore exclude large sections of the population. The employer's regulations may also be important; the person who is heard to say 'Oh dear! I have not used all this year's sick leave yet' is not unknown. The durations of illness will be affected by the tendency to return to work on particular days of the week, e.g. a Monday, and if

someone starts an illness on Friday and returns on Monday, is that one day's or three days' illness? The statistics will also be influenced by the workers themselves—how readily they choose to absent themselves for minor complaints obviously may be affected by loss of income. In other words, we are inevitably dealing with the frequency and durations of *absences* and not with the frequency and durations of pathological conditions; the former may overstate or understate the latter and may do so differently between persons, jobs, etc.

(5) *Notifications of disease* are usually limited to infectious diseases and depend upon an adequate supply of doctors to provide the information. They will be influenced by the difficulties of diagnosis, by the degree to which laboratory assistance is available and by the readiness of practitioners to notify (which may well vary with the importance of the disease to the public health, e.g. typhoid fever compared with measles). However, in spite of known omissions and imperfections, they can be important as indicators to the medical officer of health of the presence of disease in the community, of the danger of a possible epidemic and of the need for preventive action, e.g. by isolation or identification of the source of an epidemic. Such, indeed, is their primary purpose. They can, however, be of considerable importance in revealing the epidemiological features of a disease, e.g. its distribution in space and time. In this context special problems have been reported in the UK in the statistics resulting from notifications made by clinics for sexually transmitted diseases. The figures relate to *cases* seen, not *patients*. Thus the same patient can reappear several times in the statistics for a given year and may do so for a number of reasons, e.g. he may have more than one disease at the same point of time, he may contract one or more diseases on separate occasions, he may be a new re-infection or a relapse. The cases thus reported must, therefore, be an *overestimate* of a problem in terms of the number of persons involved. In other words, the sexually transmitted diseases may not be so common as the statistics would suggest but be limited to an appreciably smaller and possibly high-risk group of the population—with obvious implications for preventive medicine.

(6) *Registration* of all cases of a disease, e.g. cancer or tuberculosis, is another system by which sickness in the population may be identified and measured. The register may permit long-term studies of a chronic disease, its course and response to treatment, as well as its epidemiological features.

In spite of the problems outlined above, all these sources of data have their own value and, in their proper sphere, can contribute to administration, research and knowledge. No single one will be valid for all purposes.

Unfortunately those who, with the best of motives, put restrictions on what data may be gathered, recorded and retained, can do great harm to useful enquiries. The suggestion that it would be unethical to test for the AIDS virus *anonymous* blood samples, taken in any case for other purposes, does seem unnecessarily restrictive—at what point do ethical objections become themselves unethical?

Rates of morbidity

In deciding upon appropriate rates of morbidity, the illnesses that exist in a population during a given time interval may first be classified as follows:

(1) illnesses beginning during the interval and ending during the interval;
(2) illnesses beginning during the interval and still existing at the end of the interval;
(3) illnesses existing before the beginning of the interval and ending during the interval;
(4) illnesses existing before the beginning of the interval and still existing at the end of the interval.

For each of these categories we shall need to decide whether we take as our measure the number of *persons* sick or the number of *spells of illnesses* occurring, e.g. if in a given time interval a patient has three attacks of bronchitis, do we count one person or three attacks? (Maybe we shall need both.)

The most useful morbidity rates in the total population or at specific ages will, then, be these:

(1) The *incidence rate*, defined as the number of illnesses (spells or persons as applicable) *beginning* within a specified period of time (categories 1 and 2 above) and related to the average number of persons exposed to risk during that period (or at its mid-point). In short, the object is to show the number of cases of sickness *arising* in a given interval, e.g. how many persons fell sick with influenza in the third week of the year.

(2) The *period prevalence rate*, defined as the number of illnesses (spells or persons as applicable) *existing at any time* within a specified period (all 4 categories above) and related to the average number of persons exposed to risk during that period (or at its mid-point). In short, the object is to show the total number of cases of sickness which *existed* during a given interval, e.g. how many persons were sick with enteric fever during the month of July.

(3) The *point prevalence rate*, defined as the number of illnesses *existing at a specified point of time* (all 4 categories above) and related to the number of persons exposed to risk at that point of time. In short, the object is to show how many cases of sickness were in existence *on this day*, e.g. how many persons were sick with enteric fever on 15 July. We still need to query however what is the 'point' concerned. Is one day (15 July, say) an adequate point, or should time of day also come into the definition?

(4) The *average duration of sickness* (and the frequency distribution upon which it is based). Such an average may be in terms of (a) *the total population exposed to risk* to give the average duration of sickness per person, (b) *the number of persons sick* to give the average duration of sickness per sick person, or (c) *the number of illnesses* to give the average duration of sickness per illness. In all these measures of duration, consideration must be paid to the circumstances of the four categories set out above. Is it intended to limit the duration to that experienced *within*

the defined period or is any note to be taken of the durations *preceding* the period but extending into it (categories 3 and 4) or of durations *following* the interval of illnesses that began within it (category 2)? There can be no categorically 'right' or 'wrong' procedure. The decision must turn upon the nature of the circumstances and the availability of the data. As already stressed above, it is essential in publication that full details be given of the procedure actually in use. The Sub-Committee on Absenteeism referred to above recommended to industry (1) that, normally, *calendar* days of absence should be used but that *working* days are permissible so long as their use is clearly stated, and (2) that the record should include all absences of one day or more up to 365 days, and again, any variation from this should be clearly stated.

The terms *incidence* and *prevalence* perhaps need special reference, for they are often misunderstood. The basic definition, as shown above, is that incidence refers to *new* cases whereas prevalence refers to *all* cases, new and old. Prevalence describes, it may be said, the static situation actually existing at a specified point of time (or over a specified period), e.g. how many persons are then ill? By incidence, on the other hand, we measure a changing situation over a defined period, e.g. how many persons are falling ill in that interval? Another feature of interest may lie in the number of persons who experience in some interval of time (say a year) 0, 1, 2, 3 or more spells of sickness or absence from work. What proportion of a labour force contributes to the sickness absenteeism? What proportion makes repeated contributions?

Whatever answers are chosen to these questions, one requirement always exists. Namely that it is essential to define with precision *exactly* what is being done. Vagueness is anathema.

Summary

Statistics of morbidity are required to supplement statistics of mortality, but in definition and analysis present much greater difficulties than the latter. As the English proverb says, stone dead has no fellow. In view of these difficulties it is essential in publication that full details be given of the definitions used, the method of collection of the data, and of the analyses applied to it. The most useful measures comprise the incidence rate showing the frequency with which new cases of disease arise and the prevalence rate showing the frequency with which established cases exist, either during a specified interval of time or at one specific point of time. Measures illustrating the duration of sickness can also be usefully employed.

Chapter 23

Clinical trials

> Therapeutics is the branch of medicine that, by its very nature, should be experimental. For if we take a patient afflicted with a malady, and we alter his conditions of life, either by dieting him, or by putting him to bed, or by administering to him a drug, or by performing on him an operation, we are performing an experiment. And if we are scientifically minded we should record the results. Before concluding that the change for better or for worse in the patient is due to the specific treatment employed, we must ascertain whether the result can be repeated a significant number of times in similar patients, whether the result was merely due to the natural history of the disease or in other words to the lapse of time, or whether it was due to some other factor which was necessarily associated with the therapeutic measure in question. And if, as a result of these procedures, we learn that the therapeutic measure employed produces a significant, though not very pronounced, improvement, we would experiment with the method, altering dosage or other detail to see if it can be improved. This would seem the procedure to be expected of men with six years of scientific training behind them. But it has not been followed. Had it been done we should have gained a fairly precise knowledge of the place of individual methods of therapy in disease, and our efficiency as doctors would have been enormously enhanced (Sir George Pickering, Presidential Address to the Section of Experimental Medicine and Therapeutics of the Royal Society of Medicine)

It would be difficult to put the case for the clinical trial of new (or old) remedies more cogently or more clearly. The absence of such a trial in the past may well have led, to give one example, to the many years of inconclusive work on gold therapy in tuberculosis, while, as Pickering stressed, the grave dangers of much earlier and drastic methods of therapeutics, such as blood-letting, purging and starvation, would quickly have been exposed by comparative observations, impartially made.

The ethical problem

Before embarking upon any such trial there is, however, a fundamental problem that must be carefully studied and adequately met. The basic requirement of clinical trials is comparison. In their most exacting form they call for concurrent 'controls', in other words a group of patients corresponding in their characteristics to the specially treated group but *not* given that special treatment. The question at issue, then, is whether it is proper to withhold from any patient a treatment that might, perhaps, give him benefit. The value of the treatment is, clearly, not proven; if it were, there would be no need for a trial. But, on the other hand, there must be some basis for it—whether it be from evidence obtained in test-tubes, animals, or even in a few patients. There must be some such basis to justify a trial at all. In other words it is no rare occurrence for physicians to have hopes about a new treatment but genuine doubts as to the balance of its benefits and risks. The responsibility of the doctor to his patient and the requirements of the trial may then be at issue. The problem will clearly turn in part, and often very considerably, upon what is at stake. If, for example, it be a question of treating the common cold in young adults and seeing whether the duration of 'illness' can be effectively reduced, then the morality of a rigidly controlled trial would not be seriously in doubt—and any other form of trial would probably be uninformative and a waste of time and money. At the other extreme it might be quite impossible to withhold, even temporarily, any treatment for a disease in which life or death, or serious after-effects, were at stake.

The problem is eased more often than not by the state of our ignorance. Frequently we have no acceptable evidence that a particular established treatment does benefit patients and, whether we like it or not, we are then experimenting upon them. As F.H.K. Green has pointed out, 'Where the value of a treatment, new or old, is doubtful, there may be a higher moral obligation to test it critically than to continue to prescribe it year-in–year-out with the support merely of custom or wishful thinking.' It is certainly not always recognised that it may be unethical to introduce into general use a drug that has been poorly or inadequately tested. The ethical problem is, indeed, not solely one of human experimentation; it can also be one of *refraining from human experimentation*.

The problem is particularly clearly demonstrated in the history of a mother in the early nineteenth century, who had seven children of whom only three survived, and we read that

> The sad truth is that those who died were the victims of Ann's sense of duty and of her belief in the medical lore of the day. However young they were, and whatever was wrong with them . . . she conscientiously dosed them with calomel; if there was nothing wrong she gave a precautionary dose every few days. Calomel is a cumulative poison and few survived. (Dr Joan Evans, *The Endless Web*, Jonathan Cape, 1955)

What young mother, conscientiously doing her duty, could bear to randomise her children into those to receive the supposed elixir and those to be denied it? Yet, had she done so, perhaps she would have saved the lives of the control group, as well as learning the truth.

Even nowadays, in withholding a new treatment from some patients it is well to remember that all the risks do not lie on one side of the balance. What is new is certainly not always the best and, as the history of antibiotics and hormones and of many other modern drugs has shown, it is by no means always devoid of serious danger to the patient. It may, therefore, be far more ethical to use a new treatment under careful and designed observation, in comparison with patients not so treated, than to use it widely and indiscriminately before its dangers as well as its merits have been determined. Such an evaluation should minimise the risk that a useless or dangerous treatment will become widely adopted. It should be realised, too, that *no special treatment* certainly does not imply *no treatment*. Constantly the question at issue is: Does this particular form of treatment offer more than the usual orthodox treatment of the day?

However that may be, every proposed trial must be exhaustively weighed in the ethical balance—each according to its own circumstances and its own problems. In other words on every occasion certain general questions must be considered by the investigator and they must be answered *within the specific and particular circumstances of the trial* he wishes to undertake. These questions will invariably include the following:

(1) *Is the proposed treatment safe and therefore unlikely to bring harm to the patient?* There can very rarely be a categorical answer Yes or No to this question. Not one of the enormously beneficial treatments that have revolutionised therapeutics since 1940 is wholly free of undesired 'side-effects' or without any hazard to the patients. None could have been introduced if complete safety for every patient had been demanded.

Obviously the possible dangers of a treatment may need to be considered in relation to the dangers of the disease treated. Thus it might be concluded that the potentialities of a new treatment could be properly explored in man in a disease of some severity but not in a mild self-limiting condition. Whether the side-effects of the treatment are likely to be reversible or irreversible will certainly call for much reflection. And so on.

(2) *For the sake of a controlled trial can a treatment ethically be withheld from any patient in the doctor's care?* The basis of a controlled trial of a new treatment compared with the old is that we are ignorant of the relative value of the treatments. If we are *not* ignorant, if there is good evidence that one or other treatment is, on average, better than the other, then (subject to special circumstances such as sensitivity in the patient) the better cannot be withheld. Often, however, we shall know something of the value of the *older* treatment but nothing of the new. And so the question may well be whether the doctor can withhold the established treatment in favour of the new and unproven. Alternatively we may know that the orthodox treatment is of exceedingly little, if of any, value, e.g. the treatment of some forms of cancer. It may be impossible then for the doctor to withhold any new treatment that appears to offer hope of

success. At the other extreme, in young adults suffering from the common cold it might well seem proper, as already suggested above, to withhold the latest 'wonder drug' from a suitable number of patients to establish, or disestablish, its alleged wonders.

In short, this question must inevitably be weighed with much care in the great diversity of circumstances that prevail between these extremes. It may be well to remember that if in any trial the treatment has been proved to be effective, then the treated patients will, fortunately, have had an advantage. But meanwhile the control group will have been in a position no different from that of all such patients suffering from the disease and thereby have not been specifically disadvantaged. By being members of a carefully conducted trial they may well have had the benefit of rather more careful observation or more laboratory investigations than would be customary.

(3) *What patients may be brought into a controlled trial and allocated randomly to different treatments?* The essential feature of a controlled trial that determines the answer to this question is that it must be ethically possible to give each patient *any* of the treatments involved. The doctor accepts, in other words, that he really has no knowledge at all that one treatment will be better or worse, safer or more dangerous, than another. If for the patient's benefit he believes that he should give one treatment rather than another, then that patient cannot be admitted to the trial. Only if, in his state of ignorance, he believes the treatment to be a matter of indifference can he accept a random distribution of the patients to the different groups.

Even if there be no evidence that one treatment is better than another, there will still be need to think whether certain types of patients should be excluded from a trial, e.g. pregnant women, patients with complicating conditions or diseases, the old and frail and the very young.

(4) *Is it necessary to obtain the patient's consent to inclusion in a controlled trial?* On this critical point opinions vary somewhat. In coming to a conclusion at least two aspects should be considered: (a) whether the patient will be subjected through a treatment to a special discomfort or pain which is not an inevitable concomitant of the disease or of its orthodox treatment; (b) whether the situation implicit in the controlled trial wholly exists, namely two (or more) possible treatments and an ignorance of their relative values (and hazards). If that situation does exist then that situation would need to be described to the patient so that he can give an adequately *understanding* consent. Without an adequately understanding consent it can be argued that the ethical decision must still lie fairly and squarely with the doctor himself. And if he believes that any of the treatments involved may, in his state of ignorance, be equally well exhibited to the patient's benefit, then it can also be asked whether there is a real basis for seeking consent or refusal.

However that may be, such trials are rarely permissible nowadays without the consent of an appropriate ethical committee, which in turn will usually demand that the patient's informed consent is needed. It is sometimes suggested that the doctor performing the trial may be less interested in the welfare of the patient than in the prospect of forwarding his own career—but for some reason this argument does not apply to the self-appointed expert in medical ethics seeking to forward *his* own career.

(5) *Is it ethical to use a placebo or dummy treatment?* This question can hardly arise whenever there already exists an established treatment of proven or accepted value. In such a situation the controlled trial must necessarily be directed towards proving the *relative* value of the newly proposed treatment, i.e. relative to the known value of the old. On the other hand, there may sometimes be no orthodox treatment of known or generally accepted value. It is in that situation that the placebo may be put forward as justifiable to indicate whether the treatment under test has any value whatever. Although experience shows that the 'dummy' treatment can produce 'side-effects', it is at least unlikely that it can produce irreversible harm!

(6) *Is it proper for a trial to be 'double-blind'?* i.e. for the doctor not to know the particular treatment that he is giving to a particular patient. By the so-called double-blind procedure, when neither patient nor doctor knows the nature of the treatment given to an individual case, it is hoped that unbiased subjective judgements of the course of the illness can be obtained. There is, indeed, much to be said in favour of the 'triple-blind' procedure in which the statistician (or whoever else) while analysing the results is kept in ignorance of which group is which. Sometimes one can escape the issue of the question by arranging for one doctor to treat the patient and another, without any knowledge of that treatment, to make the required assessments of X-rays, pathological specimens or even of the patient himself. When that procedure is not possible one must consider what might conceivably happen to the detriment of the patient if the doctor does *not* know the treatment. That is what calls for reflection in the special circumstances of a trial. It may be that nothing whatever is likely to happen to the detriment of the patient; alternatively, it may be that without a knowledge of the drug being administered the doctor cannot adjust the dose finely enough to meet the needs of the individual. And it is in such terms that the answer to the question must be sought. Finally, it must always be axiomatic that at any moment the code can be broken if the doctor in charge of the patient thinks it necessary.

Indeed, looking more widely, it must be implicit at every stage of a trial that its 'rules' must be broken if it appears to the doctor that the patient's well-being requires such action. Though this may on occasions nullify the value of a trial it is nevertheless an essential proviso.

The attitude and advice of the British Medical Research Council and the World Medical Association on these very difficult problems are given in Appendix G, and all concerned with the planning and conduct of clinical trials should study these statements with the utmost care.

Publications of the Royal College of Physicians of London are also worthy of careful study (*Guidelines on the Practice of Ethics Committees in Medical Research Involving Human Subjects* (2nd edn), 1990; *Research Involving Patients*, 1990).

For trials in children, workers should also consult the report of a working party set up by the British Paediatric Association: Guidelines to aid Ethical Committees considering research involving children. *Br. Med. J.* **i**, 229 (1980) and *Archives of Disease in Childhood* **55**, 75 (1980). On the general issues, strongly recommended reading is The scientific and ethical basis of the clinical evaluation of medicines, *Euro. J. of Cl. Pharm.* **18**, 129 (1980), the report of an international conference of distinguished physicians and research workers from Canada, Denmark, Finland, Germany, Israel, the Netherlands, Sweden, Switzerland, the UK and USA.

In conclusion on this aspect of the clinical trial, sometimes the difficulties are such that a scientifically imperfect trial must be accepted, sometimes it may be impossible to carry out any trial at all. The result of this situation in clinical medicine, unique, perhaps, in scientific work, is that second-best or even more inferior 'controls' are often adopted and, sometimes, though certainly not invariably, must be adopted.

Imperfect comparisons

Thus the following ways and means have been used from time to time, and are still used:

(1) The treatment of patients with a particular disease is unplanned but naturally varies according to the decision of the physicians in charge. To some patients a specific drug is given, to others it is not. The reactions and progress of these patients are then compared. But in making this comparison in relation to the treatment the fundamental assumption is made—and must be made—that the two groups are equivalent in all respects relevant to their progress, except for the difference in treatment. It is, however, almost invariably impossible to believe that this is so. The drug may be used or not used in relation to the patient's condition when he first comes under observation and also according to the subsequent progress of his disease, e.g. it may be given to the severely ill and not to the less ill. The two groups may therefore be quite incomparable, and more often than not the group given the specific drug is heavily weighted by the more severely ill. In such circumstances no conclusion as to its efficacy can possibly be drawn.

(2) The same objections must be made to the contrasting in a trial of patients who volunteer for a treatment with those who do not volunteer, or in everyday life between those who accept and those who refuse. There can be no knowledge that such groups are comparable; and the onus lies wholly, it may justly be maintained, upon the experimenter to prove that they are comparable, before his results can be accepted. Particularly, perhaps, with a surgical operation the patients who accept may be very different from those who refuse (see also Chapter 24).

(3) The contrast of one physician, or one hospital, using a particular form of treatment, with another physician, or hospital, not adopting that treatment, or adopting it to a lesser degree, is fraught with much the same difficulty—apart from the practicability of being able to find such a

situation (with, it must be noted, the same forms of ancillary treatment). It must be proved that the patients are alike in relevant group characteristics, e.g. age, sex, social class, severity of illness at the start of therapy, before they can be fairly compared and their relative progress, or lack of progress, interpreted. That proof is clearly hard to come by.

Retrospective, or historical, controls

One approach to the required comparison of patients given a new treatment with those not so treated lies in the use of past records, i.e. the comparison of the responses of patients to whatever was the treatment of the day, before the new treatment was available, with the patients now given the new treatment. Obviously in this method there lies one great advantage, that the ethical problem is greatly reduced (though it may still remain in the selection of which patients are to be given the new treatment). Clearly, the method can be effective in certain circumstances, namely if the natural history of the disease is so consistent and well documented that we can be sure that we are comparing like with like. If, for instance, in the past a disease has invariably and rapidly led to death, there can be no possible need for controls to prove a change in the fatality rate. Thus the trial of streptomycin in tuberculous meningitis needed no concurrent control group. Given a precise and certain diagnosis of the condition the success of treatment could be measured against the past 100 per cent fatal conclusion.

Similarly, controls may be unnecessary to prove *broadly* the value of such drugs as insulin and penicillin which quickly reveal their dramatic effects. Such dramatic effects occurring on a large scale and in many hands cannot be long overlooked. Unfortunately these undeniable producers of dramatic effects are the exception rather than the rule. Most therapeutic advance occurs in modest and gradual steps which call for a precise valuation.

Can this precise valuation be provided by a comparison of documented cases of the past with the documented cases of the present? The new treatment may well be given only to certain defined patients, defined closely by certain characteristics. They will then form a selected group and are not necessarily comparable, and, indeed, unlikely to be comparable, in all respects with previous patients who were not defined solely by those characteristics. Secondly, it may be that there has been a change in time in the patients presenting themselves for treatment (or referred by their general practitioners)—a change that the very existence of the new treatment may itself promote, e.g. by bringing in the intractable case with renewed hope of cure. Thirdly, we have to be sure that there has been no secular change in the nature or severity of the disease itself.

We are invariably faced with the question: Were these two groups similar in all respects relevant to the issue? It can be very difficult to give a decisive, or even acceptable, answer. An instructive example lies in a published study of polyarteritis nodosa treated with cortisone. From 1950 onwards 17 cases were thus treated. As controls, 19 cases, all proved by biopsy, were extracted from the clinical case records of the same hospitals during the preceding years 1941–49. The two groups were alike in sex and age at onset and the severity of the constitutional illness appeared similar. But they were found to differ in one critical respect. Of the controls nearly half (8 out of the 19 patients) had hypertension, while of those treated with cortisone there was only one single patient with hypertension (1 out of the 17 patients). The unequal distribution of this serious complication makes direct comparison impossible and leads to serious doubts as to whether the two groups may not differ in other and undetected ways. The authors themselves were led to conclude that the chief importance of their labours was 'to emphasise that the assessment of therapeutic activity by the use of retrospective controls is an inherently fallacious method' (*Br. Med. J.*, **1**, 611 (1957)). It may well be thought that this is too sweeping a condemnation to be drawn from a single example (see, for instance, Cranberg, *Br. Med., J.*, **2**, 1265 (1979)). It is, indeed, argued that such *observed* disparities in important diagnostic features between past and present can be effectively taken into account by a suitable statistical analysis of the data. The operative word is *observed*. Certainly such an analysis can be used to adjust for such features; but that gives no certainty that there are not others undetected from the records for which no correction can be made.

In short, if the past and present obviously differ from one another, how do we know that we have become aware of *all* their differences? Can we be sure that differences of importance are confined to those that have been identified and measured? This is the crux of the problem—and, indeed, the main reason for concurrent controls to eliminate the effects of such variables as changes in diagnostic criteria, changes in methods and degrees of observation, the use of ancillary treatment, etc.

At a lower level there will often be practical difficulties in securing the necessary records of past patients. These difficulties will arise because certain observations or measurements are missing in some, or all, past records, or were made at different times and by different methods. If in a case/control approach each new case is matched by a corresponding old case there may be more than one of the latter that fulfils the matching criteria. Knowing the upshot in the latter case may lead to a biased selection unless special steps are taken to avoid it, e.g. by a random selection. Possibly there might be some advantage in inverting the process. In other words the worker would first construct the required group of past records over some chosen interval of time. By an analysis of these data he would define the more important prognostic features of, say, mortality, speed of recovery, etc. Then, as new cases are admitted to the trial of the new treatment each patient would be matched in these features against a specific past case. In the course of time the response of the new case would be revealed. Any new case not matched would be discarded.

There is no doubt that in certain circumstances the 'historical' approach is inevitable. Equally there should be no doubt in the user's mind of the very grave difficulties in ensuring comparability of past and present.

Finally in this connection we may note that the position has been reached in which we are not contrasting older orthodox methods of treatment with a potent modern drug, but one modern drug with another. To prove that a fatality rate of the order of 60 per cent has fallen to 15 per cent is a very much easier task than to prove that with a new treatment the 15 per cent can be reduced to 10 per cent. Even a poor clinical trial could hardly destroy the evidence of the former profound change; it may take a very efficient one to prove the latter. Yet, in the saving of life, that improvement is a very important change.

The aim of the controlled trial

Turning to the trial with concurrent treated and control patients, the first step is to decide precisely what it sets out to prove. It is essential that initially its aims should be laid down in every detail. For example, the object in the British Medical Research Council's first trial of streptomycin upon the introduction of this drug was to measure the effect upon respiratory tuberculosis. This illness, may, however, denote many different things: the minimal lesions just acquired by the patient, the advanced and progressive disease that offers a poor prognosis, or the chronic and relatively inactive state. Equally, its course, its rapidity of development or the recovery of the patient, may differ with age, whether early childhood, adolescence or old age. The question, therefore, must be made more precise. It was, in fact, made precise by restricting the trial to 'acute progressive bilateral pulmonary tuberculosis of presumably recent origin, bacteriologically proved, unsuitable for collapse therapy, age-group 15–25 (later extended to 30)'. Thus it was ensured that all patients in the trial would have a similar type of disease and, to avoid having to make allowances for the effect of forms of therapy other than mere bed-rest, that the type of disease was one not suitable for such other forms. In such cases the chances of spontaneous improvement were small, but the lesion, on the other hand, offered some prospect of action by an effective chemotherapeutic agent. In short, the questions asked of the trial were deliberately limited, and these 'closely defined features were considered indispensable, for it was realised that no two patients have an identical form of the disease, and it was desired to eliminate as many of the obvious variations as possible' (*Br. Med. J.*, **2**, 769 (1948)).

This planning, as already pointed out, is a fundamental feature of the successful trial. But to start out without thought and with all and sundry included, with the hope that the results can somehow be sorted out statistically in the end, is to court disaster.

This is not to say that a wide range of patients can never be admitted to a trial. Indeed in some circumstances, it may be very desirable that they should be. With a liberal entry we can subsequently analyse the results within narrow groups. Thus we may see whether the effects of the treatment were more, or less, favourable in patients of a defined clinical category. But so far as possible we should, with thought, define these groups *beforehand*. To be able to generalise from the particular we must precisely define the particular.

One difficulty may sometimes arise with a precise and exact definition of the cases to be included in a trial. In testing, for instance, the effects of different specific treatments on rheumatic fever, the Anglo-American co-operative trial wisely endeavoured to lay down criteria which had to be fulfilled before any patient was accepted. The patient had to have such-and-such signs and symptoms. By such means misdiagnoses would be avoided and other illnesses simulating rheumatic fever excluded. But the exhibition of these signs and symptoms, while denoting the undoubted case of rheumatic fever, may also denote that the patient is no longer in the very early stages of the disease. The very early case with highly suspicious but still indefinite signs might under these criteria be excluded until the signs had become definite—when it is no longer in those desirable early stages. That is a situation which will sometimes need close consideration and the weighing of alternatives (and was, of course, given such in the trial quoted). For instance, patients with an illness in which bacteriological confirmation of the diagnosis is required may be brought into the trial at once but subsequently discarded if the bacteriological result is negative.

The basic rule, however, on such exclusions must be that *every* patient admitted to the trial must appear in the final analysis, even if later found to be unsuitable for the trial. Otherwise biases are likely to be produced. It is, of course, permissible to exclude certain subjects from some parts of the analysis provided it is made clear that that is being done.

The construction of groups

The next step in the setting up of the trial is the allocation of the specifically defined patients to be included in the treatment and the non-treatment groups (or to more than two groups if more than one treatment is under test). As stated earlier, before admission to a trial *every* patient must be regarded as suitable for *any* of the treatments under study. If this freedom is not present, then equivalent groups cannot be constructed and comparisons are impossible. The aim, then, is to allocate those admitted to the trial in such a way that the two *groups*, 'treatment' and 'control', are initially equivalent in all respects relevant to the inquiry. Individuals, it may be noted, are not necessarily equivalent; it is a group reaction that is under study. In many trials this allocation has been successfully made by putting patients, as they present themselves, alternately into the treatment and control groups. Such a method may, however, be insufficiently random if the admission or non-admission of a case to the trial turns upon a difficult assessment of the patient and if the clinician involved knows whether the patient, if accepted, will pass to the treatment or control group. By such knowledge he may be biased, consciously or unconsciously, in his acceptance or rejection; or through fear of being biased, his judgement may be influenced. The latter can be just as important a source of error as the former but is more often overlooked. For this reason, it is better to avoid the alternating method and to adopt the use of random sampling numbers; in addition, the allocation of the patient to treatment or control should be unknown to the clinician until *after* he has made his decision upon the patient's admission to the trial. Thus he can proceed to that decision—admission or rejection—without any fear of bias. One such technique has been for the statistician (or other associated worker) to provide the clinician with a set of numbered and sealed envelopes. After each patient has been brought into the trial the appropriately numbered envelope is opened (no. 1 for the first patient, no. 2 for the second, and so on) and the group to which the patient is to go, treatment (*T*) or control (*C*), is given upon a slip inside. Alternatively a list showing the order to be followed may be prepared in advance, e.g. *T*, *T*, *C*, *T*, *C*, *T*, *T*, *T*, *C*, etc., and held confidentially, the clinician in charge being instructed after each admission has been made.

It may sometimes be thought advisable, rather than using absolutely random allocation, where each individual envelope is equally likely to be a *T* or a *C*, to enforce equality of *T*s and *C*s after every *n* envelopes for some value of *n*, so that if the trial has to be stopped early there will be enough of each group for conclusions to be attempted. Details of such methods are given in Appendix E.

When the results of treatment are likely to vary between, say, the sexes or different age groups, then a further extension of this method may be made, to ensure a final equality of the total groups to be compared (by stratified sampling as already described in Chapter 2). Separate sets of envelopes, or of lists, are provided for subgroups of the patients to be admitted e.g. for each sex separately, for specific age groups or for centres of treatment, and in each subgroup the number of *T* cases is made equal to the number of *C*. Thus we may have allocation lists, or envelopes based upon them, as shown in Table 23.1. There is, of course, no point in such separation of subgroups if pure randomisation is used.

It may be thought that these fine subdivisions are unwarranted, since the

Table 23.1 Assigning patients to groups.

Patient's number in each subgroup	*Male*		*Female*	
	20–29 years	*30–39 years*	*20–29 years*	*30–39 years*
1	*T*	*T*	*C*	*T*
2	*C*	*T*	*T*	*C*
3	*C*	*T*	*T*	*C*
4	*T*	*C*	*T*	*T*
5	*T*	*C*	*C*	*C*
6	*C*	*T*	*C*	*T*
7	*T*	*C*	*C*	*T*
8	*C*	*C*	*T*	*C*
.				
.				
.				

numbers within the subgroups may finally be too small to justify any comparisons. Whether that be true or not, this argument overlooks the fundamental aim of the technique; if a trial is not being confined to a narrowly defined group, the aim is to ensure that when the *total* groups, *T* and *C*, are compared they have within themselves equivalent numbers of persons with given characteristics. Thus in the above example when the *T* and *C* experiences as a whole need to be compared, it will be found that there are 8 males and 8 females in the *T* group and 8 males and 8 females in the *C* group; 8 *T* and 8 *C* persons are aged 20–29, 8 *T* and 8 *C* are aged 30–39. The *T* and *C* groups have been automatically equalised, a result which—and particularly with small numbers—would not necessarily have been achieved with a single allocation list which ignored age, sex and place. With large numbers, equality of characteristics in the two groups will result in the long run and this refinement is not worthwhile; with small numbers, it will be wise to ensure equality, or near equality, in such characteristics as will affect the final comparisons.

The prescribed random order must, needless to say, be strictly followed or the whole procedure is valueless and the trial breaks down. Faithfully adhered to, it offers three great advantages: (1) it ensures that our personal feelings, or judgements, applied consciously or unconsciously, have not played any part in building up the various treatment groups—from that aspect, therefore, the groups are unbiased; (2) it removes the very real danger, inherent in any allocation which is based upon personal judgements, that believing our judgements may be biased, we endeavour to allow for that bias and in so doing may 'lean over backwards' and thus introduce a lack of balance from the other direction; (3) having used such a random allocation we cannot be accused by critics of having set up personally biased groups for comparison.

It should be noted that although a certain amount of enforced equality between the groups is permissible, and may sometimes be desirable, there must always remain a random element to the allocation. We can equalise only for such features as we can measure or otherwise observe, but we also need unbiased allocation for all other features, some of which we may not even know exist. Only randomisation can give us that, and no form of equalisation can be a satisfactory substitute for it.

The treatment

In regard to treatment there are frequently, and obviously, a great many questions that can be asked of a trial. We can, for instance, choose one dose of a drug out of many; we can vary the interval of its administration; maybe we can give it by different routes; we can give it for different lengths of time; and so on. In testing a new form of treatment knowledge is at first scanty, and it may often, therefore, be best to choose one closely defined regimen which should, it is believed, reveal the potentialities and perhaps some of the dangers of the drug, or whatever may be concerned. Thus the question asked of the trial may run: If to a closely defined type of patients

2 g of drug *X* are given in four divided doses daily and orally, and for three months, what is the progress of such a group of patients compared with a corresponding group not treated with this drug? After the answer has been reached we may be able to extend our knowledge by experimenting with other dosages of the drug, etc.

Sometimes, however, it might be more informative to allow for individual idiosyncrasies during the basic trial or, in other words, to permit the clinician to vary the dosage, etc., according to his own judgement of the patient's needs as shown by the latter's responses. We have then, it is clear, deliberately changed the question asked of the trial so that it now runs: if clinicians in charge of a closely defined type of patient administer drug *X* orally in such varying amounts and for such varying durations of time, and so forth, as they think advisable for each patient, what, at the end of three months, has been the progress of such a group as compared with a corresponding group not treated with this drug?

The moot point, that must be considered in given circumstances, is which is the better question to ask. In many trials it would certainly seem most desirable to lay down a fixed schedule which, except under exceptional circumstances, e.g. for ethical reasons, must not be varied by those taking part in the trial. (The fixed schedule, can, of course, be different for patients of different ages or body-weights, etc.) If dozens of people in a co-operative trial are free to vary the dosage just as they personally think fit, and in circumstances where too little is known to give any reasonable basis for such variations, then it may be very difficult finally to extract any clear or useful answer at all from the trial. Thus, for example, in the basic trials of isoniazid in pulmonary tuberculosis it was decided to give a particular dosage for a particular length of time in comparison with a correspondingly defined schedule of streptomycin. On the other hand, when it seems clear that individuals may react very differently to some therapeutic agents, e.g. to treatment with a hormone, it may be wiser to design a trial that permits the clinicians to select their treatment over a range of dosages and to vary it from time to time in accordance with the response of the patient. This was the procedure adopted in a trial of cortisone compared with aspirin in the treatment of early cases of rheumatoid arthritis. It should be clear that such a procedure does not prohibit the use of the customary measures of progress—body-weight, erythrocyte sedimentation rate, pyrexia, etc.—even though the physician may base his treatment upon the level of these characteristics and their changes. The question asked of the trial is: what after so many months are the clinical conditions, degrees of fever, etc., of two groups of patients, one treated with, say, cortisone and the other with, say, aspirin, the treatment having varied within both groups at the will of the clinicians? Do they differ? We may thereby see whether what the doctors chose to do with cortisone was any better, or worse, than what they chose to do with aspirin. No dissection of the two groups, however, can possibly be made to see how patients fared upon different regimes of treatment. The physicians have deliberately varied these regimes in accordance with the patients' responses; it is not reasonable, then, to turn round at the end of the trial and observe the responses in relation to the regimes. That would be circular reasoning. To measure the effects of different regimes there can be no other way than the setting up of a trial to that end, randomising the patients to the different regimes and observing the responses. In some circumstances that may be a possible and very desirable procedure. Clearly, much might be learned about the action and value of a drug if a number of randomly constructed groups could be set up and given various amounts of it, ranging, perhaps, from none to some maximum value. The graded responses to these different amounts would be very informative. Too few trials appear to be set up on this basis.

Consideration must also be given to the treatment of the control group. In such a trial as that of streptomycin in respiratory tuberculosis in young adults (quoted above), involving frequent injections of the drug, it would be quite unethical to institute any corresponding procedure for the controls. They were, therefore, treated as they would have been in the past.

On the other hand, in the treatment of the common cold a dummy treatment would be essential, for one cannot invite volunteers for a trial, obviously keep half as controls, and then hope for co-operation and good records. The fact that they are taking part in a trial should be made clear to the participants, but they should not be told which treatment they receive. The importance of the control treatment is shown by the results of one such trial—of colds of under one day's duration at the start of treatment, 13.4 per cent were reported as 'cured', and 68.2 per cent 'cured' or 'improved', on the second day following administration of an antihistamine compound; but with the placebo, the corresponding figures were 13.9 and 64.7 per cent (*Br.Med.J.*, **2**, 425 (1950)).

In many cases, therefore, a placebo treatment is desirable and its practicability must be considered—as well, of course, as the ethics of such a procedure in each instance. The object is twofold. On the one hand we hope to be able to discount any bias in patient or doctor in their subjective judgements of the treatment under study. For example, a new treatment is often given a more favourable judgement than its value actually warrants. The effects of suggestibility, anticipation and so on must thus be allowed for. In addition the placebo provides a vital control for the frequency of spontaneous changes that may take place in the course of a disease and are independent of the treatment under study. In these two ways the placebo aids us to distinguish between (1) the pharmacological effect of a drug and (2) the psychological effect of treatment and the fortuituous changes that can take place in the course of time.

Finally under this heading of treatment we have to remember that patients may neglect to take the drugs prescibed for them. This may apply particularly to trials undertaken outside hospital control. Prolonged courses of treatment, as in tuberculosis, are, of course, specially likely to lead to default. We must at the very least ensure that all patients understand the instructions as to what they should do and we must take whatever steps are possible to minimise and measure default. For example, it is probably preferable to have one dose per day rather than divided doses. With widespread or considerable defaulting we may be measuring accurately our *intent* to treat in a certain way. But that is not the same thing as the actual treatment in that certain way. The drug is not given the full opportunity to display its value—or lack of value.

It should also be remembered that the double-blind procedure, however desirable, may be very difficult to achieve. To make two treatments indistinguishable to doctors and patients over a course of time is by no means as easy as it appears on paper.

Measuring the results

There is one great advantage of a placebo used in such a way that the clinician does not know to which patients it is allocated, i.e. the double-blind trial. In these circumstances the clinical impression can be included and given full weight in the analysis. If, in other words, two groups of patients are being treated, one *T* and one *C*, and the clinician does not know the components of the groups, then he can without fear or favour assess the progress and condition of every patient. And thus clinical instinct and opinion, as well as more strictly objective measures, can be used, without risk of bias, to assess the result. This may be a very valuable addition to the trial and prevent us from substituting a collection of precise qualitative, but uncoordinated, details for a 'coherent though impressionistic picture'.

Such a method was used in the Medical Research Council's clinical trial of an antihistaminic drug in the treatment of the common cold (quoted above). To make sure that no bias should enter into the assessment of the results, neither patient nor clinician was aware whether antihistamine tablets or control tablets had been given in a particular case. To ensure this result, numbered boxes of tablets and similarly numbered record sheets were issued to the centres taking part in the trial, each box to be used in conjunction with the appropriate sheet. Box No. 1, for instance, might contain antihistamine tablets, Box No. 2 the control tablets, and so on, as determined beforehand from randomly constructed lists. Neither box had any label indicating its contents. In the final analysis, therefore, record sheet No. 1 must relate to the antihistaminic group, record sheet No. 2 to the comparative group, record sheet No. 3 to the comparative group, and so on. But neither patient nor investigator in the field could know that (nor, necessarily, the worker analysing the results). This method of randomisation and presentation of the treatment was adopted in place of the more usual one of labelling one product as *X* and the other as *Y*, and presenting them in random order, because it was believed that the drug could have noticeable side-effects. If such effects were clearly observed in only one patient, e.g. on drug *X*, then the nature of *X* (and *Y*) would thereafter be known, or suspected. But with the random presentation of unlabelled boxes of tablets the identity of one box might be suspected because of the side-effects, but this would be no guide to the contents of any other box.

In the adoption of other measures of the effect of treatment, detailed planning must as usual play its part. Before the trial is set under way it must be laid down, for example, precisely when and how temperatures will be taken, when full clinical examinations will be made, and what will be specifically recorded, how often and at what intervals X-rays will be taken. Standard record forms must be drawn up, and uniformity in completing them stressed. Unless these rules and regulations are well kept and observed by the clinicians in charge of the trial, many and serious difficulties arise in the final analysis of the results, e.g. if some X-ray pictures were not taken at the required monthly interval, or, if some tests of the erythrocyte sedimentation rate (ESR) were not made. In fact, *every* departure from the design of the experiment lowers its efficiency to some extent; too many departures may wholly nullify it. The individual may often think 'it won't matter if I do this (or don't do that) just once'; he forgets that many other individuals have the same idea.

It is fundamental, too, that the same care in measurement and recording be given to both groups. The fact that some are specially treated and some are not is wholly irrelevant. Unless the reactions of the two groups are equally noted and recorded, any comparisons of them must clearly break down. For the same reason, if a follow-up of patients is involved it must be applied with equal vigour to all. In some circumstances the 'blind' assessment technique of the patient's condition can be applied. Thus in the trial of streptomycin in respiratory tuberculosis in young adults the chest radiographs of all patients were viewed, and changes assessed, by three members of a special radiological panel working separately and not knowing whether the films came from treated or control patients. The setting up of a team whose members worked independently gave an increased accuracy to the final result; the 'blind' assessment removed any possibility of bias or over-compensation for bias. Similarly in another trial one worker injected the patient with a compound or with saline, and another, not knowing the nature of that injection, measured the results of compound or saline in the patient.

What is not often recognised is that in many trials it is desirable that if any judgements or interpretations are required of the data recorded for each patient, then those judgements or interpretations should be made *before* the treatments are decoded.

Reporting the results

In reporting the results of clinical trials it is important to describe the techniques employed and the conditions in which the investigation was conducted, as well as to give the detailed statistical analysis of results. In short, a statement must be made of the type of patient brought into the trial and of the definitions governing the selection of a case; the process of allocating patients to treatment and control groups should be exactly defined; the treatment should be stated precisely; the assessments and measurements used must be clearly set out, and it must be shown whether they were made 'blind' or with a knowledge of the treatment given. In other words, the whole plan and its working out should be laid before the reader so that he may see precisely what was done.

Secondly, even if a random allocation of patients has been made to the treatment and control groups, analysis must be made to show the degree of equality of the groups at the start of the trial. With large numbers such an equality will almost invariably be present, but with small numbers the play of chance will not invariably bring it about. It is important, therefore, to see whether there is an initial equality or an inequality for which allowance must be made (e.g. by subdivision of the records or by standardisation). In the trial of the antihistamine drug in the treatment of the common cold, for example, there were 579 persons given the drug and 577 a placebo. Of the former group 34.7 per cent had had symptoms for less than a day before treatment, of the latter 30.0 per cent; 55.4 per cent of the former had a blocked nose as a presenting symptom, and 53.2 per cent of the latter; 7.6 per cent of the former and 8.0 per cent of the latter said at entry that they 'felt ill'.

With the much smaller-scale trial of streptomycin in young adult respiratory tuberculosis a good degree of equality was also reached. Of 55 patients in the treatment group 54 per cent were in poor general condition at the start of the trial; of 52 patients in the control group 46 per cent were in a similar condition. Twenty and 17 respectively were desperately ill, 32 and 30 had large or multiple cavities, 19 and 19 showed radiological evidence of segmental atelectasis. To such an extent can a carefully designed and deliberately limited inquiry bring about equality in even quite small groups. But that it has achieved that aim must first be shown.

In published reports of trials the writer will sometimes give the results of tests of significance of the initial differences between the treatment groups; but as D.G. Altman has said 'it is pointless to test whether something *might* be due to chance when, if you have randomised properly, you already *know* that it is'. Of course if the differences *are* significant one will need to consider whether (1) a relatively rare event has turned up by chance or (2) the random allocation has not, in fact, been followed, either by error or deliberately. In either case some action has to be taken. But if the differences are not significant in the statistical sense *it does not follow that the two groups are initially so equal that we can ignore the differences between them*. That is a *non-sequitur*. For example on treatment *A* there might be 10 mildly ill patients, 30 moderately ill and 15 severely ill; and with treatment *B* the numbers might be 6, 29 and 20. The difference is not statistically significant but that is irrelevant. It may yet be sufficiently large to upset the overall comparison of the 55 patients in each group. We shall need to consider that in making our analysis of the results of the trial. No test of statistical significance will take care of it.

Differential exclusions

In the protocols of a proposed trial, specifications for any exclusions from it, e.g. of the old and severely ill, should be laid down explicity in advance. They should *not* be determined after the entry of a patient and the allotment of a treatment. However, some exclusions at this latter point of time are usually inevitable.

In analysing the results of a trial we have, therefore, a vital question to consider—have any patients *after admission* to the treated or control group been excluded from further observation? Such exclusions may affect the validity of the comparisons that it is sought to make; for they may *differentially* affect the two groups. For instance, suppose that certain patients cannot be retained on a drug—perhaps through toxic side-effects. No such exclusion may occur on the placebo or other contrasting treatment, and the careful balance, originally secured by randomisation, may thereby be disturbed. Another specific example might lie in a trial of pneumonectomy versus radiation in the treatment of cancer of the lung. At operation there is no doubt that pneumonectomy would sometimes be found impossible to perform and it would seem only sensible to exclude these patients. But we must observe that no such exclusions can take place in the group treated by radiation. If we exclude such patients on the one side and inevitably retain them on the other, can we any longer be sure that we have two comparable groups differentiated only by treatment? Unless the losses are very few and therefore unimportant, we may inevitably have to keep such patients in the comparison and thus measure the *intention to treat* in a given way rather than the actual treatment. On the other hand, as earlier stated, if the diagnosis of an illness needs to be confirmed by a bacteriological, or other, test, there is less objection to excluding patients who were randomly entered but in whom the test has shown the diagnosis to be wrong. The exclusions should, except for the play of chance, occur equally in the two, or more, groups. At the very least, though, the number of such exclusions in each group must be reported so that readers can judge whether anything has gone seriously amiss. Similarly if, after randomisation, death (or some other defined event) were to take place *before* treatment could be begun there should be no objection to the removal of such patients from the trial *so long as the lapse of time between randomisation and the beginning of treatment is, on the average, the same in both groups*. In such circumstances there is, again, no reason why the numbers should differ materially between the groups—and if they did one would need to seek an explanation. In practice the lapse of time might differ between, for instance, a group allocated to, and likely to *await*, surgery and a group allocated to, and *immediately* available for, medical treatment. Possibly the only solution here lies in the *ab initio* planning, e.g. that the medical case will await the treatment under trial until the corresponding surgical case enters the trial. There can be no hard and fast rules for there is no correct answer to all situations. One thing that can be said is that whenever possible it would be wise to analyse the results of a trial (1) including and (2) omitting the patients that one proposes to exclude. What effect do the exclusions reveal? The question of the introduction of bias through exclusions for any reason (including lost sight of) must, therefore, always be carefully studied, *not only at the end of a trial but throughout its progress*. This continuous care is essential in order that we may immediately consider the nature of the exclusions and whether they must be retained in the inquiry for follow-up, measurement, etc. It will be too late to decide about that at the end of the trial.

Duration of treatment

A point of importance can also arise in trials designed to determine the most favourable duration of a form of treatment. To make the discussion specific one may take the original trials of penicillin in the treatment of subacute bacterial endocarditis. One question at issue was whether it was advisable to give *x* units to the patient within one week or the *same* number of units spaced out over four weeks. The two methods are tried and the fatality rate (or relapse rate) is observed over some specified period of time, say 1 year. The point at issue is, from what point of time are the deaths counted—from the *completion* of the course of treatment or from its *initiation*? At first sight the former seems reasonable, but, in fact, it would be wrong. If the deaths are counted only from the completion of treatment, then, artificially, the treatment of the longer duration may always show an apparent advantage, even though it has no real advantage. The reason is that by such a method all the deaths (or relapses) between day 7 and day 28 must be counted against the 7 days' treatment group (the treatment of which is complete at day 7), but no such death can be counted against the 28 days' treatment group (the treatment of which is not then complete). If the number of such deaths (or relapses) is appreciable, the answer *must* favour the long duration of treatment—indeed it would seem better to give the specified number of units over a year, for the deaths will then relate only to those who survive that length of time! The conclusion is that for a valid comparison the deaths must be counted from the *initiation* of treatment (as was indeed done in the trials upon which this comment is based). It may be argued that the specified treatment was not, in fact, then given in all cases, but the answer to that is that a specified treatment *had been chosen*. If the patient died before its completion the method of choice clearly failed, and that failure must be debited to it. We are thus led to the conclusion that the results must be measured from the initiation of treatment—though, of course, separate rates may be computed for the first month and for subsequent periods. If any exclusions are to be made, e.g. death after only one administration of the drug, they must be made uniformly from all groups, and just what is meant by 'uniformly' in this context may not be at all obvious without much thought in each particular case.

The numbers required

In the planning of a clinical trial a fundamental question is its scale. How many patients do we need to provide a valid answer? To that question there is usually no simple answer. If two groups are to be compared, a treated and a control group, then the size of the sample necessary to 'prove the case' must depend upon the magnitude of the difference that ensues.

If, to take a hypothetical example, the fatality rate (or any other selected measure) is 40 per cent in the control group and 20 per cent in the treated group, then by the ordinary test of significance of the difference between two proportions, that difference would be more than is likely to occur by chance with 50 patients in each group. In other words, with those fatality rates we must have observed at least 50 patients in each group to feel at all confident in our results. If there were 50 patients in each group and 20 died in the control group and 10 in the specially treated group, that difference is more than would be likely to occur by chance. If, on the other hand, the improvement was a reduction of the fatality rate from 40 to 30 per cent we should need at least 200 patients in each group. If we had 200 in each group and 80 died in the one and 60 in the other, that difference is more than would be likely to occur by chance. Finally, if the fatality rate was only 4 per cent in the control group and 2 per cent in the treated group, we should require as many as 650 patients in each group to be able to dismiss chance as a likely explanation. With that number in each group there would be 26 and 13 deaths, and a difference of this order on smaller numbers might well be due to chance. (In such a case the fatality rate, of course, might not be the best measure of the advantages of the treatment.)

The determination of the numbers required is based, it will be noted, upon the difference observed between the groups. In practice, we often do not know what that difference is likely to be, until at least some preliminary or pilot trials have been made. Alternatively we may have some indication from past experience of the magnitude of the rate likely to occur in the 'control' group. We may then decide how much fall in that rate we would wish to see in the 'treated' group before we would regard it as of clinical importance. On that basis we would be able to calculate the numbers required to give a statistically significant answer. The crucial feature, as shown above, is the frequency of the critical events upon which the comparisons will be based and the conclusions drawn.

On the other hand, the object of a clinical trial is not usually limited to showing that there is a significant difference in efficiency between treatments. Interest will lie in such features as the incidence and nature of adverse reactions, in the ease of administration and the relative cost of the treatments; and these features may have a bearing upon the scale of the trial.

A pilot trial should not be started without careful thought about whether it is really going to help. If it gives very little information, no purpose will have been served by doing it; but if it gives considerable information, it may not be enough to answer the question that the main trial was going to ask, and yet be suggestive enough to make that main trial unethical, so the whole plan collapses.

With the modern large-scale trial the choice may lie between (1) a large number of patients rapidly recruited from a large number of centres and a relatively short follow-up and (2) a smaller number of patients from fewer centres and a long follow-up. The former scheme can present formidable administrative problems and difficulties in ensuring the constant use of the same protocol throughout the centres, while the latter may suffer from the effects of changing clinicians, waning interest, and so on. But obviously a long follow-up is essential if long-term results of the treatment are the essence of the question under study. Many current trials are too small to give much hope of detecting the sort of effects that might plausibly be expected.

It appears sometimes to be thought that there is some necessary antagonism between the clinical assessment of a few cases and the 'cold mathematics' of the statistically analysed clinical trial dealing with a larger number. It is difficult to see how, in fact, there can be any such antagonism. The clinical assessment, or the clinical impression, must itself be numerical in the long run—that patients are reacting in a way different from the way the clinician believes was customary in the past. In the controlled trial an attempt is made to record and systematise those impressions (and other measurements) and to add them up. The result reached is, of course, a group result, namely, that *on the average* patients do better on this treatment than on that. No one can say how one particular patient will react. But that, clearly, is just as true of the approach via clinical impressions and two cases, as it is via a controlled and objectively measured trial and 100 cases. Also it may be noted that observation of the group does not prevent the most scrupulous and careful observation of the individual at the same time—indeed it demands it. 'Without doubt', said an anonymous leader writer (*Br.Med.J.*, **1**, 235 (1975)), 'clinicians who take part in trials eventually learn many new facts about the natural history of cancer, and their whole approach to the disease is influenced by the lessons of the trial. Anything that helps to convert from anecdote to accurately recorded evidence placed in its true context can only be a step in the right direction.'

Which patients respond?

Frequently the investigator should try to do much more than determine whether a new treatment has, on the average, more (or less) value than, say, the current standard treatment. He should specifically set out to determine why some patients respond to a certain treatment and others do not. What are the characteristics in the patients that lead to this difference?

To solve, or contribute to, this vital problem may call for the recording of much observation about each patient before the trial starts, and for subsequent acute observation throughout its course. It may then be possible at its conclusion to contrast the features of those patients who showed a favourable response with the features of those who did not. Thus we might narrow down the basic issue of the 'treatment of choice'. The danger here is that we may overload the trial with innumerable irrelevancies. At the end of it with the aid of a computer there may be no difficulty in sorting out all the data. But at the start of it, and throughout its course, the clinician, or some other worker, has to make and record all these observations, a suitable computer package has to be chosen for analysis and, finally, all these answers have to be studied. It would be better, perhaps, to consider initially, and with much care, what features *might* be relevant to prognosis. The recording of the effects of treatment would thus be limited to the hypotheses that we have set up to be tested. Here the trouble is that we may be too ignorant to know what could be important—what, in fact, to look for. And so, in practice, we may be led to what has been described as a 'fishing expedition'. We have a host of measurements and we sort through them to see if we can isolate some that are associated with a favourable response to the treatment. Here we need caution in interpreting 'significant'. If we seek through a large number of features we can expect to find many which satisfy the 1 in 20 significance test but are, in fact, due only to the play of chance. We cannot justifiably seize upon just that one out of many contrasts which shows an exciting difference and ignore the rest. It may need to be put to the test in a future trial. In short, any association thus arising fortuitously rather than by prior definition of a question asked of the trial should be treated with extra caution. This is, of course, not to say that the 'fishing expedition' should not take place. To seek through one's data for clues is demanded of every investigator. By accident or sagacity one may light upon something quite unthought of, but of real importance.

The patient as his own control

In some instances it may be better to design the trial so that each patient provides his own control—by having various treatments in turn. This is known as a 'cross-over' trial. The advantages and disadvantages of the procedure will need careful thought.

By such means we may sometimes make the comparison more sensitive since we have eliminated the variability that must exist *between* patients treated at the same stage of the disease in question (so far as can be judged). We have done so, however, at the expense of introducing as a factor the variability *within* patients from one time to another, i.e. we may be giving the patient treatment *A* and treatment *B* at *different* stages of the disease. For instance, with a disease that naturally declines in severity with the passage of time, e.g. the acute sore throat, a very misleading answer would be obtained if treatment *A* was invariably applied first and treatment *B* invariably applied second. The apparent advantage of treatment *B* would be attributable in reality to the natural history of the disease. Clearly we should have been much better off in a trial of the treatment of this illness by setting up two groups of different patients, one on treatment *A* and one on treatment *B*. On the other hand we could alternatively have randomly allocated the *order* in which each patient was to receive the two treatments, so that in half the patients *A* preceded *B* and in the other half *B* preceded *A*. We would then be in a position to make a comparison of the two treatments within the same patients and, at the same time, allow for the natural progression of the disease.

In general, such comparisons within patients give no advantage and are, indeed, usually impracticable, with diseases running an acute course to death or recovery, e.g. pneumonia and other fevers. We cannot easily change the treatment during the illness nor measure the relative effect of doing so. Comparisons within patients are also not likely to give an advantage even with a long-protracted disease if the disease is one which shows a trend in time, e.g. respiratory tuberculosis. Their advantage lies rather with chronic diseases with a relatively *constant* level of disability, etc., where after the use of one drug its effect will disappear and the patient will return to the *status quo*. Even here serious ethical difficulties may sometimes arise. For instance, with a patient suffering from rheumatoid arthritis and *inadequately* maintained on cortisone there would obviously be no difficulty in changing the treatment to aspirin (or vice versa) and judging the relative merits of these treatments. But suppose the patient is being *adequately* maintained on cortisone, can one then change to aspirin and perhaps persevere with it for some weeks (or months) to see whether the patient can be thus equally well maintained in good, or reasonable, health?

A simple example of an easily and effectively conducted 'within-patient' trial was, however, the assessment of a rapidly acting agent in this disease, rheumatoid arthritis. Each of 43 patients (to whom the experimental nature of the trial had been explained) was injected on one occasion with the test substance and on another occasion with normal saline. The order of injection was randomised for each patient, with the test substance given first on half the occasions and saline first on the other half. This is an essential step for, apart from the contents of the injection, it might be that the patient would react differently (subjectively or objectively) on the first or second occasion. The nature of the injection was unknown to the patient and unknown to the doctor assessing its results. There could therefore be no bias in the judgements of either. In fact, of the 43 patients 18 reported less pain in the joint after their injection of the test substance and 17 after injection of saline; 19 with the test substance and 19 with saline reported less stiffness. The average strength of grip rose by 5 mm Hg after injection of the test substance and 4 mm Hg after the injection of saline; the average speed of a step test fell by 1.3 seconds after the test substance and by 1.6 seconds after saline (*Br. Med. J.*, **2**, 810 (1950)).

An example of a more elaborate design in the same field has been reported by D.D. Reid. Patients were invited to take part in a trial of three ointments intended to relieve articular pain. The ointments were dispensed in identical containers and the contents were unknown to doctor and patient. They were administered on a pre-arranged randomised plan to the affected joint in each patient on successive visits to the clinic. Thus the three ointments were applied to the first patient in the order X, Y, and Z on his first three visits and in the reverse order, Z, Y, X, on his next three visits. With the second patient the order was Y, Z, X for his first three visits and the reverse order, X, Z, Y, for his next three visits. With the third patient the order was *Z*, *X*, *Y* and then the reverse, *Y*, *X*, *Z*. These treatment orders randomly allocated were repeated for each group of 3 patients. Thus, tabulating, the order of administration of the 3 ointments was as shown in Table 23.2.

It will be seen that each ointment has been tested twice on every patient so that any differences in response *between patients* are equally represented. Further, each ointment has been tested once at *each order* of visit to the clinic from first to sixth; so the tendency in the patients to a natural recovery in time affects all three ointments equally. Finally *each ointment has preceded any other* (e.g. *X* before *Y*) just as often as it has followed any other (*X* after *Y*) and these two sets of orders are repeated in the same patient. The tendency for a patient to make comparisons between an ointment and the one given immediately before it is thereby equalised. Possible disturbing factors in the required comparisons have thus been allowed for in this 'balanced' type of design. In fact the results showed that there was very little to choose between the three ointments, which were a standard preparation in common use, the special preparation under test, and a vanishing cream used as a placebo.

A careful experimental design, with random allocation within that design, can often produce such advantages. Experience suggests, however, that not quite enough thought is often put into making up such a design. In the case quoted above, for instance, still better balance could have been achieved by using all 6 possibilities instead of only 3 of them (Table 23.3 shows the additional 3). Had these been used as well it might have made no difference to the conclusions, but there is *nothing* to be lost, and there could have been a gain.

In these cross-over trials from one treatment to another in the same patient one must also consider whether there is likely to be a carry-over effect from drug to placebo or from one drug to another. To prevent this disturbing feature it may be necessary to have a no-treatment period between each therapy. If this is not feasible it might be better to abandon

Table 23.2

Patient	*Order of visit to clinic*					
	1	*2*	*3*	*4*	*5*	*6*
1st	*X*	*Y*	*Z*	*Z*	*Y*	*X*
2nd	*Y*	*Z*	*X*	*X*	*Z*	*Y*
3rd	*Z*	*X*	*Y*	*Y*	*X*	*Z*

Table 23.3

X	*Z*	*Y*	*Y*	*Z*	*X*
Y	*X*	*Z*	*Z*	*X*	*Y*
Z	*Y*	*X*	*X*	*Y*	*Z*

the cross-over technique and to make it a straightforward between-patients comparison.

Matched pairs

Another, and sometimes useful, form of the controlled trial of a treatment lies in the comparison of *pairs of patients*. The object is to make the members of each pair as alike as possible in such factors as are known, or thought likely, to affect the issue, e.g. sex, age, severity of presenting symptoms. One of the pair (*chosen at random*) is given the treatment under test and the other the established treatment of the day (which may sometimes amount to placebo); or, if the aim of the trial is the comparison of two drugs, one of the pair may be given drug *A* and the other drug *B*.

The pair must, of course, be treated alike in supplementary ways and, as far as possible, observed contemporaneously. This last requirement is, in fact the main difficulty of the method. If one makes the matching too precise, then one is unlikely to have two patients available at the same time. As a result the application of the method is often limited to a chronic disease where a pool of patients is readily available in a clinic or hospital and can be drawn upon at one specified time.

The factors to be matched must be thought out and laid down in advance—including definitions of the precision required (e.g. will any age between, say, 50 and 54 years be accepted as equivalent?).

A special advantage of this type of trial is that one may be able to detect in a wide distribution of pairs, which *kind* of patient it is that has favourably responded to a given treatment, e.g. is it pairs of a particular age or pairs with some specific symptoms? For instance, we might have two or three hundred patients to whom it seemed a matter of indifference which of the two drugs they were given—either both members of the pair improved or neither member improved. But to some relatively small group, it seemed to matter very much since the comparisons showed improvement on one of the drugs and no improvement in the matched patient on the other drug. Why, then, this difference in response? We can seek for it in a comparison of the characteristics of the two minority groups, where the responses of the pairs *do* differ, with the characteristics of the two majority groups where they do *not* differ.

For example, we might have the results in a trial with 200 matched pairs of patients (i.e. 400 patients in all); one in each pair was treated with drug *A* and the other with drug *B*. Table 23.2 shows the tabulation of the 400 patients *if we ignore the fact that they had originally been paired*, i.e. we insert the result for each patient as a separate individual. This comparison shows no appreciable difference between the two treatments; 57.5 per cent of the patients on drug *A* benefited and 50.0 per cent of those on drug *B*. The difference is not statistically significant (χ^2 with Yates' correction, on 1 degree of freedom, is 1.97, giving $P = 0.16$).

Table 23.3 shows the results *when we take into account the extra information given by the original pairing of the patients*, i.e. we insert the

Table 23.2 Results of treatment in 400 patients.

Result	*Drug A*	*Drug B*	*Total*
Effective	115	100	215
Ineffective	85	100	185
Total	200	200	400

Table 23.3 Figures of Table 23.2 rearranged as 200 pairs of patients.

			Drug A	
		Effective	*Ineffective*	*Total*
	Effective	100	0	100
Drug	*Ineffective*	15	85	100
B	*Total*	115	85	200

results of each *pair*. We see that in the very great majority of the pairs it is immaterial whether they had drug *A* or drug *B*. In 100 pairs the drugs were both effective, in 85 pairs they were both ineffective. Possibly in these 185 pairs nature was merely taking its course and the drugs were irrelevant. Anyway there was no apparent reason to exhibit one rather than the other.

However, there were 15 pairs of patients in which there was a distinct difference. Drug *A* was seen to be effective in the 15 patients so treated and drug *B* was seen to be ineffective in their 15 opposite numbers. Furthermore, there were no pairs in which drug *B* was effective and drug *A* was ineffective. Alerted by this result we should return to the records and endeavour to identify the characteristics of the patients who have produced it. McNemar's test now gives us $P = 0.00006$ showing that, in this case, the pairing has made a dramatic difference.

In short, for the patients at large the advantage of one drug over the other may be negligible but for a small minority it may be vital. It is in such circumstances that the paired comparison can be so important. By ignoring the pairing in an analysis of the data one might fail to discover the dominating feature.

The procedure of matched pairs has been described above in relation to a comparison *between pairs of patients*. It is equally applicable to comparisons of *pairs of observations within the same patients*. Thus two drugs might be applied in random order to each patient on different occasions. The two responses could then be compared. Or, for a specifically local effect, it might be possible to apply one treatment to one limb and another treatment to the corresponding limb, e.g. in chronic rheumatoid arthritis.

The interaction of treatments

In some clinical trials the effects of more than one treatment may be effectively and simply explored. Thus for patients with hypertension we might be concerned with (1) the value of an antihypertensive drug, (2) the value of a diuretic drug, and (3) the value of *both* these drugs when used together or, in other words, their interactions.

Such a trial is known as a *factorial* trial. We need four groups and must allocate the patients to them randomly. Thus for 100 patients admitted and randomly distributed we might have the figures in Table 23.4. To estimate the effect of the antihypertensive drug, from the results observed, it would not be correct to compare directly all the 49 given it with all the 51 not given it, for these were not all treated alike in other ways, but it is correct to take the top row of the table and compare the 26 who were given it with the 25 who were not, for these were all treated alike except for the one feature that we are investigating. Similarly we can compare the 23 with the 26 in the bottom row. It is then possible to combine the results of these two 'within row' comparisons, and thus use all 100 patients for this purpose.

Likewise, we can estimate the effect of the diuretic drug by the combination of the two 'within column' comparisons, thus using all 100 patients for this purpose too. Finally we can use all 100 patients to look for interaction between the two drugs by combination of the two 'within diagonal' comparisons. Thus we have made very efficient usage of the data and have been able to investigate points that could not have been investigated by means of the two separate trials.

The factorial design of trial can be extended to examine three or more things at a time and, properly analysed, it can be both powerful and economical. The problem may be to get enough patients to have at least some in every cell of the table, and the ethical situation may, in any case, rule this out. In the treatment of a form of cancer, for example, it might be ethical to compare the effects of surgery with the effects of radiation and with the effects of the two combined. A factorial trial, however, also requires a group treated in neither fashion, and this would almost certainly be unethical.

Table 23.4 A factorial trial.

		Antihypertensive drug		
		Given	*Not given*	*Total*
Diuretic drug	*Given*	26	25	51
	Not given	23	26	49
	Total	49	51	100

Summary

The history of medicine shows many examples of forms of treatment widely considered as effective on grounds of clinical impression which have turned out to be ineffective or even harmful. The aim of the clinical trial is to circumvent this situation by means of a carefully, and ethically, designed experiment. In its most rigorous form it demands equivalent groups of patients concurrently treated in different ways. These groups are constructed by the random allocation of patients to one or other treatment. Sometimes carefully matched pairs of patients may provide the contrast. In some instances patients may form their own controls, different treatments being applied to them in random order and the effects compared. In principle the method is applicable with any disease and any treatment. It may also be applied on any scale; it does not necessarily demand large numbers of patients. It should be designed to promote rather than hinder the traditional method in medicine of acute observation of disease by the clinician at the bedside.

In short, in the controlled clinical trial our endeavour is to measure the relative value of a defined treatment (or treatments) by the comparison of a group of patients treated in one way with a similar group not so treated. Everything turns upon the validity of that final deduction from the results of the trial—whether it be between or within patients, whether it be double-blind, whether we use a placebo, whether there be constraints upon the variability of the patients admitted. *There is no hard-and-fast 'cook-book' recipe to fit every trial. The trial itself promotes the choice of rules.* At its best such a trial shows what can be accomplished with a medicine under careful observation and certain restricted conditions. The same results will not invariably or necessarily be observed when the medicine passes into general use; but the trial has at the least provided background knowledge which the physician can adapt to the individual patient.

Chapter 24

Fallacies and difficulties: mixing non-comparable records

In this and the following three chapters are set out examples of the misuse of statistics in medicine. The present chapter is concerned with the dangers of mixing non-comparable records and, in particular, with the statistical problems that arise in the assessment of the value of forms of immunisation. Chapter 25 is concerned with one of the most fundamental of all problems in the use of statistics (of whatever kind)—the absence, or neglect, of the numbers of persons 'exposed to risk'. Without a knowledge of these basic numbers we cannot calculate a *rate* of occurrence e.g. of deaths in a population. The deaths are the numerator of the fraction and the population at risk is the denominator. In the absence of the denominator we may be forced to rely upon a proportional rate and there are considerable difficulties consequent upon that. Chapter 26 deals with the special problem of medical statistics in the use of mortality rates and particularly with relation to the causes to which the deaths are attributed. Finally, Chapter 27 takes up a few other specific problems which commonly arise in medical statistics.

In some of the examples used in these chapters the figures have been taken from published papers; in others hypothetical figures have been used to indicate the type of error which has led the worker to fallacious conclusions. No principles are involved that have not been discussed in the previous chapters. The object is merely to illustrate, at the risk of 'damnable iteration', the importance of these principles by means of simple numerical examples; in some instances—e.g. (1) below—the figures are deliberately exaggerated to make clearer the point at issue. The fact that in practice such grossly exaggerated differences rarely occur does not lessen the importance of accurate statistical treatment of data. Differences do occur very often of a magnitude to lead to erroneous conclusions, if the data are incompetently handled in the ways set out.

Perils of non-comparability

(1) Let us suppose that in a particular disease the fatality rate is twice as

high among females as it is among males, and that among male patients it is 20 per cent and among female patients 40 per cent. A new form of treatment is adopted and applied to 80 males and 40 females; 30 males and 60 females are observed as controls. The number of deaths observed among the 120 *individuals* given the new treatment is 32, giving a fatality rate of 26.7 per cent, while the number of deaths observed amongst the 90 *individuals* taken as controls is 30, giving a fatality rate of 33.3 per cent. Superficially this comparison suggests that the new treatment is of some value; in fact, that conclusion is wholly unjustified, for we are not comparing like with like. The fatality rates of the total number of individuals must be influenced by the proportions of the two sexes present in each sample; males and females, in fact, are not equally represented in the sample treated and in the sample taken as control. Tabulating the figures shows the fallacy clearly (Table 24.1).

The comparison of like with like—i.e. males with males and females with females—shows that the treatment was of no value, since the fatality rates of the treated and untreated sex groups are identical, and equal to the normal rates. Comparison of the total samples, regardless of sex, is inadmissible, for the fatality rate recorded is then in part dependent upon the proportions of the two sexes that are present. There are proportionately more females amongst the controls than in the treated group, and since females normally have a higher fatality rate than males their presence in the control group in relatively greater numbers must lead to a comparatively high fatality rate in the total sample. Equally their relative deficiency in the treated group leads to the comparatively low fatality rate in that total sample. No comparison is valid which does not allow for the sex differentiation of the fatality rates.

(2) A more extensive example of the same kind is shown in Table 24.2. In this it is presumed that the fatality rate of an illness varies with the sex of

Table 24.1 Non-comparability of sexes.

	Males	*Females*	*Males and females combined*
Normal fatality rate	20%	40%	
Number of patients given new treatment	80	40	120
Deaths observed in treated group	16	16	32
Fatality rates observed in treated group	20%	40%	26.7%
Number of patients used as controls	30	60	90
Deaths observed in control group	6	24	30
Fatality rates observed in control group	20%	40%	33.3%

Table 24.2 Non-comparability of sex, age and severity.

			Hospital A			Hospital B		
Sex	*Age (years)*	*Severity of disease*	*No of cases*	*No of deaths*	*Fatality rate (%)*	*No of cases*	*No of deaths*	*Fatality rate (%)*
Male	50+	Severe	40	10	25	160	40	25
Male	50+	Mild	100	5	5	200	10	5
Male	<50	Severe	60	6	10	100	10	10
Male	<50	Mild	200	4	2	50	1	2
Female	50+	Severe	40	6	15	100	15	15
Female	50+	Mild	100	4	4	50	2	4
Female	<50	Severe	60	3	5	40	2	5
Female	<50	Mild	200	2	1	100	1	1
Total			800	40	5	800	81	10

the patient (males are more vulnerable than females), with the age of the patient (the old are more vulnerable than the young), and with the severity of the attack (the severely ill are more vulnerable than the mildly ill), all very reasonable assumptions. The comparison of 800 patients in Hospital A with 800 patients in Hospital B suggests a serious state of affairs—a fatality rate in the latter *twice* as high as that in the former (10 per cent compared with 5 per cent). A test of significance shows that the difference is highly 'significant': $\chi^2 = 10.9$ on 1 degree of freedom, $P = 0.0010$. Yet the difference is entirely spurious and the comparison is not valid. Dissection of the data shows that Hospital B has twice as many severely ill patients as Hospital A (400 to 200), nearly 30 per cent more men (510 to 400), and nearly twice as many persons above the age of 50 (510 to 280), and these are all features associated with a high fatality rate. Comparison of like with like in these respects shows that the hospitals have, in fact, identical fatality rates. It is the amalgamation of the different groups in different proportions that has led to the impressively 'significant' but fallacious conclusion. The importance of trying to unravel the chain of causation is apparent.

Proposals are sometimes seen that hospitals' fatality rates should be published, to let patients make their own choice between them. Should such proposals be adopted, it would be hard to avoid either publishing such thoroughly misleading figures on the one hand, or being accused of 'cooking' the figures on the other.

(3) The following (hypothetical) figures show the attack rates of a disease upon an inoculated and an uninoculated population (Table 24.3).

Table 24.3 Changing exposure to risk and attack rate over time.

	Number of persons		Number of persons attacked		Attack rates (per cent)	
Year	*Inoculated*	*Uninoculated*	*Inoculated*	*Uninoculated*	*Inoculated*	*Uninoculated*
1981	100	1000	10	100	10	10
1982	500	600	5	6	1	1
Total	600	1600	15	106	2.5	6.6

In each calendar year the attack rate of the inoculated is equal to the attack rate of the uninoculated. Between 1981 and 1982 there has, however, been a large change in the size of the inoculated and uninoculated populations and also a large change in the level of the attack rate. Summation of the results for the two years leads to the fallacious conclusion that the inoculation afforded some protection. The large uninoculated population in 1981 when the attack rate was high leads to an absolutely large number of cases—though in relation to their numbers the uninoculated are at no disadvantage compared with the inoculated. The inoculated cannot contribute an equal number of cases, for the population at risk in that year (1981) is far smaller. Thus amalgamation of the unequal numbers of persons exposed to different risks in the two years is unjustified. No fallacy would have resulted if the attack rate had not changed or if the proportions exposed to risk had not changed, as the figures of Table 24.4 show.

When the populations at risk and the attack rates *both* vary, the calendar year becomes a relevant factor, and must be taken into account by the calculation of rates within the year. Such a problem has arisen quite frequently in practice in assessing the incidence of various infectious diseases in immunised and unimmunised children.

Table 24.4 Changing exposure to risk or attack rate, while the other remains constant.

	Number of persons		*Number of persons attacked*		*Attack rates (per cent)*	
Year	*Inoculated*	*Uninoculated*	*Inoculated*	*Uninoculated*	*Inoculated*	*Uninoculated*
(1) Constant attack rates						
1981	100	1000	10	100	10	10
1982	500	600	50	60	10	10
Total	600	1600	60	160	10	10
(2) Constant exposure to risk						
1981	500	600	50	60	10	10
1982	500	600	5	6	1	1
Total	1000	1200	55	66	5.5	5.5

Neglect of the period of exposure to risk

A further fallacy in the comparison of the experiences of inoculated and uninoculated persons lies in neglect of the time during which the individuals are exposed first in one group and then in the other. Suppose that in the area considered there were, on 1 January 1982, 300 inoculated persons and 1000 uninoculated persons. The numbers of attacks are observed within these groups over the calendar year and the annual attack rates are compared. This is a valid comparison *so long as the two groups were subject during the calendar year to no additions or withdrawals*. But if, as often occurs in practice, individuals are being inoculated *during* the year of observation the comparison becomes invalid unless the point of time at which they enter the inoculated group is taken into account.

Suppose on 1 January 1982, there are 5000 persons under observation, none of whom is inoculated; that 300 are inoculated on 1 April, a further 600 on 1 July, and another 100 on 1 October. At the end of the year there are, therefore, 1000 inoculated persons and 4000 still uninoculated. During the year there were registered 110 attacks amongst the inoculated persons and 890 amongst the uninoculated. If the ratio of recorded attacks to the population *at the end of the year* is taken, then we have rates of $110/1000 = 11.0$ per cent amongst the inoculated and $890/4000 = 22.3$ per cent amongst the uninoculated, a result apparently very favourable to inoculation. Such a result, however, *must* be reached even if inoculation is completely valueless, for no account has been taken of the unequal lengths of time over which the two groups were exposed. None of the 1000 persons in the inoculated group were exposed to risk for the *whole* of the year but only for some fraction of it; for a proportion of the year they belong to the uninoculated group and must be counted in that group for an appropriate length of time.

The calculation should be as follows (presuming, for simplicity, that one attack confers no immunity against another):

All 5000 persons were uninoculated during the first quarter of the year and therefore contribute $(5000 \times \frac{1}{4})$ years of exposure to that group. During the second quarter 4700 persons belonged to this group—i.e. 5000 less the 300 who were inoculated on 1 April—and they contribute $(4700 \times \frac{1}{4})$ years of exposure to the uninoculated group. During the third quarter 4100 persons belonged to this group—i.e. 4700 less the 600 who were inoculated on 1 July—and they contribute $(4100 \times \frac{1}{4})$ years of exposure. Finally in the last quarter of the year there were 4000 uninoculated persons—i.e. 4100 less the 100 inoculated on 1 October—and they contribute $(4000 \times \frac{1}{4})$ years of exposure. The 'person years' of exposure in the uninoculated group were therefore $(5000 \times \frac{1}{4}) + (4700 \times \frac{1}{4}) + (4100 \times \frac{1}{4}) + (4000 \times \frac{1}{4}) = 4450$, and the attack rate was $890/4450 = 20$ per cent—i.e. the equivalent of 20 attacks per 100 persons per annum. Similarly the person-years of exposure in the inoculated group are $(0 \times \frac{1}{4}) + (300 \times \frac{1}{4}) + (900 \times \frac{1}{4}) + (1000 \times \frac{1}{4}) = 550$, for there were no persons in this group during the first three months of the year, 300 persons during the second quarter of the year, 900 during the third quarter, and 1000 during the last quarter. The attack rate was, therefore, $110/550 = 20$ per cent and the inoculated and uninoculated have identical attack rates. Neglect of the durations of exposure to risk must lead to fallacious results and must favour the inoculated. The figures are given in tabulated form (Table 24.5).

Fallacious comparison. Ratio of attacks to final population of group. Inoculated $110/1000 = 11.0$ per cent. Uninoculated $890/4000 = 22.3$ per cent.

True comparison. Ratio of attacks to person-years of exposure. Inoculated $110/[(300 \times \frac{1}{4}) + (900 \times \frac{1}{4}) + (1000 \times \frac{1}{4})] = 20$ per cent. Uninoculated $890/[(5000 \times \frac{1}{4}) + (4700 \times \frac{1}{4}) + (4100 \times \frac{1}{4}) + (4000 \times \frac{1}{4})] = 20$ per cent.

In real-life figures, each individual person will have an individual date of

Table 24.5 Inoculations during a year.

		Inoculated		*Uninoculated*	
	Inoculated at each point of time	*Exposed to risk in each quarter of the year*	*Attacks at 5 per cent per quarter*	*Exposed to risk in each quarter of the year*	*Attacks at 5 per cent per quarter*
Jan 1	0	0	0	5000	250
Apr 1	300	300	15	4700	235
Jul 1	600	900	45	4100	205
Oct 1	100	1000	50	4000	200
Total at end of the year		1000	110	4000	890

inoculation and the above simple style of analysis needs to be replaced by a life-table approach.

This example is an exaggerated form of what may (and does) happen in practice if the time factor is ignored. Clearly even if the time factor is allowed for, interpretation of the results must be made with care. If the inoculated show an advantage over the uninoculated it must be considered whether at the point of time they entered that group the incidence of the disease was already declining, due merely to the epidemic swing. Strictly speaking, the experience of the vaccinated should be compared with that of the unvaccinated only from the time that the former entered the vaccinated group. It would seem necessary in these circumstances to select randomly at the proper time a comparable control person for each person vaccinated, the former to remain unvaccinated. But this procedure, while an epidemic is in progress, would be impracticable and unethical. Comparisons based upon vaccination *during* an epidemic should therefore be regarded with caution, if not sceptism.

Neglect of numbers exposed to risk

In certain circumstances the doctor treating cases of a disease against which a vaccine is in use may be easily, but grossly, misled by the patients whom he sees. Let us suppose that a vaccine against whooping cough is not 100 per cent effective in conferring protection upon its recipients (and this, for unknown reasons, is undoubtedly true of such vaccines). We may, therefore, have the figures of Table 24.6. Of the total of 1000 children in the population 10 per cent, say, i.e. 100, are not inoculated at all (because the parents do not choose), and 100 were inoculated but the vaccine conferred no protection. The cases of whooping cough seen by the general practitioner will, therefore, be about equally divided between these two groups, the uninoculated and inoculated, say 30 in each 100. The observer might conclude that the vaccine is useless—'I see just as many attacks in the inoculated as I do in the uninoculated.' But thereby he is forgetting the

Table 24.6 A comparison of inoculated and uninoculated groups (hypothetical figures).

		Attacked by whooping cough		
		Yes	No	Total
Inoculated	Yes	30	870	900
	No	30	70	100
	Total	60	940	1000

large number of children who were effectively inoculated and whom, therefore, he never sees. The true comparison of incidence is between 30 in 100 and 30 in 900.

Volunteers for inoculation or survey

Another very frequent source of error in the assessment of a vaccine lies in the comparison of persons who volunteer, or choose, to be inoculated (or volunteer on behalf of their children) with those who do not volunteer or choose. The volunteers may tend to come from a different age group and thus, on the average, be older or younger; they may tend to include more, or fewer, males than females; they may tend to be drawn more from one social class than another. Thus, in the early days of vaccination against poliomyelitis one survey in the USA showed that the vaccinated children came more from the white families, more from mothers with a high standard of education behind them, and more from fathers in 'white collar' and relatively well-paid occupations. Given adequate records these, and similar easily defined characteristics, can be identified. The comparisons of the inoculated and uninoculated can then be made within like categories. But far more subtle and undetectable differences may well be involved. The volunteers for inoculation may be people who are more careful of their health, more aware of the presence of an epidemic in the community. They may on such occasions take other steps to avoid infection, e.g. the avoidance of crowded places. It may (as in the example quoted above) be the more intelligent mothers who bring their children to be inoculated; and more intelligent mothers may also endeavour to protect their children from infection in other ways. They may, too, have smaller families, perhaps, so that the inoculated child is automatically less exposed to infection from siblings. Thus in one pioneer trial of vaccines against whooping cough it was found that 47 per cent of the inoculated children were 'only' children, i.e. with no siblings, compared with 20 per cent of the control children.

Lastly, with diseases that do not have a uniform incidence and do not themselves confer a lasting immunity, e.g. influenza and the common cold, there may well be in the act of volunteering a grave element of self-selection. People who rarely suffer from such minor illnesses are unlikely to volunteer, those who are constantly troubled by them may well seek relief through inoculation. In other words, self-selection may tend to bring the susceptibles into the inoculated class and leave the resistants as 'controls'. The comparison is, then, valueless. We are not comparing like with like. Such dangers are not imaginary; they have been demonstrated often in carefully documented trials.

Similarly problems may arise in making a survey. Thus in a survey of respiratory illness it was reported that three groups were specially studied: (1) unco-operative persons who had to be visited at home, (2) co-operative persons who would come to the clinic, and (3) enthusiastic persons who readily volunteered to be examined. It was found that the unco-operative group had, on average, more morning phlegm and lower vital capacity than the co-operative group. In other words, an examination limited to those coming to the clinic would have led to an underestimate of the amount of respiratory disease in the community.

Self-selection of patients

An interesting example of the effects of self-selection by patients is to be found in the statistics of migraine. It has often been stated that the patients are more intelligent than the average; and a study of patients seen in general practice showed that the prevalence was relatively high in the professional classes. On the other hand an epidemiological study of a random sample of a whole population, well or ill, showed no such relationship with intelligence or social class. It did, however, show a tendency for more of the intelligent individuals and more of those of the higher social classes to consult a doctor because of their headaches (Waters, W.E., *Br. Med. J.*, **2**, 77 (1970)). In other words the clinical evidence may well be true—migraine patients *who attend their doctors* are more intelligent and of a higher social class. But this may be the result of their behaviour, their own self-selection in seeking medical aid. We cannot safely project the observations of the patients seen to the population of patients as a whole, seen or not seen.

Associations in time and space

Standing alone, associated movements in time of two characteristics may be very poor evidence of cause and effect—particularly if the trend be merely a steady upward, or downward, movement. Thus in the last 40 years the emission of exhaust gases from motor vehicles has risen in the large cities of Great Britain. There has also been a rise in the death rate of cancer of the lung. This concurrent movement in time, together with the obvious possibility of a causal relationship, might well justify a carefully planned investigation. But without quite other evidence we have no sure basis for deduction. Many factors move together in time, e.g. in these 40 years the increase in television and the decline in long skirts in women. The movements of some may be directly associated, the movements of others be only indirectly associated. The problem is to sort them out.

Associations in space may also be difficult to interpret. The fact that in areas of Scandinavia the level of the birth rate varies directly with the prevalence of storks is not likely to mislead us. But what of the fact that the incidence of a specific disease appears to be rather less in, say, Wales than in England? Half a dozen factors might be thought of as relevant—genetic and environmental—but none can take us further than a useful basis for further investigation.

Even a planned investigation, or experiment, revealing an association in time or space can be quite unconvincing without the introduction of a control group.

(1) For example, a well-planned trial was made of vaccination against the common cold (*J. Amer. Med. Assoc.*, **2**, 1168 (1938)). University students who believed themselves to be particularly susceptible were invited to participate and were allocated at random to a vaccine or control group (vaccinated with normal saline). An extract from the results is given in Table 24.7.

Whatever may have been the explanation (e.g. exaggerated memories of the past year's experience), the most striking feature is the enormous reduction in colds observed in the year of the trial compared with those of the previous year. *But suppose no control group had been used*; would one not have been impressed by so great a reduction as 73 per cent in the vaccinated group? It is only alongside the control group that it falls into its proper perspective and relieves us of the danger of deducing cause and effect merely from a time relationship. In fact, the vaccine, it appears, achieved nothing very remarkable for its recipients. Nevertheless some of them were well satisfied, for the authors report that from time to time physicians would write to them saying, 'I have a patient who took your cold vaccine and got such splendid results that he wants to continue it. Will you be good enough to tell me what vaccine you are using?' It must have been embarrassing to reply 'water', yet in that reply lies all the fallibility of the *individual* judgement *post hoc ergo propter hoc*.

(2) As an instance of a study in both space and time, the possible effect of bacteriophage was measured by comparing the incidence of cholera in two areas, one in which bacteriophage was distributed, the other serving as a standard of comparison. In the former the incidence was found, over an observed period of time, to be at a lower level than the incidence in previous years or in the area observed as a control (the question of

Table 24.7 A study of vaccination against the common cold.

	Vaccine group	*Control group*
Number of persons vaccinated	272	276
Average number of colds per person		
Previous year	5.9	5.6
Current year	1.6	2.1
Reduction in current year	73%	63%

duration of exposure to risk, dealt with above, having been properly observed). It is clear that there is an association both in time and space between the incidence of cholera and the administration of bacteriophage. Is that association one of cause and effect? The answer must be that the results are perfectly *consistent* with that hypothesis, but that consistency is not the equivalent of proof. The incidence of epidemic disease fluctuates both in time and space for unknown reasons, and the abnormally low attack rates in the area in which bacteriophage was administered *may* be the results of the influence of those undetermined natural causes operating at the same time as the experiment was carried out. Repetition of the experiment in another area with equivalent results would strengthen the hypothesis that bacteriophage was beneficial. With observations of this kind, limited in time and space, it is well to reflect upon the fact that 'If when the tide is falling you take out water with a twopenny pail, you and the moon together can do a great deal.' The history of scarlet fever may well be remembered, in this connection, as illustrated by the testimony of R.J. Graves (*A System of Clinical Medicine*, Dublin, 1843). In the first years of the nineteenth century the disease 'committed great ravages in Dublin' and was 'extremely fatal'. After the year 1804 it assumed a 'very benign type' and was 'seldom attended with danger until the year 1831'. In 1834 it again took the form of a 'destructive epidemic'. The low fatality rate after 1804 was 'every day quoted as exhibiting one of the most triumphant examples of the efficacy' of new methods of treatment. But Graves candidly admits that 'the experience derived from the present [1834–35] epidemic has completely refuted this reasoning, and has proved that, in spite of our boasted improvements, we have not been more successful in 1834–5 than were our predecessors in 1801–2' (quoted from Charles Creighton's *History of Epidemics in Britain*, Cambridge University Press, **ii**, 722–725, 1894).

Chapter 25

Fallacies and difficulties: 'exposure to risk' and proportional rates

Absence of 'exposed to risk' and standard of comparison

It often happens that an investigation is confined to individuals marked by some characteristics.

(1) For example, a detailed inquiry is made into the home conditions of each infant dying in the first year of life in a certain area over a selected period of time, and it is found that 15 per cent of these infants lived under unsatisfactory housing conditions. Do such conditions, or factors associated with them, lead to a high rate of infant mortality? The limitation of the inquiry to the dead makes it quite impossible to answer this question. We need information as to the proportion of *all* infants who were born in that area over that period of observation who live under unsatisfactory housing conditions. If 15 per cent of *all* infants live under such conditions, then 15 per cent of the deaths may reasonably be expected from those houses and unsatisfactory housing appears unimportant. If, on the other hand, only 5 per cent of *all* infants are found in these conditions but 15 per cent of the deaths come from such houses, there is evidence of an excess of mortality under the adverse conditions. In practice, it may be impossible for financial or administrative reasons to investigate the home conditions of all infants. It should be possible, however, to inquire into a random sample of them. Without some such standard of comparison no clear answer can be reached. Such limited investigations have been made into the problems of both infant and maternal mortality. Where they are valuable is not in measuring the risk of unexplored features but in pinpointing the presence of some distinct features in the pregnancy or delivery or in the care of the infant, which are known from past experience to be highly undesirable or dangerous and ought, therefore, to be eliminated.

(2) After very careful inquiry it is shown that of motor-car drivers involved in accidents a certain proportion, say three-quarters, had consumed alcohol during some period of hours previous to the accidents,

and one-quarter had not. There is, of course, convincing evidence of the dangers of drinking and driving but the deduction that alcohol contributes to the risk of accident is not justified from these figures alone. It is well recognised that white sheep eat more than black sheep—because there are more of them. Before the ratio of 3 'alcoholics' to 1 'non-alcoholic' amongst the accident cases can be interpreted, information is also required as to the comparable ratio amongst drivers *not involved in accidents*. Suppose, for example, there are 1000 drivers on the roads, and 48 accidents are recorded. Of the 48 drivers involved in these accidents three-quarters are found to have consumed alcohol (i.e. 36) and one-quarter (i.e. 12) have not. If three-quarters of all the 1000 drivers have consumed alcohol within a few hours of driving and one-quarter have not, then the populations 'exposed to risk' of accidents are 750 and 250. The accident rates are, then, identical—namely, 36 in 750 and 12 in 250, or 4.8 per cent in each group. A knowledge of the exposed to risk, or at least of the ratio of alcohol consumers to non-consumers in a random sample of all drivers, is essential before conclusions can be drawn from the ratio in the accident cases.

Careful inquiry into the destination of drivers involved in accidents on a Sunday morning might show that a larger proportion was driving to golf than to church. The inference that driving to golf is a more hazardous occupation is not valid until we are satisfied that there are not, in this case, more black sheep than white sheep. Lest it should be thought that undue stress is being laid upon the obvious, the following quotation from a debate in the House of Lords in 1936 may be of interest. A noble Lord is reported to have said that 'only 4 per cent of the drivers involved in fatal accidents were women, and that was because they drove more slowly.' Without evidence of the hours of driving endured (perhaps a fitting word nowadays) by each sex—and perhaps of the type of area—that conclusion cannot be justified.

The continuation of such errors from one generation to the next is shown by a report in 1969 of the *colours* of cars involved in accidents. But no such information was available of the numbers of cars of different colours that were, in fact, in existence; and so the relative accident rates could not be calculated. From merely the numbers involved in an accident it would seem that orange might be a safe bet! However observation and reflection would suggest that there were very few orange cars around to have accidents.

Another 21 years later, we read that, of 200 deaths from heart attacks during or soon after taking part in sports, 93 occurred among squash players, 27 among swimmers, etc. Perhaps the original report did consider the question of exposure to risk, but if so such essential information failed to be reported by the newspapers, which reacted as if the numbers of deaths were directly proportional to the relevant risks.

As said before, and it can hardly be said too often, this neglect of the 'exposed to risk' is one of the commonest elementary mistakes in the everyday use of statistics and the reader should be constantly on his guard against it.

Proportional rates

Without a knowledge of the number of persons exposed to risk, rates of mortality, morbidity, etc., cannot be calculated. In their place refuge is often taken in *proportional* rates.

It is not uncommon for some confusion to be shown over the difference between such absolute and relative figures. For instance, take the figures in Table 25.1 (which relate to a study of the trend of the mortality from tuberculosis):

From the estimated populations and the numbers of deaths registered the death rates experienced in the two age groups can be calculated. At ages 5–9 years the death rate from tuberculosis was 243 per million persons and at ages 10–14 it was 247, a negligible difference. In other words, at this date, the risk of dying from tuberculosis did not vary materially between these two age groups. Suppose, however, we had no means of estimating the populations exposed to risk, and the sole information at our disposal consisted of the deaths registered from tuberculosis and from all causes. In that event the only rate that it is possible to calculate is a proportional rate—i.e. the deaths from tuberculosis as a percentage of the deaths from all causes. At ages 5–9 this rate is $774 \times 100/6601 = 11.7$ per cent and at ages 10–14 it is $827 \times 100/4643 = 17.8$ per cent. The *proportional death rate* from tuberculosis is very much higher at ages 10–14 than at ages 5–9. It is necessary to be quite clear on the meaning of this rate. The information it gives is that *in relation to other causes of death* mortality from tuberculosis was, in these data, more frequent at ages 10–14 than at ages 5–9. This is certainly an important aspect of the figures; it shows that any action which reduced deaths from tuberculosis (without changing the death rate from other causes) would have more effect on the total death rate at ages 10–14 than at ages 5–9, because more of the total deaths were attributable to tuberculosis at the higher ages. But obviously we cannot deduce from the proportional rates that the *absolute* risk of dying from tuberculosis differs in the same ratios; we have, in fact, seen that the absolute risks were virtually equal. The proportional rate is dependent upon the level of two factors—namely, the deaths from tuberculosis and the deaths from all causes—which form the numerator and denominator of the ratios. Differences in *either* factor will influence the ratio. In the case cited the

Table 25.1 Illustration of proportional death rates.

Age-group (years)	*Estimated population in age-group*	*Number of deaths registered*		*Death rates per million*	
		All causes	*Tuberculosis (all forms)*	*All causes*	*Tuberculosis (all forms)*
5–9	3 181 900	6601	774	2075	243
10–14	3 349 100	4643	827	1386	247

numerators (deaths from tuberculosis) are nearly equal, 774 at ages 5–9 and 827 at ages 10–14, but the denominators (deaths from all causes) differ appreciably, 6601 at 5–9 and 4643 at 10–14. The proportion of deaths due to tuberculosis at ages 10–14 is higher than the proportion at ages 5–9, not because of a rise in incidence of tuberculosis but because between these ages there is a fall in the incidence of other causes of death. While other causes of death, frequent at that time, e.g. diptheria, scarlet fever, measles—became less at ages 10–14, tuberculosis remained nearly steady. Relatively, therefore, its importance increased, absolutely it was unchanged.

Those who deplore the fact that some particular cause of death is a high proportion of all deaths should always be asked which other causes they wish to see proportionately increased. The total for all causes must add to 100 per cent whatever happens.

Changes in proportional rates

To take another example, in 1917 the deaths at age 0–1 year formed 12.9 per cent of all the deaths registered in England and Wales, whereas in 1918 the proportion was only 10.5 per cent. It does *not* follow that the infant mortality rate of 1918 was lower than that of 1917; in fact, it was very slightly higher, 97 per 1000 live births compared with 96 in 1917. The lower *proportion* in 1918 was due to the great increase in deaths *at ages over* 1 as a result of the influenza pandemic of that year, one of the great epidemics of history. The numerator of the ratio did not change appreciably (64 483 deaths under 1 in 1917 and 64 386 deaths in 1918); the denominator did change appreciably (from 498 922 deaths at all ages in 1917 to 611 861 deaths in 1918).

Similarly an absence of change in a proportional rate does not necessarily denote an absence of change in the absolute rate. If in 1950 there were 5 per cent of all deaths due to cause X and in 1983 the proportion was still 5 per cent, this is not incompatible with an absolute decline in the mortality due to cause X. If the death rates due to cause X and to all other causes have *both* been halved, the proportion must remain the same.

Proportional death rates in occupations

Similarly a lower proportion of deaths due to a particular cause amongst, say, coal miners than amongst bank clerks does not necessarily denote an absolute lower death rate from that cause amongst the miners. Expressing the number of deaths due to each various cause as percentages of the total number due to all causes gives a series of figures that must add up to 100. Differences between any one proportion amongst miners and bank clerks must, to some extent, influence all other differences. For instance, if the proportion of deaths amongst miners which is due to accidents is very much higher than the corresponding proportion amongst bank clerks, it follows that other causes of death amongst miners must be *proportionately* decreased. The point is clear if we suppose only two causes of death to exist, say accidents and infectious diseases. Amongst populations of clerks and miners of equal size there are, let us suppose, in the clerks 50 deaths of each kind, while amongst miners there are 100 deaths from accidents and 50 deaths from infectious diseases. The actual mortality rates from infectious diseases (deaths divided by the number in the population) are identical, for the populations are of equal size; but proportionately those deaths form only one-third of the total amongst miners and one-half amongst clerks. The relative excess of deaths from infectious diseases amongst clerks is not due to an absolutely greater risk but to a lower risk from accidents.

The cardinal rule in the interpretation of proportional rates is to pay equal attention to the numerator and to the denominator of the ratios. Departure from that rule may be illustrated in one or two examples culled from published figures.

Proportional rates in infancy

Table 25.2 shows the infant mortality experienced at different stages of the first year of life in a developing country, differentiated betweeen its urban and rural areas. Expressing these figures as proportions, it was concluded that, 'It is significant that while the proportion of infants succumbing in the first month after birth was 31 per cent in urban areas, it rose to 41 per cent in rural areas, which points to the lack of institutional facilities in connection with confinements in such districts and to the difficulty often experienced of summoning skilled medical assistance in time.' Clearly that conclusion does not necessarily follow from the proportions. The *absolute* rate under 1 month is *lower* in the rural areas than in the urban areas (23.77 to 29.67)—though possibly it might be lower still given better institutional facilities. But it will remain *proportionately* higher just as long as the urban environment shows a relatively excessive death rate at ages 1–12 months. If 69 per cent of the deaths are at ages 1–12 months in the urban areas and only 59 per cent in the rural areas, then whatever the absolute rates at ages 0–1 month the proportions *must* be 31 in the urban and 41 in the rural areas at that age—i.e. a *relatively* unfavourable state of affairs in the rural areas. Given the best possible institutional facilities, that state of affairs will

Table 25.2 Deaths per 100 live births at different months of age.

		Months				
	Under 1 month	*1–*	*2–*	*3–*	*6–12*	*Under 1 year*
Urban	29.67	12.06	8.73	22.77	22.14	95.37
Rural	23.77	7.44	5.18	10.98	11.29	58.66

prevail so long as the urban areas show higher rates during the remainder of the first year.

From the absolute rates, in fact, two arguments may be advanced. The smaller advantage in the rural areas at 0–1 month compared with later ages may be the result of inferior institutional facilities there, or it may be due to the fact that the urban environment does not exert its maximum unfavourable effects until after the age of one month.

Proportional rates in hospital statistics

In hospital statistics the population at risk—i.e. from which the recorded cases are drawn—is often not known. Mortality or incidence rates cannot, therefore, be calculated, and one falls back on proportional rates.

To take a published example, cases of pernicious anaemia were expressed as a proportion of all cases of illness admitted to hospital, the hospital being situated in the United States. Of the 47 203 admissions of individuals born in the United States, 291, or 0.62 per cent, were cases of pernicious anaemia; of the 2814 admissions of individuals born in England, 25 were cases of pernicious anaemia or 0.89 per cent; and of the 7559 admissions of individuals born in Russia, 14 were cases of pernicious anaemia or 0.19 per cent. It is impossible to accept these figures as adequate evidence of racial differences in the liability to pernicious anaemia. At that time the mortality from the disease was correlated both with age and with sex, the death rate rising with age and falling more heavily on women than on men. There is no doubt that the incidence varied similarly. The numerator of the proportional rate (the number of cases of pernicious anaemia) is therefore influenced by the sex and age composition of the population from which the cases are drawn, and it is unlikely that the composition is the same in native-born and immigrant populations.

Equally, the age and sex compositions of the populations at risk in the area may influence the admission rates for other causes of illness—i.e. the denominators of the ratios—since many causes of illness are correlated with age and sex. In the absence of any knowledge of the constitution of those populations it is impossible to draw reliable conclusions from these different proportions of pernicious anaemia.

It is clear that in the clinical records compiled at a hospital (or in general practice) many characteristics of sick persons can be studied—age, sex, race, occupation, family and personal history, etc. From these records much can be learned about the symptoms, the pathology, the course and the prognosis of the disease, either in the total group of sick persons or in subgroups. But in the absence of a knowledge of the exposed to risk we must be cautious in deducing its epidemiology. We are unable to answer the basic questions, '*who* in particular succumbs to this disease, *when*, *where*, and *why*?' We have the numerators of the fractions but no denominators. Proportional rates are usually quite inadequate substitutes. The fact for instance that 20 per cent of certain cases are aged, say, between 40 and 49 years may mean that this age group is particularly vulnerable to the disease in question; but it may mean merely that 20 per cent of the population from which the hospital draws its patients are aged 40–49. In medical literature many comparisons are annually made between the characteristics of a series of cases collected in Hospital A by one worker and in Hospital B by another. Yet, the differences may reflect no more than the characteristics of the populations served by these two hospitals. It is obvious, for instance, that the occupations followed by the patients with peptic ulcer will differ between hospitals in a Lancashire town and those in a nearby rural district of Derbyshire. It is not obvious that similar, if less pronounced, differences in the populations at risk may always underlie the contrasts we make and the conclusions we seek to draw from an enumeration of cases only.

An illuminating example can be found in a study of the incidence of spontaneous rupture of the ventricle in London made by Margaret Crawford and J.N. Morris (*Br.Med.J.*, **2**, 1624 (1960)). This sudden cause of death, they found, led most victims to the coroner's mortuary and relatively very few to hospitals (less than 15 per cent of the total). One result of this stringent selection is that the hospital statistics alone give quite a false picture of the age and sex distribution of the condition, seriously under-stating the number of cases in old age and over-stating the number involving middle-aged men. Thus, only by a knowledge of all the cases occurring and of the population at risk of dying could the true age and sex specific rates be reached.

Berkson's fallacy

A peculiar fallacy in the use of statistics in hospital admissions was noted by a distinguished American statistician after whom it is named. It may occur as follows.

Let us suppose that we have a population served by a particular hospital (or hospitals) and that in that population over some period of time there occur 1000 cases of rheumatic diseases and 1000 cases of cancer. In this population there appears a somewhat rare skin disease; it attacks, during the time referred to above, 5 per cent of the population. It is quite unrelated to rheumatic diseases and cancer. Thus in the population as a whole we shall have the incidence of the illnesses shown in Table 25.3.

Now let us suppose that 10 per cent of the patients with rheumatic diseases are admitted to hospital, and 80 per cent of the patients with cancer. So what we see in the hospital is shown in Table 25.4.

So far so good. But in the population as a whole there still remain 45

Table 25.3 Illustration of Berkson's fallacy (1).

Rheumatic diseases			*Cancer*		
Total cases	*95% without skin condition*	*5% without skin condition*	*Total cases*	*95% without skin condition*	*5% with skin condition*
1000	950	50	1000	950	50

Table 25.4 Illustration of Berkson's fallacy (2).

Rheumatic diseases			*Cancer*		
Total cases	*95% without skin condition*	*5% with skin condition*	*Total cases*	*95% without skin condition*	*5% with skin condition*
100	95	5	800	760	40

Table 25.5 Illustration of Berkson's fallacy (3)

Rheumatic diseases		*Cancer*	
Without skin condition	*With skin condition*	*Without skin condition*	*With skin condition*
95	(1) 5 (2) 9	760	(1) 40 (2) 2
Total 109		*Total* 802	

patients with rheumatic diseases and the skin condition (i.e. the original 50 minus the 5 who were admitted for their rheumatic disease); and 10 patients with cancer and the skin condition (i.e. the original 50 minus the 40 who were admitted for their cancer). So let us suppose that 20 per cent of patients with the skin condition come to hospital, solely because of that skin condition; i.e. 20 per cent of the remaining 45 with rheumatism and 20 per cent of the remaining 10 with cancer. Then what we finally see in the hospital admissions is shown on Table 25.5.

The observed frequency of the skin disease is now 14/109, nearly 13 per cent, in the rheumatic patients and 42/802, or only 5 per cent, in the cancer patients. It therefore appears that the skin condition is associated with rheumatic complaints. The fallacy has arisen because (1) the two illnesses, rheumatism and cancer, have quite different admission rates and (2) the skin condition itself brings patients to hospital independently of the other illnesses.

In this example the fallacy has deliberately been made to show a very considerable effect upon the final statistics. None the less in any such comparison of hospital statistics the possibility of its presence should not be overlooked.

Statistics of post-mortems

Much attention has been paid during the last 60 years to the extent to which cancer of the lung has increased. The recorded mortality statistics have been regarded doubtfully, for their rise is certainly due in part to more accurate certification of death now than in the past. The number of post-mortems at which this form of malignant disease is found has, therefore, sometimes been held to be a more certain basis. This number can be taken as the numerator of the rate. But in the absence of a knowledge of the population from which the hospital draws its inmates, an incidence rate cannot be calculated. In its place a proportional rate is used, taking, for instance, as the denominator either the total number of post-mortems carried out, or the total of all cases of disease admitted to the hospital. Thus we have the cases of cancer of the lung observed at post-mortems as a proportion of all post-mortems or of all cases admitted. Calculating this ratio year by year shows its secular trend. But in interpreting this trend we have to consider possible changes both in the numerator and the denominator. The numerator—cases of cancer of the lung found at post-mortems—may be influenced by an increasing interest in the condition; its victims may be more frequently submitted to post-mortem than in past years. If the denominator were unchanged this would give an apparently rising incidence. But it is equally important to inquire whether the denominator has changed—i.e. all post-mortems or all cases admitted. If the criteria upon which post-mortem examination or admission to hospital are based have changed with time, this will also inevitably influence the trend of the ratio. An addition to the denominator of cases or post-mortems that would not have been present in earlier years must result in a declining ratio, even though the actual incidence rate of cancer of the lung in the population exposed to risk has not changed, or, indeed, has risen.

For example, suppose the incidence of cancer of the lung in the unknown population is 50 cases per annum, all of which come to the hospital, and are recognised, and this number is unchanged between 1950 and 1990. If in 1950 there were 500 post-mortems carried out, then the proportional rate is 50 in 500; if in 1990 a laudable attempt to determine more accurately the causes of death of patients led to an increase of the post-mortem examinations to 600, the ratio falls to 50 in 600, without any change in the real incidence. Alternatively, we may suppose that the cases of cancer of the lung have actually gone up threefold, i.e. to 150. At the same time all post-mortem examinations have doubled to 1000. The proportional rates are 50 in 500, or 10 per cent, and 150 in 1000 or 15 per cent. The real rise in cancer of the lung is masked by the general rise.

It may, perhaps, be reiterated that the denominator needs just as much consideration as the numerator.

Ignoring censored observations

In statistics a *censored* observation is one where we cannot tell its precise value; we know only that it is greater than (or less than) some value. Ignoring the fact that censored observations exist can produce fallacious results very similar to those produced by ignoring exposure to risk.

A good example is to be found in comments that are sometimes made concerning life-imprisonment for murder. This does not literally mean imprisonment for life, but only until such time as release is thought to be safe and reasonable, and statements are sometimes made to the effect that 'life' means only $8\frac{1}{2}$ years on average (or some such figure). What has usually been done is to take the average time served by those *who have been released*, ignoring the censored observations of those who, for example, have done 10 years so far but are still confined. Also ignored may be any who died in prison.

To make a good estimate of a meaningful figure is clearly difficult, but the figure given is useless nonsense.

Chapter 26

Fallacies and difficulties: incidence and causes of mortality

Statistics of causes of death

In making comparisons between death rates from different causes of death at different times or between one country and another, it must be realised that one is dealing with material which the distinguished American statistician Raymond Pearl long ago described as 'fundamentally of a dubious character', though of vital importance in public health work. Although much progress has been made in certifying and tabulating causes of death and thus in international comparability, the recorded incidence of a particular cause may still be influenced by differences in nomenclature, by differences in interpreting the entries on the death certificate and thus in the resulting tabulation, by medical fashions, and by the frequency with which the cause of death is certified by medically or non-medically qualified persons or follows a post-mortem examination.

Striking differences between reported death rates for particular diseases may, therefore, be true in fact or merely due to lack of uniformity in methods of registration, diagnosis or coding. Looking, for instance, to the influences of coding practices, the World Health Organisation (in 1962–65) took a sample of one thousand completed death certificates and asked five European countries to code them according to their own normal national practice. The results from each country were compared with the coding adopted at the WHO's own Centre for Classification of Diseases, in London. Exact agreement varied between 85 per cent and 66 per cent; broad agreement lay between 90 and 76 per cent; but material differences in allocation of the certificate to the cause-of-death group was found in 5 per cent at the best and as many as 22 per cent at the worst. A few specific examples of such risks of comparison may therefore be given.

Mortality from cancer

The crude death rate from cancer (all forms) in Eire 60 years ago was well below that registered in England and Wales. Part of this difference may have been due to a more favourable age distribution of the population in Eire—i.e. standardised rates should be used in the comparison—but it is also likely that it arose from differences in the certification of causes of death. In Eire at that time considerably more deaths were ascribed to senility than in England and Wales—for example, 15 per cent in the former in 1932 against 4 per cent in the latter. Such a difference cannot inspire confidence in the death rate from such a disease as cancer, in which the majority of deaths fall at advanced ages. In general, therefore, in comparing the cancer death rates of different countries or of the different areas of the same country—e.g. rural and urban—it is not sufficient to pay attention to the cancer rubric; other headings such as 'uncertified', 'senility', and 'ill-defined causes' must be taken into consideration, and an attempt made to determine whether transferences between these rubrics are likely to play a part.

The kind of indirect correlation that one may observe is this. It is stated that the death rate from some cause is associated with the consumption of, say, sugar and the level of the death rate is compared with some measure of sugar consumption in different countries. It is found that the countries with a low consumption of sugar have relatively low death rates from the specified cause. But it is at least possible that those countries which have a high standard of living have a relatively higher sugar consumption, and also a higher standard of vital statistics, and therefore more accurate death rates, than countries with a low standard of living and less accurate vital statistics. Other 'causes' of death—e.g. ill-defined and old age—would need study as well as those attributed directly to the specific cause.

Maternal mortality

It has been recognised that the maternal death rates of different countries may be affected by varying rules of tabulation in vogue. Some years ago a sample of deaths associated with pregnancy and childbirth that took place in the USA was assigned by different statistical offices of the world to puerperal and non-puerperal groups according to the rules of those offices (Children's Bureau Publication No. 229). The variability was considerable. In the USA 93 per cent were tabulated to puerperal causes, in England and Wales 79 per cent, in Denmark 99 per cent. Such differences may still make international comparisons difficult.

Mortality from respiratory causes

In England and Wales bronchitis and pneumonia have shown pronounced differences in their incidence in different parts of the country at certain ages. It appears that the 'bronchitis' of one area might include deaths which would be attributed to pneumonia in another. This problem is equally prominent in international comparisons of death rates from these respiratory causes. The term used by doctors in death certification may vary widely.

Infant mortality rates

The international comparison of mortality in infancy can be particularly difficult for it may be materially affected by the laws regulating the registration of births and deaths. In many countries, e.g. England and Wales, if the infant shows any signs of life but then dies within a short period of time, the law requires the registration of a live birth and then of a death. In some countries, on the other hand, a baby born alive but dying before the end of the 2–3 days allowed for registration of the birth is not included in the statistics of infant mortality but is treated as a stillbirth. Very early infant mortality is thereby excluded from the rate and the comparison of the figures for the various countries is not strictly valid. Apart from this the comparisons will also be affected by the criteria adopted in defining a stillbirth. When is it that the infant is to be regarded as having been born alive and when as a stillbirth? The definitions vary in different countries and affect the relative levels of their infant mortality and stillbirth rates. It is for this reason that the perinatal death rate has been widely adopted, i.e. the sum of the stillbirths and the infant deaths occurring in the first week of life.

The average age at death

The average age at death is not often a useful measure. Between one occupational group and another it may be grossly misleading. For instance, as William Farr, the medical statistician of the General Register Office in the nineteenth century, pointed out, the average age at death of bishops is much higher than the corresponding average of curates. But making all the curates bishops will not necessarily save them from an early death. The average age at death in an occupation must, of course, depend in part upon the age of entry to that occupation and the age of exit from it—if exit takes place for other reasons than death. Bishops have a higher age at death than curates because few men become bishops before they have passed middle life, while curates may die at any age from their twenties upwards.

The following misuse of this average is taken from a report on hospital patients.

It is stated that in 31 cases of renal hypertension which came to autopsy the average age of death was 45. 'Thus the common fate of the renal hypertensive is to die in the fifth decade of life.' This may be a true statement of fact, but it clearly cannot be deduced from the average age; the average might be 45 years without a single individual dying in the fifth decade. The report continues, 'In 86 cases of essential hypertension which came to autopsy the average age at death was found to be 60, while in 20 cases seen in private practice the average age at death was nearly 70. Thus, the fate of the non-renal hypertensive is very different from that of the renal. The subject of uncomplicated essential hypertension may reasonably expect to live into the seventh and even the eighth decade.'

The first deduction is probably valid, though obviously information regarding the *variability* round those averages is required. The frequency distributions of the age at death for the two groups should be given. What the patient may 'reasonably expect' has no foundation in the figures given. If the subjects of uncomplicated essential hypertension mainly live into the seventh or eighth decade such wording might be reasonable. But if the average age is derived from individual ages at death varying between, say, 40 and 90, one has no justification for using that average in such a way. The statistical meaning of 'expectation' is clearly not intended, for in that usage 'reasonably expect' has no meaning.

The author regards statistics as 'dull things' and therefore refers to them as briefly as possible—so briefly that in his hands they are of very little use.

To take one further example, a difference in the average ages at death from, say, silicosis in two occupations may imply that in one occupation the exposure to risk is more intense than in the other and thus leads to earlier death; but this interpretation can only hold, as is pointed out above, so long as the employed enter the two occupations at the same ages and give up their work at the same ages and to the same extent. It is usually very difficult to secure good evidence on these points, and the average ages at death must be regarded with great caution.

The expectation of life

It was stated in Chapter 21 that the 'expectation of life' is the average number of years lived beyond any age by the survivors at that age when exposed to some selected mortality rates. For instance, at the age-specific mortality rates of England and Wales in 1910–12 the average length of life of males after birth was 51.5 years; at the rates of 1980–82 it was 71.0. It is sometimes wrongly deduced that those figures imply an increase of some 20 more years of life to enjoy *when we approach retiring age*. Actually the expectation of males at age 60 was 13.8 years in 1910–12 and 16.9 years in 1980–82, an increase of only 3.1 years. Apart from the fact that the calculation involves the assumption that current mortality rates will continue to prevail, it must be remembered that the average length of life after birth will be influenced by reductions in the death rate at *any* stage of life. The increase in the expectation at birth over this period has been largely due to the fall in infant mortality and deaths in early childhood, which implies much fewer very short durations of life than formerly. Such a fall would increase the average length of life even if not a soul survived age 60 then or now. After the age 60 male mortality rates did not fall sufficiently between 1910–12 and 1980–82 to increase the average length of survival substantially.

To clarify the point, suppose there were a population in which 50 per cent of all those born died on their first day of life, the other 50 per cent all surviving to the age of 70 years and dying on their 70th birthday. The expectation of life at birth in that population would be 35 years. If a method were then discovered to halve the infant mortality, so that only 25 per cent died on their first day, the remaining 75 per cent continuing to 70 years, the expectation would increase to $52\frac{1}{2}$ years, yet those reaching age 30 would have a further 40 years in prospect, just like their ancestors.

Problems of inheritance

Many disorders or derangements in mankind have been recorded as showing evidence of hereditary factors. Sometimes the evidence consists merely of the appearance of the disease or disability in a more or less orderly fashion among related individuals. In many instances, of course, there is no doubt that hereditary factors are important, but in others their presence is difficult to prove, in the inevitable absence of controlled breeding experiments and the impossibility sometimes of distinguishing genetic from environmental influences. Cases are reported, for example, of a familial incidence of cancer; a man whose father died of cancer of the stomach died himself of cancer in the same site, while his wife died of cancer of the breast and their six children and one grandchild all died of various forms of cancer. This is a very striking family history, but it is not necessarily evidence of an inherited factor. If each of these individuals had been known to have passed through an attack of influenza we should not deduce a particular family susceptibility to influenza, since we know that influenza in the whole population is so widespread that a familial incidence is bound to occur very frequently. Similarly we want to know the probability of observing a series of familial cases of cancer merely by chance. Even if that probability is small it must be remembered that the field of observation amongst medical men is enormously wide and a few isolated instances of multiple cases cannot be adequate evidence. Usually, too, only one part of the field is reported in medical literature, for notice is taken of the remarkable instances and no reference made to the cases in which no inheritance is apparent. This is also, and particularly, true of reports in the lay press of such family events, e.g. of the long life of members of the same family. A father dying at the age of 85 and four children all dying at ages above 80 is 'good copy' and worthy of notice. But if the children had died at the ages of 20, 65, 69, and 90 the occurrence would not have found a place in the paper. Indeed, the evidence presented through the 'media' can be incredibly selective and in no way representative of the population at large. In fact, scientific evidence of the inheritance of length of life in man is extremely hard to come by.

With specific diseases the data required in such a problem are reasonably large numbers of family histories, so that, if possible, it may be seen whether the distribution of multiple cases differs from the distribution that might be expected by chance, or whether the incidence in different generations suggests a Mendelian or other specific form of inheritance. In considering family histories that rely only or largely on memory it must also be kept in mind that a patient with a specific disease may well have become aware of other instances of that disease in relatives whereas a person without the disease would not have necessarily done so, having no particular interest in it.

Even if the distribution of multiple cases differs from that expected on a chance hypothesis, the question of a common family environment cannot be ruled out—e.g. multiple cases of tuberculosis may occur more frequently in families of a low social level not through an inherited diathesis but through undernourishment.

Family histories

As stated above, possible errors in family histories, taken from patients or healthy persons, must not be overlooked. Thus in the USA, Tecumseh Community Health Study, 208 persons were identified with probable coronary heart disease. Each gave a family history and, where appropriate, a cause of death or disability for their parents, siblings and children. The relatives who had died were subsequently identified on death certificates. The agreement between history and death certificate was 73 per cent for heart disease, 83 per cent for cancer but only 50 per cent for stroke. The disagreements were spread over many other causes of death on the certificates. Of living relatives 243 were traced and interviewed by the same interviewers using the same family history form. The discrepancies were considerable. Thus, family history gave 7 per cent with stomach trouble whereas the direct interview gave 43 per cent; the corresponding figures were 9 and 26 per cent for hypertension, 4 and 27 per cent for kidney troubles.

Not all the errors need be due to the family history. Some, of course, may come from the death certificates and interviews. We cannot tell but, obviously, the limitations of the family history must be kept in mind when such a history is the only available source of information.

Chapter 27

Fallacies and difficulties: various problems

'The errors balance out'

It is sometimes reasonable to suppose that errors in statistical data will balance out and therefore are of no importance. For instance, in calculating the arithmetic mean of grouped data we presume that the observations in a frequency distribution are spread across each of the frequency groups in such a way that we may accept, as a sufficiently accurate approximation, the centre of the group as applying to all the observations in it. Spread over the whole of the distribution, the errors we make are likely to balance out. Similarly there may sometimes be errors in observations which will balance out. For example, deaths attributed to a form of cancer may be overstated by the attribution to it of deaths really due to other causes; they may be understated by the omission of deaths attributed to other causes which were in fact due to it. The errors could balance out (though we should need evidence to that effect).

On the other hand, biased errors cannot possibly balance out. If in taking case histories of patients, the majority deliberately—or merely through wishful thinking—understate their normal daily consumption of alcohol, there can be no balancing out of errors. The final answer, and the figures derived from it, *must* be an understatement. Whether errors are likely to balance out or to lead to an erroneous picture often requires, therefore, careful thought in planning an inquiry or in analysing its results. The answer is not always so clear-cut as in the examples above.

Similar reflection may be required in the study of the *relative* position of two groups in which the observations are liable to error. To take a specific example, in which the incidence of influenza in a vaccinated and an unvaccinated group we are certain, owing to the difficulties of diagnosis, to be including illnesses due to other respiratory causes against which the vaccine would be expected to be powerless. We must by this dilution weaken the evidence in favour of the vaccine. Instead of comparing the two relative attack rates of influenza (x_1 against x_2) we are comparing the relative rates of influenza *plus in each group a constant amount of other diseases* (x_1+y against x_2+y). If the other diseases are relatively infrequent, as they may well be when influenza is sharply epidemic, then clearly the assessment of the vaccine will not be seriously at fault (y is a small and unimportant component). But if other diseases included are relatively frequent, as they may well be when influenza is merely sporadic, then the assessment of the vaccine may be in serious jeopardy (y dominates the comparison). There can be no question of errors balancing out.

The 'all or none' syndrome

A widespread misconception of scientific evidence of cause and effect, which arises in even well-educated minds, lies in the argument that some personally observed exception gravely weakens, or even destroys, the more general and, often, very extensive evidence of that relationship. Exceptions to an apparent rule are, of course, important in all circumstances of observation or experiment and we must never lightly dismiss them. The fact, for instance, that a non-smoker contracts cancer of the lung, or that a man who never drinks water sickens with typhoid fever, is both interesting and important. We are by such observations informed, *whatever we may believe about smoking and polluted water supplies*, that there must certainly be other means of acquiring these diseases. Similarly the fact that heavy smokers or drinkers of a contaminated water supply do *not* fall sick is also important. *However much we believe in the relationship*, we are thereby informed that there must be further undetected environmental or constitutional factors to sway the balance upwards or downwards in the individual. But, clearly, none of the exceptions is incompatible with the observation that *on the average* drinkers of polluted water supplies have more typhoid fever than those who have a pure supply or that *on the average* those who smoke most die at a higher rate from lung cancer. Whether we deal with the typhoid baccilus, a carcinogenic agent, or an industrial dust hazard, there are very few things in life that are 100 per cent effective or have a simple all-or-none reaction.

The 'normal' value

Many errors of clinical interpretation have arisen (and no doubt may continue to arise) from a failure to recognise the variability that is a characteristic of man in all his measurable features. Such variability between one person and another is clearly a necessary and distinctive attribute of life. How then in this situation can we define a 'normal' value? What value, or values, can we accept as normal in the sense that they are compatible with good health? On what point of a scale should we begin to regard them as abnormal, i.e. pathological?

There is indeed no easy answer to these questions. Obviously we can say that the average value is a wholly unacceptable guide. The very fact that there *is* such variation as referred to above, i.e. in persons in good health, at least shows that we must accept as 'normal' some *range of values*, e.g. of the systolic blood pressure at a given age, of the weight of the new-born baby, of the pulse rate after strenuous exercise, of body temperature.

If, however, we have adequate figures of the frequency distribution in an apparently healthy population, and thus of the variations in the measured characteristic which prevail, then it will be reasonable to regard with suspicion, or as 'abnormal', values that lie far from the average. Much greater difficulty will, of course, lie, as J.A. Ryle once wrote (*Lancet*, **1**, 1, (1947)) in less measurable characteristics such as the heart sounds, the strength of the knee jerk, the size of the tonsils, and so on.

There is another problem. Although we may have observed that a range of values is compatible with *present* good health, we have very little information indeed as to their being compatible with *continuing* good health. In many circumstances the observations have just never been made. Further, values on a numerical scale that we accept as 'normal' (in the sense of healthy) will often shade imperceptibly into the borderline and then imperceptibly into the pathological (in the sense of 'unhealthy'). There will not necessarily be any sharp demarcation, any decisive line of division between 'normal' and 'abnormal'. There is, without doubt, a decisive line between a sound femur and a fractured femur. It does not follow that there is an equally decisive line between 'normotension' and 'hypertension'.

Values quoted as 'normal ranges' are often chosen by examining samples from suitable populations and setting the limits so as to include the central 95 per cent, either as observed or after fitting an appropriate theoretical frequency curve, with $2\frac{1}{2}$ per cent outside the limits at each end. Such limits then tend to be quoted and used as if everybody should come within them, whereas it is obviously to be expected that 5 per cent should continue to fall outside without that indicating anything wrong.

Finally, if we regard the average of some population as 'normal' we must remember that it may well represent a situation that is really abnormal. It may be *usual* in western countries to be obese in mid-life and edentulous in old age. But one can hardly argue from that that the usual is normal in any physiological sense.

In short the problems of defining the normal are such that we should be very careful in our acceptance and use of the term.

The problem of attributes

In medical statistics subdivision of the patients treated have often to be made in order to ensure that like is being compared with like. It is obviously idle to compare the fate of two groups of patients with, say, cancer of the cervix uteri treated by different methods, if one group contains 50 per cent of persons whose disease was far advanced when treatment was begun and the other has but 10 per cent in the same category. To avoid this fallacy an attempt is made to classify the patients according to the stage, or severity, of the disease when first seen. Such a division, it is recognised, may be influenced by the different interpretation of what constitutes degrees of severity by the observers, and such statistics must always be regarded with that difficulty in mind.

For instance, one hospital classifies two thirds of its patients to stages of disease 1 and 2 and one third to stages 3 and 4; another hospital has only one quarter of the patients at stages 1 and 2 and three quarters at stages 3 and 4. Do patients really present themselves at these hospitals in such widely differing proportions or is the difference due to differing standards of classification? The answer to that question is unknown, but in comparing the fatality rates following treatment the problem must not be disregarded. If the stage of the disease is, in fact, related to the prognosis, then it must be remembered that different standards of classification may favourably influence one hospital as compared with another at *every* alleged stage of the disease. This may be shown diagrammatically by supposing the patients are set out in a line in order of severity, from mildest to most severe.

```
Hospital A    Stage 1 | Stage 2 | Stage 3 | Stage 4              |
              ------------------------------------------------------
Hospital B              Stage 1 | Stage 2 | Stage 3 | Stage 4    |
```

Hospital B includes in Stage 1 a proportion of patients which Hospital A relegates to Stage 2; the latter hospital, therefore, has only the very best of the patients in Stage 1 and will in consequence have a more favourable result than Hospital B at that stage. In the Stage 2 group Hospital A includes favourable cases which B had called Stage 1, while B includes less favourable cases that A classifies to Stage 3. Hospital A will again show a better result. The same difference is apparent in the subsequent stages, Hospital B having only the extreme cases in Stage 4 and A including some that B would include in Stage 3. At each stage Hospital A will, therefore, compare favourably with Hospital B, though in the total there may be no difference whatever.

In any series of statistics based upon division by attributes by different observers this possible fallacy must be considered with care. The difficulty has occurred acutely in the assessment of the state of nutrition among school-children where different observers have used different criteria of determination and placed a different meaning on such terms as 'good', 'fair', 'poor', etc. This can lead to an incidence differing unbelievably from one place to another, unless some simple objective measure can be devised to replace the subjective assessment. In the same way, differing interpretations by different people made the statistics of the blind and dumb collected at the census enumerations in the past of doubtful value, and led (in England and Wales) to the questions being abandoned. Faced with such statistics the reader must always ask himself: Could these classifications have been made by means of objective measurements, which, given equal skill in making those measurements, would not vary appreciably from observer to observer? If not, how far may the presence of subjective influences affect the results and in what way?

Standardising a scale

In trying to overcome the problem discussed in the previous section, the approach is often made of picking out typical cases for comparison with future ones. For example, in X-ray pictures of the lungs the classification can be much improved by making the assessment in comparison with a number of selected examples. However the mistake is generally made, in devising such schemes, of having the standard cases as 'Typical of stage 1', 'Typical of stage 2', etc. It would be much better to seek for standard pictures that are 'Borderline between stages 1 and 2', 'Borderline between stages 2 and 3', etc. There is then a genuine point of comparison with the decision that is actually to be made.

Prospective and retrospective data

It was pointed out in Chapter 5 that with data gathered retrospectively it may be impossible, and is always difficult, to calculate actual rates of incidence, etc. One possible and serious error is worthy of mention. Suppose, for example, 12 women all suffer during the first 3 months of pregnancy from an attack of rubella and that of the babies subsequently born to them 4 have specified congenital defects and 8 are found to be normal. Thus, tabulating, we have the following results:

A	Normal
B	Congenital cataract
C	Normal
D	Normal
E	Normal
F	Congenital deafness
G	Normal
H	Congenital cataract
I	Normal
J	Normal
K	Congenital disease of the heart
L	Normal

It is clear that these data viewed *prospectively*, and correctly, give the risk of a baby suffering from a congenital defect after rubella occurring in the mother during the first trimester as 4 in 12 or, in other words, 1 in 3. But suppose that the problem is approached *retrospectively* (as is so often done with hospital data), i.e. by observing merely the 4 affected children and inquiring of the mother's history. *All* 4 mothers will then be found to have had rubella during the first 3 months and from this it may be quite erroneously concluded that the risk is 100 per cent. Once more the original number exposed to risk is unknown, and its paramount importance has been forgotten.

Difficulties with maps

Where death rates, or sickness rates, vary from place to place, setting them out as a map with various colours representing the rates may be helpful in raising questions—why does that particular form of cancer occur so much more frequently in that particular country? Such maps can be beautifully produced, and totally honest in their intentions, but as with all diagrams we have to ask whether the impression gained by the eye is the truth or not.

The difficulty is that the sparsely inhabited large areas are likely to dominate the picture, while densely populated small areas, which may contain the majority of the people concerned, are virtually unnoticeable. A distorted map that gives equal areas to equal numbers of people, but nevertheless retains the correct geographical relationship so far as possible, is always worth considering.

A further difficulty with maps occurs when the eye notices something like Figs. 27.1 and 27.2. Surely the fact that the cancer rate and the density of population both have the same boundary, complete with a 'blip' at the same point, must mean something—indeed, it does; it means that that is the shape of the boundary between the local govenment administrative areas by which the statistics are collected, and it is likely to have little, if any, more meaning that that.

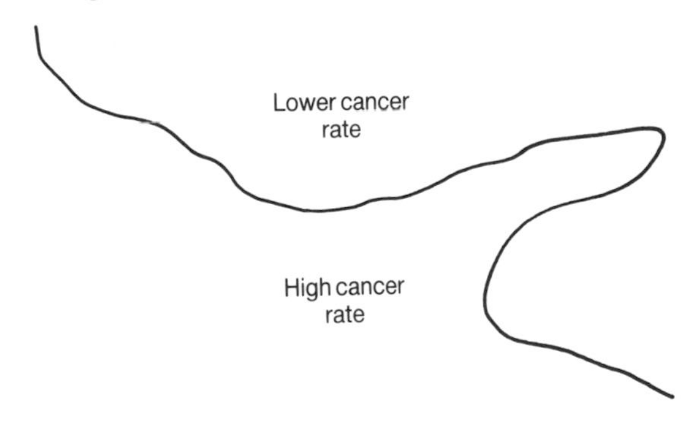

Fig. 27.1 Hypothetical boundary between two cancer rates shown on a map.

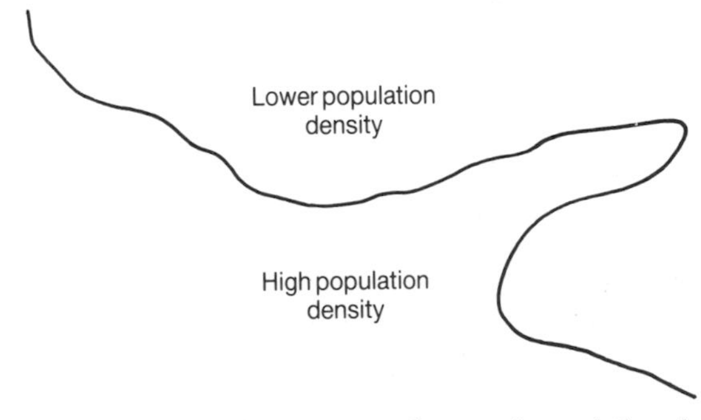

Fig. 27.2 Corresponding boundary between two figures of population density shown on another map.

Chapter 28

Statistical evidence and inference

The aim of this book has been to make clear in non-mathematical terms some of the techniques that the statistician employs in presenting and in interpreting figures. Much attention has been paid to two basic problems:

1 the *significance*, or reliability in the narrow sense, of a difference which has been observed between two sets of figures—be those figures averages, measures of variability, proportions, or distributions over a series of groups;
2 the *inferences* that can be drawn from a difference which we are satisfied is not likely to be due to chance.

A secure foundation for argument

The discussion of the first problem led to the development of tests of significance—the standard errors of individual values, the standard errors of the differences between values, and the χ^2 test. The object of such tests is to prevent arguments being built up on a foundation that is insecure owing to the inevitable presence of sampling errors. Medical literature is full of instances of the neglect of this elementary precaution. Illustration is hardly necessary, but the following is just one quotation: 'A mere list of the treatments which have been tried in thrombo-angiitis obliterans would be of formidable length and there is little point in mentioning many of them—they have only too often fallen by the way after an introduction more optimistic than warranted by results.'

In general, worker A, who is at least careful enough to observe a control group, reports after a short series of observations that a particular method of treatment gives him a greater proportion of successes than he secures with patients not given that treatment, and that therefore this treatment should be adopted. Worker B, sceptically or enthusiastically, applies the same treatment to similar types of patients and has to report no such advantage. The application of the simple probability tests previously set out would have (or should have) convinced A that though his treatment *may* be valuable, the result that he obtained *might* quite likely have been due to chance. He would consequently have been more guarded in his conclusions and stressed the limitations of his data.

'Significant' and 'not significant'

On the other hand it is important to keep in mind the meaning of statistically 'significant' and 'not significant'. The calculation of probabilities is undoubtedly of value in preventing us from 'overcalling a hand' and from attributing too freely to some cause what might very well be due only to chance. The calculation also has the advantage of setting a standard of judgement that is constant from one person to another. Beyond that it cannot go. A 'significant' answer does not *prove* that the difference is real—chance is still a possible, though unlikely, explanation; it certainly does not prove that the difference is of any material importance—it could be significant on large enough numbers but utterly negligible; inspection and consideration of the results themselves must be (and *always* should be) made for that; it never proves (or even indicates) the cause of the difference. On the other hand a 'not significant' answer does not tell us that Group *A* does *not* differ from Group *B*—the figures in front of us show, in fact, that it does, though maybe immaterially. But it does tell us that chance may easily be the reason for that difference, and it is clearly important to know that, before we seek some more recondite explanation or before we take action. When a material difference *is* apparent between two groups, but, with the numbers involved, is insufficient to pass the formal probability test, it is better (particularly in a matter of importance) to take 'statistically not significant' as 'non-proven' rather than as 'not guilty'.

The second problem referred to above is usually very much more difficult than that presented by the formal test of significance, namely what inferences can we reasonably draw from a difference which we are satisfied is *not* likely to be due to chance? In the clinical trial, for instance, we would ask ourselves: Were the groups of differently treated patients equivalent in other relevant characteristics that might have a bearing on the progress of the illness? Was treatment the only respect by which they were differentiated? This question immediately emphasises the importance of the initial planning of clinical trials, as discussed at length in Chapter 23. Only by planning and foresight can we ensure comparable groups of patients, and only in such circumstances can we with reason infer that the significant difference between groups is more likely to be due to the specific treatment than to any other factor. For in the well-planned trial such other factors are likely to be equally present in the various groups and therefore to contribute equally to the upshot.

In short, if we can experiment we can usually conduct our experiment in such a way as to justify the conclusions we draw from it. The problem of inference is thereby kept within bounds.

The observational approach

On the other hand, the problem is very greatly increased when we are unable to experiment and must rely upon our observations of the relationships that occur in nature before our eyes. This was, of course, the classical method of the early epidemiologists who, faced for example with an epidemic of enteric fever, found by painstaking inquiry that, in comparison with *unaffected* persons, appreciably more of the *patients* had consumed a particular foodstuff or had drunk from a particular milk or water supply (i.e. they adopted the retrospective, case/control approach which is, of couse, still fundamental in the epidemic setting). One advantage that such observers had, and still have, was the relatively short interval that elapsed between exposure and illness, i.e. between cause and effect. On the other hand, with the chronic diseases, to which so much attention is being paid today, it is possible that the cause may precede the effect by many years. The true relationship of one to the other is thereby made most difficult to detect and, at the same time, spurious, or indirect, associations will confuse the issue.

Faced then with a clear and significant association between some form of sickness or cause of death and some feature of the environment, what ought we to consider specifically in drawing conclusions about the nature of the relationship—*causation* or merely *association*?

The strength of association

First upon this list of features to be closely considered one might well put the *strength* of the association. In other words, we must consider the relative incidence of the condition under study in the populations contrasted. Thus, to take a specific example, prospective inquiries into smoking have shown that the death rate from cancer of the lung in cigarette smokers is nine to ten times the rate of non-smokers, while the rate in heavy cigarette smokers is twenty to thirty times as great. On the other hand the death rate from coronary thrombosis in smokers appears to be no more than twice, possibly less than twice, the death rate in non-smokers. Though there is indeed good evidence to support causation it is much easier in this latter case to think of some features of life that may go hand-in-hand with smoking—features that might conceivably be the real underlying cause or, at the least, an important contributor, whether it be lack of exercise, nature of diet or other factors. But to explain the pronounced excess in cancer of the lung in any other environmental terms requires some feature of life which is so closely correlated with cigarette smoking, and with the amount of smoking, that such a feature should be easily detectable. In the absence of any such features we can more readily accept direct causation as the likely explanation.

Whether we consider the absolute difference between the death rates of the various groups or the ratio of one to the other depends upon what we want to know. If we want to know how many extra deaths from cancer of the lung will take place through smoking (i.e. presuming causation), then obviously we must use the absolute differences between the death rates—for example 0.07 per 1000 per year in non-smoking doctors, 0.57 in those smoking 1–14 cigarettes daily, 1.39 for 15–24 cigarettes daily and 2.27 for 25 or more daily. But it does not follow that this measure of the effect upon mortality is also the best measure in relation to *aetiology*. In this respect the ratios of the smokers' rates to that of non-smokers, 8, 20 and 32 to 1, are far more informative. It does not, of course, follow that the differences thus revealed by ratios are of any practical importance. Maybe they are, maybe they are not; but that is another point that needs study.

In this connection we may recall John Snow's classic analysis of the opening week of the cholera epidemic of 1854. The death rate that he recorded in the customers supplied with the grossly polluted water of the Southwark and Vauxhall Company was in one sense quite low—71 deaths in each 10 000 houses. What stands out vividly is the fact that this small rate is *14 times* the figure of 5 deaths per 10 000 houses supplied with the sewage-free water of the rival Lambeth Company. Thus the association with polluted water is very strong.

Consistency

Next on the list of features to be specially considered we may place the *consistency* of the observed association. Has it been repeatedly observed by different people, in different places, different circumstances and times?

With much research work in progress today many an environmental association may be thrown up. Some of them on the customary tests of statistical significance will appear to be unlikely to be due to chance. Nevertheless whether chance is the explanation or whether a true hazard has been revealed may sometimes be answered only by a repetition of the circumstances and the observations.

Returning to the example taken above, an Advisory Committee to the Surgeon-General of the United States Public Health Service found the association of smoking with cancer of the lung in 29 retrospective and 7 prospective inquiries. The important lesson here is that broadly the same answer has been reached in quite a wide variety of situations and techniques. In other words we can justifiably infer that the association is not due to some constant error or fallacy that permeates every inquiry. And we have indeed to be on our guard against that.

Thus we have the somewhat paradoxical position that the *different* results of a *different* inquiry certainly cannot be held to refute the original evidence; yet the *same* results from *precisely the same* form of inquiry will not invariably greatly strengthen the original evidence. Much weight should be put upon similar results reached in quite different ways, e.g. in prospective and retrospective inquiries.

On the other hand there will be occasions when repetition is absent or impossible and yet we should not hesitate to draw conclusions. The experience of nickel refiners in South Wales is an outstanding example. The population at risk, workers and pensioners, numbered about one thousand. During the ten years 1929–38, sixteen of them had died from cancer of the lung, eleven of them had died from cancer of the nasal sinuses. At the age-specific death rates of England and Wales at that time, one might have anticipated one death from cancer of the lung (to compare with the 16), and a fraction of a death from cancer of the nose (to compare with the 11). In all other bodily sites cancer had appeared on the death certificate 11 times and one would have expected it to do so 10–11 times. There had been 67 deaths from all other causes of mortality and over the ten years' period 72 would have been expected at the national death rates. Finally, division of the population at risk in relation to their jobs showed that the excess of cancer of the lung and nose had fallen wholly upon the workers employed in the chemical processes.

A later analysis showed that in the nine years 1948–56 there had been 48 deaths from cancer of the lung and 13 deaths from cancer of the nose. The numbers expected at normal rates of mortality were respectively 10 and 0.1.

In 1923, long before any special hazard had been recognised, certain changes in the refinery had taken place. No case of cancer of the nose was observed in any man who first entered the works after that year, and in these men there was no excess of cancer of the lung. In other words, the excess in both sites is uniquely a feature in men who entered the refinery in, roughly, the first 23 years of the twentieth century.

No causal agent of these neoplasms has been identified. Until much later no animal experimentation had given any clue or any support to this wholly statistical evidence. Thus minds had to be made up on a unique event; and there was no difficulty in doing so.

This situation very clearly makes nonsense of the assertion that if the evidence is 'only statistical' we cannot accept it for action. Action would be right even if no laboratory in the world had been able to produce malignant changes in the noses and lungs of animals. We know too little of species differences in relation to carcinogenic agents to be influenced by a negative answer in laboratory animals when the epidemiological picture in man is overwhelming.

Specificity

One reason, needless to say, is the *specificity* of the association, the third characteristic to be considered. If, as in this instance, the association is limited to specific workers and to particular sites and types of disease and there is no association between the work and other modes of dying, then clearly that is a strong argument in favour of causation.

We must not, however, over-emphasise the importance of the characteristic. Even in the present example there is a cause-and-effect relationship with two different sites of cancer—the lung and the nose. Looking at a wider field, milk as a carrier of infection and, in that sense, the cause of disease, can produce such a disparate galaxy as scarlet fever, diphtheria, tuberculosis, undulant fever, sore throat, dysentery and typhoid fever. Before the discovery of the underlying factor, namely the bacterial origin of disease, great harm would have been done by pushing too firmly the need for specificity as a necessary feature before convicting the dairy.

In modern times the prospective investigations of smoking and cancer of the lung have been criticised for not showing specificity—in other words the death rate of smokers is higher than the death rate of non-smokers from a number of causes of death. But here one must look also at the characteristic, the strength of the association. If some other causes of death are raised 10, 20 or even 50 per cent in smokers whereas cancer of the lung is raised 900–1000 per cent we have specificity—a specificity in the magnitude of the association.

We must also keep in mind that diseases may have more than one cause. Few diseases are confined to a single occupation and, in general, one-to-one relationships are not frequent.

The relationship in time

A fourth characteristic to be considered is the temporal relationship of the observed association—which is the cart and which the horse? This is a question which might be particularly relevant with diseases of slow development. Does a particular diet lead to disease or do the early stages of the disease lead to particular dietetic habits? Does a particular occupation or occupational environment promote infection by the tubercle bacillus or are the men and women who select that kind of work more liable to contract tuberculosis whatever the environment—or, indeed, have they already contracted it? This temporal problem may not arise often but it certainly needs to be remembered.

The biological gradient

Fifthly, if the association is one which can reveal a *biological gradient*, or dose–response curve, then we should look most carefully for such evidence. For instance, the fact that the death rate from cancer of the lung has been shown to rise linearly with the number of cigarettes smoked daily adds a very great deal to the simpler evidence that cigarette smokers have a higher death rate than non-smokers. That comparison would be weakened, though not necessarily destroyed, if it depended upon, say, a much heavier death rate in light smokers and a lower rate in heavier smokers. We should then need to envisage some much more complex relationship to satisfy the cause-and-effect hypothesis. The clear dose–response curve admits of a simple explanation and obviously puts the case in a clearer light.

The same would clearly be true of an alleged dust hazard in industry. The dustier the environment the greater the incidence of disease we would expect to see. Often the difficulty is to secure some satisfactory quantitative measure of the environment which will permit us to explore this dose response. But we should invariably seek it.

Biological plausibility

It will be helpful if the causation we suspect is *biologically plausible*, though this is a feature we cannot demand. What is biologically plausible depends upon the biological knowledge of the day. Thus there was no biological knowledge to support (or to refute) Pott's observation in the eighteenth century of the excess of cancer in chimney sweeps. It was lack of biological knowledge in the nineteenth century that led to a prize essayist writing on the value and the fallacy of statistics to conclude, amongst other 'absurd' associations, that it was ridiculous for the stranger who passed the night in the steerage of an emigrant ship to ascribe the typhus, which he there contracted, to the vermin with which bodies of the sick might be infected. And in the twentieth century there was no biological knowledge to support the evidence of the effects upon the foetus of rubella in the pregnant woman.

In other words, the association recorded may be one new to science or medicine and must not therefore be too readily dismissed as implausible or even impossible.

Coherence of the evidence

On the other hand the cause-and-effect interpretation of an association should not seriously conflict with the generally known facts of the natural history and biology of the disease. It should have *coherence*.

Thus it could be argued that the association of lung cancer with cigarette smoking was coherent with the temporal rise that had taken place in the two variables and with the sex difference in mortality.

Contributing to coherence is the histopathological evidence from the bronchial epithelium of smokers and the isolation from cigarette smoke of factors carcinogenic for the skin of laboratory animals. Nevertheless, while such laboratory evidence can enormously strengthen the hypothesis of causation and may even determine the actual causative agent, the lack of such evidence cannot nullify the epidemiological observations in man. Arsenic, it is known, can cause cancer of the skin in man but it has never been possible to demonstrate such an effect on any other animal. In a wider field John Snow's epidemiological observations on the conveyance of cholera by the water from the Broad Street pump would have been put almost beyond dispute if the existence of the cholera vibrio had been known and its isolation possible. Yet the fact that Robert Koch's work was to be awaited another thirty years did not really weaken the epidemiological case, though it made it more difficult to establish against the criticisms of the day.

The experiment

Occasionally, it is possible to appeal to *experimental*, or semi-experimental, evidence. For example, because of an observed association some preventive action is taken. Does it in fact prevent? The dust in the workshop is reduced, lubricating oils are changed, people stop smoking cigarettes. Is the frequency of the associated events affected? Here the strongest support for the causative hypothesis may be revealed.

We need to remember though that nothing can properly be called an experiment unless a control group is available for comparison. If to have such a group is unethical, or if it cannot be formed at random, we have a less than complete experiment and must make judgements with that in mind.

Reasoning by analogy

Finally, in some circumstances, it would be fair to judge by *analogy*. With the known effects of the drug thalidomide and the disease rubella we would be ready to accept slighter but similar evidence with another drug or another viral disease in pregnancy.

Clearly none of these viewpoints can bring indisputable evidence for or against a cause-and-effect hypothesis and equally none can be required as a *sine qua non*. What they can do, with greater or less strength, is to help us to answer the fundamental question—*is there any other way of explaining the set of facts before us? Is there any other answer which is more likely than cause and effect?*

Common sense and figures

This interpretation of statistical data turns, it should be seen, not so much on technical methods of analysis as on the application of common sense to figures and on elementary rules of logic. The common errors discussed in Chapters 24–7 are due not to an absence of knowledge of specialised statistical methods or of mathematical training, but usually to the tendency of workers to accept figures at their face value without considering closely the various factors influencing them—without asking themselves at every turn: What is at the back of these figures? What factors may be responsible for this value? In what possible ways could these differences have arisen? That is constantly the crux of the matter. Group *A* is compared with Group *B* and a difference in some characteristic is observed. It is known that Group *A* differed from Group *B* in one particular way—e.g. in treatment. It is therefore concluded too readily that the difference observed is the result of the treatment. To reject that conclusion in the absence of a full discussion of the data is *not* merely an example of armchair criticism or of the unbounded scepticism of the statistician. Where, as in all statistical work, our results may be due to more than one influence, there can be no excuse for ignoring the fact. And it has been said with truth that the more anxious we are to prove that a difference between groups is the result of some particular action that we have taken or observed, then the more exhaustive should be our search for an alternative and equally reasonable explanation of how that difference has arisen.

It is clearly necessary to avoid the reaction to statistics which leads an author to give only the flimsiest statement of his figures on the grounds that they are dull matters to be passed over as rapidly as possible. They may be dull—often the fault lies in the author rather than in his data—but if they are cogent to the thesis that is being argued they must inevitably be discussed fully by the author and considered carefully by the reader. If they are not cogent, then there is no case for producing them at all. In both clinical and preventive medicine, and in much laboratory work, we cannot escape from the conclusion that they are frequently cogent, that many of the problems we wish to solve *are* statistical and that there is no way of dealing with them except by the statistical method.

This, however, does not mean a blind or exaggerated faith in, or reliance upon, techniques. For example a regression analysis of our observations may be unnecessary. A simple scatter diagram may tell us all we want to know. No doubt we shall have to learn by experience when and when not to turn to the more elaborate procedure. But we should at the very least constantly cultivate the habit of thinking about it.

Similarly we should remember that the fact that some drug has value in the hands of half-a-dozen independent investigators can be more illuminating than the same result in the hands of a single worker—and this *whatever* his statistical test of significance may indicate. At the present time, indeed, far too much emphasis is placed upon the use of statistical tests without a full and thoughtful consideration of the actual results.

In the last analysis it is always the results that matter. Are the observations well and fairly made? Are they good or bad? If they are bad no tests of significance can compensate—they merely provide a spurious air of respectability to meaningless results. If they are good the application of a test of significance may be quite unimportant. The difference observed may be so great, or so consistent, that it would be absurd to attribute it to chance. It may be so small that it is a matter of indifference whether it is technically significant or not.

In between these two extremes the calculation has, of course, the great advantage of putting us on our guard or of encouraging us in an interpretation. And it has the advantage that, with it, all workers will be able to make the same deductions. They are unlikely to do so if they judge merely 'by eye'—at least without long training and experience in the use of such tests. This, indeed, is part of the value of tests, that they give the worker a 'feeling' for figures and their meaning.

Finally, one of the cardinal sins in the presentation of results is to give the results of tests of significance and to omit the data—means, proportions, rates, whatever they may be. Similarly one sometimes sees regression lines given, with no indication of the data that led to those lines, and the reader has no way of judging their validity. To give an example (based upon reality) it is reported that for purposes of prophylaxis a drug and a placebo were randomly allocated to a large number of individuals. The result of the trial shows, it is claimed, an incidence of subsequent illness in the treated group 'significantly less at the 0.02 level' than the incidence in the placebo group. *But in the absence of the two rates of incidence this statement is entirely useless and may be grossly misleading to the unwary.*

For instance the attack rate of subsequent illnesses may have been 18 per cent in the group given the drug and 20 per cent in the group given the placebo. Given large enough groups the difference will be 'significant'. But it is clearly of little practical importance. On the other hand the attack rate may have been 2 per cent in the group given the drug and 20 per cent in the group given the placebo. It is not only significant but clearly of great practical importance.

This is what matters. This is what we want to know and this is what we should be told. For it is upon this difference that we shall base our future actions whether it is worth while exhibiting the drug or not. The test of significance is of quite secondary importance *after* we have studied (1) the validity of the trial and the data and (2) the results the data show.

Technical skills, like fire, can be an admirable servant and a dangerous master.

Appendix A

Hand calculation of mean and median

Calculations from grouped figures: the mean

With a large number of observations it would be very laborious to calculate the mean by summing all the individual values. With no serious loss of accuracy this can be avoided by working from the frequency distribution. So long as the classes into which the observations have been grouped are not too wide we can presume that the observations in that group are located at its centre. Taking the distribution in Table A1, we presume that the 96 children whose height lies between 127 and 129 cm were all 128 cm, that the 120 children whose height lies between 129 and 131 cm were all 130 cm. With small class intervals and a distribution that is not grossly asymmetrical that will approximate quite closely to the truth. Thus to reach the mean stature we estimate the sum of all the 565 statures to be 74 742 cm, and the mean stature is, therefore, this sum divided by 565, or 132.29 cm.

With a modern pocket calculator the arithmetic involved presents no problem. For those who lack this aid the work can be further eased by a simple device. Suppose 7 children were measured and their statures were found to be, in centimetres, 116, 119, 124, 127, 132, 136 and 145. The mean stature, summing the values, is 899/7 = 128.4 cm. But instead of summing

Table A1 Calculation of the mean, using true units of measurement.

Height of children (cm)	*Number of children*	*Mid-point of group*	*Sum of statures of the children measured*
127–	96	128	96 × 128 = 12 288
129–	120	130	120 × 130 = 15 600
131–	145	132	145 × 132 = 19 140
133–	83	134	83 × 134 = 11 122
135–	71	136	71 × 136 = 9 656
137–	32	138	32 × 138 = 4 416
139–141	18	140	18 × 140 = 2 520
Total	565		74 742

these values we might proceed, if it saved time, by seeing how far each of these children differed from, say 125 cm. Originally, indeed, we might merely have measured these differences from a mark 125 cm high, instead of taking the statures from ground level, so that we would finally have the average level of stature above or below 125 cm instead of above 0, or ground level. These differences from 125 cm are −9, −6, −1, +2, +7, +11, +20, and their sum, taking sign into account, is 24; their mean is then 24/7, or 3.4. This is the mean difference of the children from 125, and their mean stature from ground level will, therefore, be 125 + 3.4, or 128.4 cm as before.

The same process can be applied to the frequency distribution. A central group can be taken as base line and its stature called 0 in place of its real value. The groups below it become −1, −2, −3, etc., and those above become +1, +2, +3, etc. Using these smaller multipliers, we can more simply find the mean stature in these units and then convert the answer into the original real units. The figures previously used then become as set out in Table A2.

The sum of the statures measured from 0 in working units is 393 − 312 = 81; their mean is 81/565, or 0.143. The base line 0 was placed, it will be noted, against the group 131–3 cm, or, in other words, against the mid-point 132. We have found, then, that in *working units* the average difference of the children's stature from this central point of 132 is +0.143. Their real mean stature from ground level must therefore be 132 plus *twice* 0.143 = 132.29, the same value as was found previously from using the real values directly. The multiplier 2 is arrived at thus: the mean is found to be 0.143 of a unit above the 0 value when the groups differ in their distances from one another's centre by unity, e.g. from 0 to 1 to 2. But in the real distribution their distance from one another's centres is 2 cm, e.g. from 132 to 134. Therefore the mean in real units must be 2 times 0.143 above 132 cm.

While the method has been demonstrated on a particular distribution it is, of course, perfectly general. A further example is shown in Table A3.

The mean age of the patients in working units is −4015/7292, or

Table A2 Calculation of the mean, using arbitrary units.

Height of children (cm) (real units)	*Number of children*	*Height (working units)*	*Sum of statures (working units)*	
127–	96	−2	96 × −2 = −192	
129–	120	−1	120 × −1 = −120	
131–	145	0	145 × 0 = 0	−312
133–	83	+1	83 × 1 = + 83	
135–	71	+2	71 × 2 = +142	
137–	32	+3	32 × 3 = + 96	
139–141	18	+4	18 × 4 = + 72	+393
Total	565			+ 81

Table A3 Calculation of the mean, using arbitrary units.

Age of patient at onset of disease (years)	*Age working units*	*Number of patients*	*Sum of ages (working units)*	
15–	−5	14	− 70	
20–	−4	163	− 652	
25–	−3	861	−2583	
30–	−2	1460	−2920	
35–	−1	1466	−1466	
40–	0	1269	0	−7691
45–	1	953	+ 953	
50–	2	754	+1508	
55–	3	221	+ 663	
60–	4	103	+ 412	
65–69	5	28	+ 140	+3676
Total		7292		−4015

−0.5506, and we have to translate this value into the real units. It lies, it will be seen, 0.5506 of a working unit *below* the 0 which was placed against the group 40–5. The centre of this group is 42.5 and the real mean is, therefore, $42.5 - 5(0.5506) = 39.75$ years.

The formula is, then:

Mean in real units = *centre value* in real units of the group against which 0 has been placed, plus (the mean in working units × width of class adopted in the frequency distribution in real units).

Points especially to be remembered, on which beginners often go wrong, are these:

(1) The 0 value corresponds in real units to the *centre* of the group against which it is placed and *not* to the start of that group.
(2) In translating the mean in working units into the real mean the multiplier, or width of the groups in real units, must be brought into play.
(3) The groups in real units should be, throughout, of the same width; if a group contains no observations, nevertheless the appropriate number in the working units must be allotted to it. Otherwise the size in working units of observations farther away from the 0 is not correctly defined.
(4) Values smaller in real units than the group taken as 0 should always be taken as minus in working units and higher values always taken as plus. Otherwise confusion arises in passing finally from the working mean to the real mean.
(5) The exact value for the centre of the group against which the 0 has been placed must be carefully considered. For simplification it has been taken above as the mid-point of the apparent group, but

sometimes that will not be strictly correct. For instance, in taking the age at onset of disease the age of the patient might be recorded either as age last birthday or as age to nearest birthday. With age last birthday persons placed in the group 40–5 may be any age between 40 years and 45 years minus 1 day, and the centre point will be 42.5. But with age to nearest birthday a person aged 39½ is called 40 and one aged 44½ is called 45. The group 40–5 therefore runs from 39.5 to 44.5 and its centre point is 42 years.

The working can be checked by changing the position of the 0 on the working unit scale—this is a better check than a repetition of the previous arithmetic.

Calculation from grouped figures: the median

For calculation of the median from the frequency distribution we may return to the statures of children, repeated in Table A4. As pointed out in Chapter 7, the median stature will be the height of the 283rd child when the observations are listed in order. In practice, however, we do not trouble to list them in order when dealing with large numbers but make an *estimate* of the median from the grouped values in the frequency distribution. The value to be found is that of the point which divides the distribution into exactly two halves, i.e. with 282.5 observations below and 282.5 above the mid-point. In other words, the value needed lies at the mid-point 565/2 or 282.5. Adding the numbers from the start of the distribution we have $96+120=216$ and $216+145=361$. The mid-line at 282.5, therefore, falls in the group with 145 children whose height lies between 131 and 133 cm. By simple proportion it will lie at a point 66.5/145 of 2 cm beyond the 131 cm at which this group of children starts (since $282.5-216=66.5$). The median may, therefore, be estimated as 131 cm + (66.5/145 of 2 cm) = 131.9 cm.

The main point to recall in calculation is that the median must be computed from the *opening* point of the group in which it is located and *not* from the mid-point of that group. Again the actual starting point of the group may have to be closely considered.

Table A4 Calculation of the median.

Height of children (cm)	*Number of children*	*Cumulative number*
127–	96	96
129–	120	216
131–	145	361
133–	83	444
135–	71	515
137–	32	547
139–141	18	565
Total	565	

Appendix B

Hand calculation of variance and standard deviation

Twenty observations of systolic blood pressure are shown in Table B1. Their mean value is 128 mm. The variability of these observations can be measured by means of the standard deviation, calculated in Table B1, by (1) finding by how much each observation differs from the mean, (2)

Table B1 Calculation of deviations, sum of squares and standard deviation.

Twenty observations of systolic blood pressure (mm) x (1)	*Deviation of each observation from mean (mean = 128)* $(x-\bar{x})$ (2)	*Square of each deviation* $(x-\bar{x})^2$ (3)
98	−30	900
160	+32	1024
136	+ 8	64
128	0	0
130	+ 2	4
114	−14	196
123	− 5	25
134	+ 6	36
128	0	0
107	−21	441
123	− 5	25
125	− 3	9
129	+ 1	1
132	+ 4	16
154	+26	676
115	−13	169
126	− 2	4
132	+ 4	16
136	+ 8	64
130	+ 2	4
Total 2560	0	3674

Standard deviation = $\sqrt{(3674/19)}$ = 13.91 mm

squaring each of these differences, (3) adding up these squares, and dividing this total by the number of observations minus one, (4) taking the square root of this number. This method of calculation would have been much more laborious if the mean blood pressure had not been a whole number—e.g. if it had been 128.472—and if each of the original observations had been taken to one decimal place (presuming such a degree of accuracy to be possible)—e.g. the first had been 98.4. The differences between the observations and their mean, and the squares of these values, would then have been less simple to calculate. But in such cases the necessary arithmetic can still be kept simple by a slight change of method.

The ungrouped series

If we call each individual observation x and the mean of all 20 $\bar{x}$ (pronounced x-bar), then by the method of Table B1, we must first find each separate deviation from the mean, $(x-\bar{x})$, as in Column (2), and then we must calculate the square of each of these deviations, $(x-\bar{x})^2$, as in Column (3). The required sum of all the squared deviations in Column (3)

Table B2 Calculation of sum of squares and standard deviation.

Twenty observations of systolic blood pressure (mm) x (1)	*Square of each observation* x^2 (2)
98	9 604
160	25 600
136	18 496
128	16 384
130	16 900
114	12 996
123	15 129
134	17 956
128	16 384
107	11 449
123	15 129
125	15 625
129	16 641
132	17 424
154	23 716
115	13 225
126	15 876
132	17 424
136	18 496
130	16 900
Total 2560	331 354

Standard deviation = $\sqrt{((331\,354-2560^2/20)/19)}$ = 13.91 mm can then be computed (3674) and may be described as $\text{Sum}(x-\bar{x})^2$. We can, however, reach this sum *without calculating any deviations at all* by means of the relationship

$$\text{Sum}(x-\bar{x})^2 = \text{Sum}(x^2) - (\text{Sum } x)^2/n$$

Thus in practice we square each observation, x, as it stands, as in Column (2) of Table B2, and we find the sum of these squares; thus $\text{Sum}(x^2) = 331\,354$. In calculating the mean we have already found the sum of the 20 observations themselves; thus, from Column (1), Sum $x = 2560$. Our required sum of squared deviations round the mean, $\text{Sum}(x-\bar{x})^2$, is therefore $331\,354 - (2560)^2/20 = 3674$. The standard deviation is then, as before $\sqrt{(3674/19)} = 13.91$ mm.

Thus to calculate the standard deviation in a short ungrouped series of figures we have five steps: (a) square the *individual observations* themselves and find the sum of these squares; (b) square the *sum* of the observations themselves and divide this by the total numbers of observations available; (c) subtract (b) from (a) to give the required sum of the squared deviations of the observations *around their own mean*; (d) divide the value by $n-1$ to reach the variance and (e) take the square root of the variance to reach the standard deviation.

It is worth reiterating here the warning given in Chapter 8 that the method using

$$\text{Sum}(x-\bar{x})^2 = \text{Sum}(x^2) - (\text{Sum } x)^2/n$$

should not be used in computer programs. The methods exhibited in this appendix are intended for hand calculation, when it is hoped that the person doing the calculations will be sufficiently awake as to notice (and start again using more decimal places) if most of the significant figures are lost when making the subtraction.

The grouped series

With a large number of observations this method of squaring each observation separately would be very laborious. A shorter method which will give very nearly the same result can be adopted. The observations must first be grouped in a frequency distribution. As an example we may take the distribution given in Table 8.1 of the ages at death from diseases of the Fallopian tube. This distribution is given again in Column (2) of Table B3.

The simplest way to calculate the required sum of the squared deviations of these observations from their mean is by use of the formula

$$\text{Sum}(x-\bar{x})^2 = \text{Sum}(x^2) - (\text{Sum } x)^2/n$$

Multiplying the central age of death, Column (3), by the numbers in each age group, Column (2), we reach the figures of Column (4). Their total gives $\text{Sum}(x) = 7670.0$. For the formula we need the square of this value divided by the number of observations, i.e. $(7670)^2/206 = 285\,577.18$.

Table B3 Calculation of standard deviation from grouped observations.

Age last birthday (years)	*Number of deaths in age-group*	*Central age*	*(2) × (3)*	*(3) × (4)*
	n	*x*	*nx*	nx^2
(1)	(2)	(3)	(4)	(5)
0–	1	2.5	2.5	6.25
5–	0	7.5	0.0	0.00
10–	1	12.5	12.5	156.25
15–	7	17.5	122.5	2 143.75
20–	12	22.5	270.0	6 075.00
25–	35	27.5	962.5	26 468.75
30–	42	32.5	1 365.0	44 362.50
35–	33	37.5	1 237.5	46 406.25
40–	24	42.5	1 020.0	43 350.00
45–	27	47.5	1 282.5	60 918.75
50–	10	52.5	525.0	27 562.50
55–	6	57.5	345.0	19 837.50
60–	5	62.5	312.5	19 531.25
65–	1	67.5	67.5	4 556.25
70–74	2	72.5	145.0	10 512.50
Total	206		7 670.0	311 887.50

Standard deviation = $\sqrt{((311\,887.5 - 7670^2/206)/205)}$ = 11.3 years

We now need $\text{Sum}(x)^2$ and to reach this we multiply Column (4) by Column (3) to give the figures of Column (5) with a total of 311 887.50. Inserting these two values in the formula we have:

$$311\,887.50 - 285\,577.18 = 26\,310.32$$

The variance is therefore 26 310.32/205 = 128.34 square years and the standard deviation is the square root of this value or $\sqrt{128.34}$ = 11.3 years. So we have finally a mean age at death of 7670/206 or 37.2 years, and a standard deviation around it of 11.3 years.

An arithmetically simpler method

For those possessing a calculator (with sufficient digits) the above method is not onerous. For those without such an aid a simpler method arithmetically can be adopted, the principle of which is merely an extension of that used in Appendix A for finding the mean, i.e. instead of working in the real, and cumbersome, units of measurement we translate them arbitrarily into smaller and more convenient units, and translate the results back again into the real units at the end.

Let us, for instance, in this case replace 32.5 by 0, 27.5 by −1, 22.5 by −2, and so on, 37.5 by +1, 42.5 by +2, and so on. (The original groups must have, it will be remembered, intervals of equal width; they were all 5-yearly in our example.) Now instead of having to multiply 27.5 by 35, for

Table B4 Calculation of standard deviation from grouped observations using working units.

Age last birthday (years) (1)	*Number of deaths in age-group* (2)	*Age in working units* (3)	*(2) × (3)* (4)	*(3) × (4)* (5)
0–	1	−6	− 6	36
5–	0	−5	0	0
10–	1	−4	− 4	16
15–	7	−3	− 21	63
20–	12	−2	− 24	48
25–	35	−1	− 35	35
30–	42	0	0	0
35–	33	+1	+ 33	33
40–	24	+2	+ 48	96
45–	27	+3	+ 81	243
50–	10	+4	+ 40	160
55–	6	+5	+ 30	150
60–	5	+6	+ 30	180
65–	1	+7	+ 7	49
70–74	2	+8	+ 16	128
Total	206		+195	1237

Standard deviation $= 5 \times \sqrt{((1237 - 195^2/206)/205)} = 11.3$ years

example, we have the simpler task of multiplying −1 by 35. These multiplications are made in Column (4) of Table B4. Their sum (taking the sign into account) is 195. The mean in working units is therefore 195/206 = 0.947.

The standard deviation can be found in these same small units, measuring, for simplicity, the deviations of the observations from the 0 value instead of from the mean. The squares of the deviations in these units are merely 1, 4, 9, 16 etc., and these have to be multiplied by the number of individuals with the particular deviation—e.g. 7×9 for the −3 group, 24×4 for the +2 group, and so forth. A still simpler process of reaching the same result is to multiply Column (4) by Column (3), i.e. instead of multiplying 7 by 9 we multiply (7×-3) by −3. This gives the figures of Column (5). The sum of these squared deviations is, then, 1237.

These deviations in working units have, however, been measured round the 0 value, whereas they ought to have been measured round the mean (in working units) of +0.947. The correction is again made by the formula $\text{Sum}(x - \bar{x})^2 = \text{Sum}(x^2) - (\text{Sum } x)^2/n$. Therefore from the values in Table B4 we can calculate $\text{Sum}(x - \bar{x})^2$ to be $1237 - (195)^2/206 = 1052.41$. The standard deviation in working units is, therefore, $\sqrt{(1052.41/205)} = 2.266$.

We have now to translate the mean, 0.947, and the standard deviation, 2.266, back into the real units. This is simply done. The mean in working units is 0.947—i.e. 0.947 working units above our 0. In real units our 0 is equivalent to 32.5, for that is the substitution we made (note, once more, the *centre* of the group against which we placed the 0, not its beginning). The real mean must therefore be $32.5 + 5\ (0.947)$ which equals 37.2 years, the same as the value we found by the long method using real units throughout. (The multiplier 5, it will be remembered, comes from the size of the interval of the original group.)

To reach the real standard deviation, all that has to be done is to multiply the standard deviation as found in working units by the original units of grouping—in this case by 5. For if this measure of the scatter of the observations is 2.266 when the range is only 14 units (from -6 to $+8$) it must be 5 times as much when the range is really 70 units (from 2.5 to 72.5). The real standard deviation is therefore $5 \times 2.266 = 11.3$ years. (It should be noted that if the original units are *smaller* than the working units then the standard deviation will be smaller in the real units, e.g. the multiplier will be 0.25 if that is the original group interval.)

Checking the arithmetic

As regards the final result for the standard deviation, as well as the mean, it is immaterial where the 0 is placed; the same answers in *real* units must be reached. From the point of view of the arithmetic it is usually best to place it centrally so that the multipliers may be kept small. For the sake of demonstration the calculations for Table B4 are repeated in Table B5,

Table B5 Repeat of Table B4 using different working units.

Age last birthday (years) (1)	*Number of deaths in age-group* (2)	*Age in working units* (3)	*(2) × (3)* (4)	*(3) × (4)* (5)
0–	1	−8	− 8	64
5–	0	−7	0	0
10–	1	−6	− 6	36
15–	7	−5	− 35	175
20–	12	−4	− 48	192
25–	35	−3	−105	315
30–	42	−2	− 84	168
35–	33	−1	− 33	33
40–	24	0	0	0
45–	27	+1	+ 27	27
50–	10	+2	+ 20	40
55–	6	+3	+ 18	54
60–	5	+4	+ 20	80
65–	1	+5	+ 5	25
70–74	2	+6	+ 12	72
Total	206		−217	1281

Standard deviation $= 5 \times \sqrt{((1281 - (-217)^2/206)/205)} = 11.3$ years taking another position for 0. This, in practice, is a good method of checking the arithmetic.

From the calculations in Table B5 we have

$$\text{mean in working units} = -217/206 = -1.053,$$
$$\therefore \text{ mean in real units} = 42.5 - 5\,(1.053) = 37.2 \text{ years}$$

(42.5 is the centre of the group against which the 0 was placed; note that the correction has now to be subtracted, for the sign of the mean in working units is negative.)

Sum of squared deviations in working units round the mean is $1281 - (217)^2/206 = 1052.41$ and, therefore, as before, the standard deviation in working units is $\sqrt{(1052.41/205)} = 2.266$ and in true units $2.266 \times 5 = 11.3$ years.

These values agree with those previously found.

Appendix C

Hand calculation of covariance

The covariance is needed in finding regression and correlation coefficients. It is defined as

$$\text{Sum}\{(x-\bar{x})(y-\bar{y})\}/(n-1)$$

The ungrouped series

For a small number of observations, the above formula can be used directly as in Table C1, giving pulse rate and height for each of 12 individuals. We find the sum of the values of [(observation of x minus mean of the observations of x) × (corresponding observation of y minus mean of the observations of y)]. For this purpose the deviation of each individual's pulse rate from the mean pulse rate of the twelve individuals is given in

Table C1 Calculation of a sum of products.

Individual number (1)	*Resting pulse-rate (beats per min)* (2)	*Height (cm)* (3)	$x-\bar{x}$ (4)	$y-\bar{y}$ (5)	*(4) × (5)* (6)	*(2) × (3)* (7)
1	62	170	−10	− 5	+50	10 540
2	74	165	+ 2	−10	−20	12 210
3	80	185	+ 8	+10	+80	14 800
4	59	178	−13	+ 3	−39	10 502
5	65	175	− 7	0	0	11 375
6	73	168	+ 1	− 7	− 7	12 264
7	78	175	+ 6	0	0	13 650
8	86	178	+14	+ 3	+42	15 308
9	64	183	− 8	+ 8	−64	11 712
10	68	180	− 4	+ 5	−20	12 240
11	75	173	+ 3	− 2	− 6	12 975
12	80	170	+ 8	− 5	−40	13 600
Total	864	2100	0	0	−24	151 176
Mean	72	175				

Column (4) and of each height from the mean height in Column (5). If there is any substantial (and direct) correlation between the two measurements, then a person with a pulse rate below the mean pulse rate ought to have a stature below the mean height, one with a pulse rate above the mean rate ought to have a stature above the mean height. (If the association is inverse, positive signs in one will be associated with negative signs in the other.) Inspection of the figures suggests very little correlation between the characteristics and a scatter diagram would show that more clearly. We need the product of the two deviations shown by each person. These are given in Column (6) and their sum is −24. The covariance is therefore $-24/11 = -2.18$ cm beats per minute.

In the calculations used in Table C1 the actual deviations of each x and y from their respective means were first found (Columns (4) and (5)). They were then multiplied together to give the required products (Column (6)). This method was used for demonstration. In practice (and particularly with a calculator) it is simpler to multiply directly the pulse rate and the stature of each person, as in Column (7), applying a correction at the end to the resulting sum of the values. The correction is given by the formula:

$$\mathrm{Sum}(x-\bar{x})(y-\bar{y}) = \mathrm{Sum}(xy) - \mathrm{Sum}(x)\mathrm{Sum}(y)/n$$

Thus in the example this gives $151\,176 - (864)(2100)/12 = -24$, or the same value as was previously reached by working with the deviations from the means.

With a larger series of observations, finding the individual squares and products becomes progressively more laborious and it is better to construct a grouped correlation table.

The grouped series

As an example, let us presume that we have data for each of a number of large towns, showing (a) a measure of the amount of overcrowding present in a given year, and (b) the infant mortality rate in the same year; we wish to see whether in towns with much overcrowding the infant mortality rate tends to be higher than in towns with less overcrowding. We must first construct a table which shows not only how many towns there were with different degrees of overcrowding but also their associated infant mortality rates.

Table C2 gives this information. The town with least overcrowding had only 1.5 per cent of its population living more than 2 persons to a room (this being used as the criterion of overcrowding); the percentage for the town with most overcrowding was 17.5. The lowest infant mortality rate was 13 deaths under 1 year per 1000 live births and the highest was 35. Reasonably narrow groups have been adopted to include those maximum and minimum values and each town is placed in the appropriate 'cell'—e.g. there were 5 towns in which the overcrowding index lay between 1.5 and 4.5 and in which the infant mortality rates were between 12 and 15, there were 2 in which the overcrowding index lay between 10.5 and 13.5 and in

Table C2 Example of a correlation table (overcrowding and infant mortality).

Infant mortality rate (per 1000)	*Percentage of population in private families living more than two persons per room*						
	1.5–	4.5–	7.5–	10.5–	13.5–	16.5–19.5	Total
12–	5	—	—	—	—	—	5
15–	9	1	—	—	—	—	10
18–	10	4	1	—	—	1	16
21–	4	7	5	2	—	—	18
24–	2	5	4	1	1	—	13
27–	—	2	2	2	—	1	7
30–	—	1	2	2	1	1	7
33–36	—	1	—	1	—	—	2
Total	30	21	14	8	2	3	78

which the infant mortality rates were between 27 and 30. If working by hand and a very large number of observations is involved it is best to make a separate card for each town, person, or whatever may have been measured, putting the observed measurements on the card always in the same order; the cards are first sorted into their proper groups for one characteristic (overcrowding), and then each of those packs of different (overcrowding) levels is sorted into groups for the other characteristic (infant mortality). The cards in each small pack then relate to a particular cell of the table.

Table C2 shows at once that there is some association between overcrowding and the infant mortality rate, for towns with the least overcrowding tend, on the average, to show relatively low mortality rates, while towns with much overcrowding tend to show high mortality rates. The table is, in fact, a form of scatter diagram.

We never need a covariance without needing the variance of at least one of the measures also. So we shall calculate the two variances as well while we are at it. We need (a) the squared deviations of the observations from their mean for the overcrowding index; (b) similar figures for the infant mortality rate; and (c) for each town the product of its two deviations from the mean—i.e. (overcrowding index − mean overcrowding index) × (infant mortality rate − mean infant mortality rate). In other words, we wish to see whether a town that is abnormal (far removed from the average) in its level of overcrowding is also abnormal in the level of its infant mortality rate. In calculating the means and the deviations from them we can entirely ignore the centre of the table; we have to work on the totals in the horizontal and vertical margins. For the former (infant mortality rate) we need Sum(x) and Sum(x^2) and for the latter (overcrowding index) we need Sum(y) and Sum(y^2). The calculations are shown in Tables C3 and C4.

$$\begin{aligned}\text{Sum}(x-\bar{x})^2 &= \text{Sum}(x^2) - (\text{Sum } x)^2/n \\ &= 42\,295.50 - (1770.0)^2/78 \\ &= 2130.1 \text{ (per cent)}^2\end{aligned}$$

Table C3 Infant mortality. Calculation of Sum(x) and Sum(x^2).

Mid-point of group x (1)	*Number in group* n (2)	*(1) × (2)* nx (3)	*(1) × (3)* nx^2 (4)
13.5	5	67.5	911.25
16.5	10	165.0	2 722.50
19.5	16	312.0	6 084.00
22.5	18	405.0	9 112.50
25.5	13	331.5	8 453.25
28.5	7	199.5	5 685.75
31.5	7	220.5	6 945.75
34.5	2	69.0	2 380.50
Total	78	1770.0	42 295.50

Table C4 Overcrowding. Calculation of Sum(y) and Sum(y^2).

Mid-point of group y (1)	*Number in group* n (2)	*(1) × (2)* ny (3)	*(1) × (3)* ny^2 (4)
3	30	90	270
6	21	126	756
9	14	126	1134
12	8	96	1152
15	2	30	450
18	3	54	972
Total	78	522	4734

Similarly Sum$(y-\bar{y})^2$ is calculated to be 1240.6 (per 1000)^2.

There is nothing new in this. The process was shown in full in Appendix B. What is new is the requirement of Sum(xy). Table C2 shows that there were 5 towns with an infant mortality rate between 12 and 15 and an overcrowding index between 1.5 and 4.5. Using the mid-points of these groups the $x \times y$ value is $13.5 \times 3.0 \times 5 = 202.5$. Similarly there were 4 towns with an infant mortality rate between 18 and 21 and an overcrowding index between 4.5 and 7.5. Their $x \times y$ value is therefore $19.5 \times 6.0 \times 4 = 468.0$.

All these values are set out in the cells of Table C5, and their total value, Sum(xy), is 12 843.0.

Applying the Sum(x)Sum(y)/n correction, we have

$$\begin{aligned} \text{Sum}(x-\bar{x})(y-\bar{y}) &= 12\,843.0 - (1770)(522)/78 \\ &= 12\,843.0 - 11\,845.4 \\ &= 997.6 \text{ (per cent)(per 1000)} \end{aligned}$$

and the covariance is 997.6/77 = 12.96 (per cent)(per 1000).

Table C5 Infant mortality and overcrowding. Calculation of Sum(xy)

Mid-points of groups	3	6	9	12	15	18
13.5	202.5	—	—	—	—	—
16.5	445.5	99.0	—	—	—	—
19.5	585.0	468.0	175.5	—	—	351.0
22.5	270.0	945.0	1012.5	540.0	—	—
25.5	153.0	765.0	918.0	306.0	382.5	—
28.5	—	342.0	513.0	684.0	—	513.0
31.5	—	189.0	567.0	756.0	472.5	567.0
34.5	—	207.0	—	414.0	—	—

Table C6 Calculation of sum of products using working units.

Infant mortality (working units)	*Percentage of population in private families living more than two persons per room (working units)*						
	−2	−1	0	+1	+2	+3	Total
−3	5(+30)	—	—	—	—	—	5
−2	9(+36)	1(+2)	—	—	—	—	10
−1	10(+20)	4(+4)	1(0)	—	—	1(−3)	16
0	4(0)	7(0)	5(0)	2(0)	—	—	18
+1	2(−4)	5(−5)	4(0)	1(+1)	1(+2)	—	13
+2	—	2(−4)	2(0)	2(+4)	—	1(+6)	7
+3	—	1(−3)	2(0)	2(+6)	1(+6)	1(+9)	7
+4	—	1(−4)	—	1(+4)	—	—	2
Total	30	21	14	8	2	3	78

If, however, we want a regression coefficient, or a correlation coefficient, instead of the covariance as such, there is no need to divide by $n-1$. The two regression coefficients are

$$997.6/2130.1 = 0.468 \text{ per 1000/per cent}$$

and $$997.6/1240.6 = 0.804 \text{ per cent/per 1000}$$

while the correlation coefficient is

$$997.6/\sqrt{(2130.1 \times 1240.6)} = 0.614$$

In practice, how much of these figures one writes down and how much one can sum continuously on a calculator is a matter of choice.

The arithmetic can be simplified by the use of working units. The method is shown in detail in Table C6. The sums of squares can be found just as was done in Appendix B giving, in squared working units, 137.85 for the overcrowding index and 236.68 for the infant mortality rate. The new requirement is for the sum of products—to find this we use the figures in parentheses in Table C6. These are the product in each case of the number in the cell, the working units measure for overcrowding and the working units measure for infant mortality. For example, in the first cell $5 \times (-2) \times (-3) = 30$.

Adding all these parenthesised figures we get 107, and applying the correction we get

$$107 - (5 \times (-60))/78 = 110.85$$

where the 5 and −60 figures are the sums of observations on each scale in working units. To get the covariance, we must multiply by 3 (because 1 working unit represented 3 real units on the overcrowding scale), and by 3 again (because 1 working unit represented 3 real units on the infant mortality scale)—*note that these two figures will not in general be identical, though they happen to be so here*—and divide by 77, to get

$$110.85 \times 3 \times 3/77 = 12.96 \text{ (per cent)(per 1000)}$$

as before.

Appendix D

Calculation of 2 × 2 table probabilities

Taking as an example a 2×2 table with a total of 19 observations, of which the 19 are divided into 4 and 15 in one categorisation and into 9 and 10 in the other, we first arrange the table so that the smaller of the first pair is the total of the first column and the smaller of the second pair is the total of the first row thus:

a	b	9
c	d	10
4	15	19

or

a	b	4
c	d	15
9	10	19

It does not matter which.

The value in cell a can then vary between 0 and 4, giving 5 possible tables in all, of which we wish to know the probability of each. The first table is

0	9	9
4	6	10
4	15	19

and its probability is

$$\frac{10! \quad 15!}{6! \quad 19!}$$

(The other factors, in the expression for the probability given in Chapter 15, cancel out where $a = 0$.)

Now $10! = 10 \times 9 \times 8 \times 7 \times 6!$
and $19! = 19 \times 18 \times 17 \times 16 \times 15!$

so the expression reduces to

$$\frac{10 \times 9 \times 8 \times 7}{19 \times 18 \times 17 \times 16} = 0.054179567$$

If working without a calculator, it is worth cancelling common factors before making this calculation (e.g. 9/18 and 8/16 can each be replaced by ½), but with a calculator it is hardly worth it. The best way of making the calculation is by alternately dividing and multiplying: starting with 10, divide by 19, multiply by 9, divide by 18, multiply by 8 and so on. The result should be taken to considerable accuracy as the rest of the calculations will depend on it, the remaining probabilities being found as:

a	b	c	d	Probability	
0	9	4	6		0.054179567
1	8	3	7	$0.054179567 \times \frac{9 \times 4}{1 \times 7}$	= 0.278637771
2	7	2	8	$0.278637771 \times \frac{8 \times 3}{2 \times 8}$	= 0.417956656
3	6	1	9	$0.417956656 \times \frac{7 \times 2}{3 \times 9}$	= 0.216718266
4	5	0	10	$0.216718266 \times \frac{6 \times 1}{4 \times 10}$	= 0.032507740
					Total = 1.000000000

each being found from the previous probability, multiplied by the previous b and the previous c, divided by the new a and the new d.

Four decimal places (as in Table 15.1) are usually enough in the figures as finally used. The extra places shown here are merely for the purposes of the calculation.

Appendix E

Use of random sampling numbers

Finding a starting place

It is often suggested that a table of random numbers should be entered at a random point, but how to do that is left unexplained. If you have a watch that can be put into stop-watch mode, the following method is suggested— start the stop-watch going and, after a minute or two without looking at it, stop it and note the hundredths-of-a-second figure: two digits anywhere from 00 to 99. This gives the column of the table to enter.

Do the same again to get another two-digit number. This time, if the number is 50 or greater, subtract 50 from it. This gives the row number of the table.

Example 1 In a controlled trial the aim is to give about half the patients treatment X and the other half treatment Y. To allocate the treatments at random let an *even* number denote treatment X and an *odd* number denote treatment Y. If the starting place is Column 32, Row 22, the first digit is 3, and continuing down the column we get 2, 6, 9, 6, etc. So the treatments are as shown in Table E1.

Table E1

Random number		*Treatment*	*Patient number*
3	odd	Y	1
2	even	X	2
6	even	X	3
9	odd	Y	4
6	even	X	5
0	even	X	6
0	even	X	7
3	odd	Y	8
2	even	X	9
0	even	X	10
7	odd	Y	11
6	even	X	12
etc.		etc.	etc.

Example 2 A simple expansion of the above method allows for three, or more, treatments. Thus numbers 1, 2, and 3 can be taken to denote treatment X, numbers 4, 5 and 6 treatment Y, and numbers 7, 8 and 9 treatment Z. Number 0 will be ignored. Starting perhaps at Column 06, Row 46, and continuing downwards to the end of Column 06 and then down Column 07 we have Table E2.

Table E2

Random number	*Treatment*	*Patient number*
1	X	1
9	Z	2
2	X	3
4	Y	4
5	Y	5
7	Z	6
7	Z	7
9	Z	8
7	Z	9
3	X	10
3	X	11
8	Z	12
etc.	etc.	etc.

Example 3 In the short run a disadvantage of the above methods may lie in too many instances of one treatment and too few of the other treatment

Table E3 Randomising in groups.

Groups of 2		*Groups of 6*	
0– 4	*TC*	00–04	*TTTCCC*
5– 9	*CT*	05–09	*TTCTCC*
		10–14	*TTCCTC*
		15–19	*TTCCCT*
Groups of 4		20–24	*TCTTCC*
		25–29	*TCTCTC*
00–15	*TTCC*	30–34	*TCTCCT*
16–31	*TCTC*	35–39	*TCCTTC*
32–47	*TCCT*	40–44	*TCCTCT*
48–63	*CTTC*	45–49	*TCCCTT*
64–79	*CTCT*	50–54	*CTTTCC*
80–95	*CCTT*	55–59	*CTTCTC*
96–99	*try again*	60–64	*CTTCCT*
		65–69	*CTCTTC*
		70–74	*CTCTCT*
		75–79	*CTCCTT*
		80–84	*CCTTTC*
		85–89	*CCTTCT*
		90–94	*CCTCTT*
		95–99	*CCCTTT*

occurring. Thus by chance the first 7 patients might all fall to treatment *X* and none to treatment *Y*. In Example 2 there are, in fact, in the first 12 patients, 6 on treatment *Z*, 4 on treatment *X*, and only 2 on treatment *Y*. To prevent this it may be wise to equalise the numbers of patients who will be on each treatment at short intervals.

If there are only two treatments, and it is desired to equalise in groups of 2, 4 or 6, Table E3 will help. Here the two treatments are called *T* and *C*, for treated and control, but if those descriptions are not relevant either treatment can be called *T* and the other *C*. Suppose we are equalising in groups of 4, and starting in Column 12, Row 40. We need two-digit random numbers, so we read both Column 12 and Column 13 to get Table E4.

If such equalisation is used it is better to keep not only the detailed allocation, but also the *fact* of equalisation, as part of the 'blindness'.

Table E4

Random number	*Treatments*	*Patients*
66	*CTCT*	1, 2, 3, 4
36	*TCCT*	5, 6, 7, 8
95	*CCTT*	9, 10, 11, 12
96	—	
14	*TTCC*	13, 14, 15, 16
11	*TTCC*	17, 18, 19, 20
45	*TCCT*	21, 22, 23, 24
30	*TCTC*	25, 26, 27, 28
24	*TCTC*	29, 30, 31, 32
13	*TTCC*	33, 34, 35, 36
30	*TCTC*	37, 38, 39, 40
73	*CTCT*	41, 42, 43, 44
etc.	etc.	etc.

Example 4 Table E3 is not adequate for all purposes. Suppose that we have 3 treatments, and have decided that in each 12 patients, 4 shall be on treatment *X*, 4 on *Y*, and 4 on *Z*. In principle it would be possible to prepare a table similar to Table E3 for this case, but it would have to have 34 650 rows to allow for every possible order, which is hardly practical.

There are a number of alternative ways of proceeding, some more efficient than others. A fairly quick and simple one is this: write down 4 *X*s, 4 *Y*s and 4 *Z*s and beside each a two-digit random number read in succession from the tables, starting say at Column 22, Row 20, to get Table E5. (Had any two-digit number been repeated, we should have ignored it and passed on to the next one.)

Now rearrange the random numbers in numerical order, taking the corresponding treatments with them, as in Table E6. It will be seen that a run of 4 *Y*s has occurred by chance. This must be expected to happen sometimes and it should not be interfered with; randomness is not, and is not intended to be, a smooth process in the short run.

Table E5

Treatment	*Random number*
X	14
X	65
X	64
X	59
Y	35
Y	50
Y	29
Y	42
Z	60
Z	25
Z	70
Z	74

Table E6

Treatment	*Random number*	*Patient number*
X	14	1
Z	25	2
Y	29	3
Y	35	4
Y	42	5
Y	50	6
X	59	7
Z	60	8
X	64	9
X	65	10
Z	70	11
Z	74	12

Example 5 It is desired to take a random sample of 600 of the records of 6780 hospital patients numbered in serial order from 1 to 6780. Starting perhaps at Column 56, Row 10, read the numbers in 4s and mark down those between 0001 and 6780, ignoring all higher numbers. Thus the records to be included in the sample are those of patients numbered 3147, 5184, 3792, 3677, 4893, etc. If the same number appears a second time it is ignored and the list continued until 600 separate numbers have been drawn. When the total population numbers less than 1000 the figures can be read in threes, e.g. starting at Row 00, Columns 00, 01 and 02 we get 136, 568, 812, etc.

Note While it is best to keep to the same method throughout the drawing of a sample, it is immaterial whether the numbers are read forwards or backwards, down the columns or up the columns. For instance, the first line can be read as 13, 61, 37, 95, etc. or proceeding backwards as 07, 33, 03, 60, etc. Thus from the 5000 digits set out many more can be generated if required.

Random sampling numbers—1

	00	01	02	03	04	05	06	07	08	09	10	11	12	13	14	15	16	17	18	19
00	1	3	6	1	3	7	9	5	7	3	9	6	0	6	3	0	3	3	7	0
01	5	6	8	8	3	6	6	7	8	3	3	5	4	5	7	3	5	8	7	6
02	8	1	2	5	2	5	1	7	5	0	4	9	7	9	7	8	9	4	5	8
03	1	9	3	8	3	7	5	9	0	3	0	9	7	4	4	2	6	9	4	0
04	4	1	6	4	1	2	6	7	8	2	5	6	0	6	0	1	5	4	3	2
05	1	5	6	2	0	9	4	3	9	8	5	8	8	5	6	6	1	4	8	5
06	0	1	6	5	2	6	9	3	1	0	3	9	9	5	1	9	3	5	8	7
07	0	8	9	8	8	0	5	8	9	6	1	4	3	6	3	9	1	7	6	6
08	4	3	9	6	5	0	1	0	6	0	9	5	1	3	1	3	3	5	4	9
09	6	7	7	6	4	7	8	7	6	9	9	7	7	7	3	2	8	2	7	1
10	9	1	2	2	6	4	4	2	9	5	6	4	7	0	4	9	6	3	1	7
11	2	5	2	6	1	3	2	5	6	2	6	9	0	9	2	7	8	0	2	9
12	4	2	1	8	7	9	3	2	6	7	4	5	1	6	3	7	5	6	2	8
13	1	0	7	9	9	8	8	7	4	6	3	2	5	0	8	0	5	5	5	8
14	7	2	2	1	1	4	5	2	8	2	7	1	1	8	2	1	8	4	9	3
15	8	2	3	4	9	3	4	1	7	8	6	3	5	2	6	6	9	3	7	3
16	7	7	7	1	2	6	8	2	2	5	7	0	6	2	2	1	8	0	6	2
17	2	0	0	6	7	6	3	6	1	8	6	7	9	7	2	8	0	5	9	8
18	6	4	0	2	9	2	1	1	8	3	1	6	0	3	3	5	4	3	2	1
19	7	2	9	3	3	5	6	3	7	0	7	5	9	1	5	2	8	7	9	4
20	1	2	0	8	8	8	3	7	1	9	8	3	6	4	0	6	1	2	6	2
21	1	9	4	7	3	5	4	3	8	8	0	5	1	5	8	2	2	5	9	3
22	8	0	0	8	8	4	8	9	6	7	3	3	0	3	7	4	0	0	7	5
23	0	0	5	6	4	9	4	2	1	7	4	4	7	8	7	2	7	0	7	7
24	4	3	4	4	9	6	3	2	8	2	2	9	2	9	4	8	0	3	5	5
25	3	5	2	4	8	0	3	8	3	9	0	3	2	3	3	2	2	1	1	6
26	2	7	8	0	6	7	0	1	1	2	7	2	5	8	0	3	4	7	4	9
27	4	7	7	2	3	7	9	4	8	6	8	6	6	0	8	9	7	4	6	0
28	0	1	6	7	3	4	7	1	9	2	9	8	6	8	1	0	1	1	6	3
29	2	8	4	9	3	4	1	3	3	0	2	3	7	4	5	3	2	3	1	5
30	1	9	5	3	5	3	5	3	2	5	4	1	6	9	7	0	8	7	3	1
31	3	0	4	6	4	7	9	8	9	2	6	8	6	5	9	8	7	0	7	5
32	1	1	8	4	0	8	7	6	4	4	2	4	2	7	6	5	8	3	9	9
33	1	5	1	6	2	5	9	8	7	9	7	1	5	4	1	9	6	3	3	4
34	3	0	4	0	3	9	5	2	1	4	8	9	6	5	3	9	4	6	8	2
35	4	7	9	9	8	7	4	9	1	9	4	8	7	8	4	2	4	3	2	8
36	5	3	2	6	0	5	6	1	3	1	4	1	2	7	5	3	5	1	2	3
37	7	7	1	6	5	6	6	7	9	9	4	8	6	6	2	1	4	5	8	1
38	4	4	2	9	5	6	5	5	9	8	1	1	9	3	5	7	3	7	6	0
39	0	6	7	9	0	5	2	1	5	6	2	6	5	1	7	5	2	7	4	5
40	0	5	1	3	3	9	7	8	1	7	3	8	6	6	2	3	0	3	9	6
41	2	5	8	6	2	3	0	7	9	8	5	8	3	6	3	2	4	6	3	4
42	1	5	9	7	4	2	3	0	0	6	0	0	9	5	3	8	9	8	7	8
43	5	9	4	9	7	1	1	3	8	4	2	1	9	6	4	1	9	1	9	4
44	1	2	5	9	7	8	4	4	7	8	5	8	1	4	4	6	8	4	8	1
45	9	2	4	5	9	0	7	5	0	7	1	4	1	1	2	5	1	4	1	4
46	7	1	1	1	2	0	1	0	5	1	7	6	4	5	1	5	3	1	6	3
47	6	4	1	2	1	2	9	2	7	9	1	0	3	0	5	7	3	0	6	3
48	8	6	1	2	9	7	2	8	0	0	3	5	2	4	0	5	4	0	0	4
49	5	4	9	4	8	2	4	1	2	6	3	5	1	3	7	3	9	5	3	3

Random sampling numbers—2

	20	21	22	23	24	25	26	27	28	29	30	31	32	33	34	35	36	37	38	39
00	6	2	6	5	2	9	0	3	5	4	2	0	6	0	4	2	8	4	0	4
01	5	6	0	7	9	0	1	6	3	4	4	2	6	0	9	3	3	6	7	9
02	7	5	0	1	3	2	5	2	2	8	4	7	9	1	8	2	2	8	8	8
03	3	3	2	9	5	6	7	2	5	6	1	1	2	9	1	1	5	5	5	0
04	5	8	0	1	5	1	8	4	2	8	4	8	5	3	0	0	4	2	0	4
05	4	5	0	4	5	5	2	2	4	9	4	8	6	0	8	3	0	8	8	1
06	6	5	0	7	0	6	9	9	0	4	2	4	3	0	4	1	4	5	8	5
07	8	4	2	0	6	0	9	6	4	1	2	0	0	5	1	8	2	4	2	6
08	8	1	5	6	7	4	5	3	1	7	4	9	1	9	2	2	4	2	1	2
09	7	3	3	0	8	1	4	2	0	6	5	9	4	4	9	5	4	5	1	0
10	1	8	9	2	8	8	1	1	1	6	9	6	4	5	8	0	6	1	0	3
11	6	0	7	6	5	4	1	6	3	6	8	4	3	9	8	5	1	3	6	2
12	7	3	1	1	7	1	4	5	0	2	2	6	5	4	6	6	4	4	6	4
13	4	1	3	4	7	5	9	1	0	6	6	2	3	1	2	4	8	2	5	0
14	0	8	1	9	2	5	1	3	9	2	2	8	3	2	5	5	9	0	8	8
15	7	4	4	6	8	6	3	4	6	1	8	8	1	9	8	1	7	4	2	5
16	6	8	0	4	7	4	9	2	5	3	5	2	7	6	8	0	7	9	9	8
17	0	8	0	4	5	7	7	6	0	5	4	4	8	8	0	7	0	4	4	6
18	6	1	1	8	4	0	4	4	1	2	7	4	8	2	2	3	0	1	0	1
19	7	5	1	8	1	2	1	3	7	2	2	6	2	3	6	9	1	0	6	3
20	0	3	1	4	5	2	5	6	1	7	0	2	1	7	1	5	4	2	1	7
21	7	4	6	5	9	0	6	0	5	5	0	4	0	8	4	3	3	5	3	6
22	4	8	6	4	4	8	4	3	3	6	1	1	3	4	1	9	5	7	7	4
23	9	6	5	9	6	8	0	9	8	0	9	9	2	4	0	9	3	3	1	5
24	1	9	3	5	5	5	1	4	0	1	0	0	6	5	9	2	4	0	9	1
25	0	5	5	0	0	6	8	0	4	2	7	3	9	2	6	3	6	5	6	6
26	4	8	2	9	3	6	4	6	8	2	2	5	6	8	0	3	8	0	1	3
27	5	2	4	2	4	2	4	3	9	8	4	1	0	9	5	3	6	0	2	9
28	7	0	6	0	8	6	2	7	3	7	2	5	0	3	2	8	3	9	9	1
29	9	0	2	5	1	3	8	0	3	8	3	2	3	3	8	3	4	7	9	8
30	1	1	7	0	4	6	3	7	6	8	0	0	2	4	3	4	2	8	4	2
31	4	7	7	4	7	8	5	3	0	6	9	9	0	8	2	4	9	2	3	3
32	8	1	8	8	4	2	5	7	6	0	0	3	7	9	0	0	5	8	2	3
33	7	1	8	3	7	9	6	6	6	6	9	0	6	7	9	2	8	8	5	4
34	9	4	1	5	1	2	0	9	6	0	5	4	6	8	3	8	3	4	2	9
35	4	1	5	1	0	8	3	9	5	0	8	4	8	2	3	2	9	5	3	6
36	8	6	8	3	0	8	9	6	5	6	5	8	0	5	8	4	1	9	2	4
37	3	6	9	5	3	8	4	2	9	2	8	6	8	7	8	3	2	4	0	3
38	6	6	5	7	3	5	6	2	9	0	0	3	1	0	1	1	1	3	4	0
39	2	7	1	2	3	6	1	9	0	8	1	2	8	3	8	4	4	9	8	3
40	3	4	0	1	7	0	9	9	8	1	4	1	6	6	4	1	4	6	4	5
41	4	2	7	1	1	9	8	9	4	7	0	4	0	4	1	6	2	0	4	5
42	8	5	6	5	9	0	0	7	5	4	6	8	1	2	3	9	8	8	2	9
43	6	1	8	3	6	3	5	7	1	5	3	7	3	8	6	2	7	3	2	1
44	9	1	3	6	5	4	9	1	2	3	1	0	7	6	7	9	7	5	4	1
45	1	9	0	1	8	1	8	2	7	9	5	8	2	8	8	9	1	7	7	2
46	7	2	5	4	8	6	8	5	8	8	4	1	0	0	8	7	5	5	5	3
47	5	1	8	9	0	5	9	5	0	3	1	3	9	5	7	8	0	7	7	8
48	7	6	2	6	0	1	6	0	0	0	5	6	7	9	1	9	9	5	6	8
49	2	9	7	2	1	7	9	5	5	9	0	7	2	1	8	2	8	7	7	5

Random sampling numbers—3

	40	41	42	43	44	45	46	47	48	49	50	51	52	53	54	55	56	57	58	59
00	5	1	4	9	4	6	4	7	5	8	9	6	4	1	8	6	4	6	2	9
01	1	6	9	2	6	4	0	5	0	7	7	1	6	8	7	2	4	5	3	3
02	0	6	6	1	4	3	5	0	7	5	3	1	9	6	4	7	2	5	9	7
03	4	5	4	5	2	7	6	0	6	7	1	7	4	9	1	3	6	7	9	8
04	9	3	3	0	3	1	6	7	7	8	3	1	5	2	1	9	2	4	4	0
05	9	0	5	4	1	8	3	2	7	6	3	6	4	0	1	4	9	9	1	8
06	9	9	0	3	8	1	6	4	6	5	7	1	7	1	7	9	8	4	5	5
07	5	0	6	2	7	1	3	8	2	6	1	1	0	8	6	4	6	4	8	5
08	0	0	1	1	3	8	5	2	3	0	4	9	0	7	4	5	5	1	7	5
09	9	6	4	5	2	3	5	3	1	4	6	3	9	0	0	1	3	7	0	5
10	2	3	8	2	5	8	1	7	7	8	2	1	2	2	2	1	9	2	7	8
11	1	1	4	5	4	3	7	8	3	5	0	5	2	1	2	8	3	1	4	7
12	4	1	2	0	4	5	2	4	6	8	5	1	1	6	6	5	5	1	8	4
13	2	3	1	4	9	6	7	2	5	6	4	9	2	7	7	5	9	8	2	1
14	9	4	2	2	5	7	9	7	0	6	2	8	0	7	1	8	3	7	9	2
15	9	8	9	4	1	8	3	4	8	0	7	8	7	3	5	8	3	6	7	7
16	5	0	5	4	3	5	4	4	9	8	3	1	1	2	2	4	4	8	9	3
17	0	0	9	1	8	6	9	4	8	8	2	2	0	6	0	1	2	5	5	3
18	4	2	5	6	8	8	3	8	3	0	8	2	4	9	2	9	5	1	3	2
19	0	3	2	0	1	5	7	6	3	8	4	3	4	5	3	7	8	1	7	9
20	3	4	6	3	2	9	7	5	8	5	1	8	8	6	1	5	0	9	3	2
21	5	8	1	9	6	9	3	9	1	9	7	0	9	0	8	3	9	8	3	2
22	0	9	0	5	6	0	9	1	7	2	5	0	5	2	3	6	5	8	6	9
23	8	5	9	1	0	0	8	2	9	5	5	1	9	0	4	0	3	4	7	7
24	2	6	3	2	6	5	5	6	7	6	8	8	6	5	0	5	0	0	7	1
25	4	3	7	3	8	8	6	5	4	8	7	2	8	2	1	1	0	4	1	3
26	5	7	4	5	4	3	1	4	9	1	6	4	9	8	6	5	9	1	9	2
27	7	6	9	0	6	8	8	3	3	8	5	4	3	8	5	7	4	3	8	1
28	8	7	1	4	6	1	3	8	4	7	3	3	6	4	3	4	7	6	7	1
29	9	6	1	1	6	6	1	6	0	9	6	1	9	1	0	8	4	3	3	6
30	6	2	8	8	3	3	1	4	2	3	6	2	4	9	6	2	4	5	6	0
31	3	1	2	2	6	4	8	5	7	5	3	9	7	7	9	1	3	4	9	8
32	7	8	0	6	8	4	2	4	0	2	6	7	5	1	4	9	8	0	0	3
33	7	7	7	7	9	4	5	5	2	1	6	7	0	2	7	9	0	4	5	7
34	5	3	2	0	8	8	3	3	2	2	7	0	6	8	3	0	1	0	2	3
35	8	9	0	4	8	9	1	2	8	7	9	9	5	7	4	6	2	6	8	3
36	8	6	1	7	9	4	4	5	5	2	8	9	1	7	3	9	6	7	3	2
37	0	0	4	9	2	0	6	8	0	2	3	6	2	3	7	4	9	0	9	9
38	7	5	3	5	1	1	3	0	5	7	8	3	1	4	5	2	4	3	1	7
39	0	5	5	8	8	8	5	8	8	9	5	1	8	7	6	2	3	7	0	8
40	7	8	7	7	4	3	1	7	7	5	8	0	4	0	8	3	0	2	7	0
41	7	0	8	3	9	2	6	6	3	6	0	6	5	1	6	4	9	7	2	5
42	8	8	3	3	4	9	8	1	6	1	0	2	4	6	2	2	5	0	5	3
43	2	8	2	3	5	9	8	9	9	9	1	8	9	2	1	5	2	9	8	6
44	0	2	3	0	7	6	7	6	8	2	3	2	8	3	7	3	9	6	5	0
45	7	9	0	0	8	5	4	6	1	2	3	7	9	7	4	0	2	6	0	3
46	5	8	0	7	0	5	6	8	6	8	9	5	1	0	2	0	7	3	7	0
47	4	5	9	3	1	8	1	4	4	0	9	4	9	7	5	9	6	4	8	2
48	9	6	8	3	2	2	2	3	5	7	2	0	7	7	8	4	1	4	2	9
49	7	5	2	5	7	4	1	1	2	2	9	6	0	4	9	6	1	2	6	3

Random sampling numbers—4

	60	61	62	63	64	65	66	67	68	69	70	71	72	73	74	75	76	77	78	79
00	4	6	5	1	5	3	3	4	9	8	8	1	0	0	3	6	1	5	4	4
01	4	7	9	1	5	5	0	1	8	8	8	2	3	0	1	7	0	1	0	4
02	8	8	2	7	4	5	3	3	0	7	0	9	9	9	4	9	0	6	6	2
03	5	6	9	1	9	8	7	2	8	8	2	3	2	9	8	9	5	8	5	0
04	1	7	5	8	3	7	1	1	1	9	0	9	5	7	8	8	1	0	2	0
05	3	4	9	2	1	1	4	2	1	9	2	7	6	4	1	2	5	8	2	7
06	9	1	9	0	2	4	8	5	8	1	6	3	5	6	6	7	1	8	3	0
07	0	9	7	0	6	4	4	4	4	8	5	8	7	2	3	2	0	2	5	0
08	9	8	2	5	4	3	7	8	8	3	7	9	0	7	7	1	9	2	8	8
09	0	7	2	3	3	6	2	0	6	4	6	9	1	4	2	7	7	4	9	5
10	7	0	9	3	2	5	2	9	9	6	5	4	3	2	3	9	3	7	4	9
11	8	6	4	0	4	5	0	9	4	9	8	1	9	1	4	5	6	4	6	0
12	9	2	0	7	0	0	0	2	1	3	0	0	6	2	0	7	8	8	4	9
13	2	6	1	0	9	2	2	0	4	4	9	9	7	6	9	0	8	9	6	9
14	5	4	9	9	4	9	8	6	3	6	4	3	3	0	9	5	3	5	2	2
15	7	7	0	1	3	7	2	8	6	6	2	7	5	6	4	0	8	0	0	1
16	3	9	8	0	0	9	7	1	3	0	0	2	6	8	1	2	2	5	9	2
17	7	4	7	3	9	3	2	8	3	8	2	3	2	1	2	8	5	0	7	0
18	1	7	4	1	3	3	9	9	7	9	0	8	9	6	4	7	9	6	5	2
19	7	4	3	3	0	6	9	5	6	9	4	8	9	3	8	2	1	1	9	4
20	4	0	7	5	0	0	7	1	8	5	9	2	3	7	1	7	4	8	3	5
21	7	9	6	9	7	1	1	1	1	3	9	6	6	2	9	2	6	9	1	3
22	7	5	1	5	0	6	2	1	6	7	9	9	9	0	1	4	9	7	8	7
23	0	0	9	8	5	8	3	2	5	7	5	0	7	8	8	0	8	1	3	3
24	2	7	2	1	3	6	5	4	6	8	6	1	9	0	7	1	4	1	9	4
25	3	0	6	2	5	4	8	3	4	5	5	0	8	3	8	7	7	9	5	2
26	5	1	3	5	7	9	4	0	4	5	7	1	0	3	4	4	6	9	8	0
27	4	1	4	3	2	8	7	9	7	9	6	0	8	6	5	8	1	7	1	9
28	5	8	3	6	6	3	0	5	2	1	9	7	4	5	8	2	7	7	7	3
29	0	0	3	3	5	4	2	3	9	5	7	8	7	4	5	8	2	7	4	7
30	2	5	9	9	4	2	3	2	8	2	1	8	2	7	9	4	2	5	8	5
31	6	6	9	9	8	9	4	7	2	9	5	0	0	7	2	4	4	4	1	1
32	1	9	7	7	1	5	8	4	1	8	2	4	3	6	6	5	2	0	1	1
33	8	2	1	6	3	6	5	7	9	7	9	0	5	0	4	9	2	0	4	1
34	5	5	4	0	4	6	2	9	9	8	0	5	2	3	6	3	9	2	7	5
35	2	8	1	8	7	0	9	7	8	2	3	3	4	4	3	5	8	3	7	2
36	1	3	8	4	5	4	3	1	1	3	1	0	5	9	0	9	8	1	0	7
37	9	9	8	7	7	4	8	6	1	1	3	7	0	4	3	5	6	0	9	0
38	7	7	9	4	3	6	6	0	4	5	4	1	4	3	0	9	0	2	1	2
39	2	3	3	7	2	7	2	9	7	9	6	1	6	2	6	3	2	0	6	6
40	6	7	1	8	1	6	4	4	3	4	0	3	2	4	7	8	7	1	0	1
41	0	0	5	7	3	3	3	2	8	6	5	3	4	6	2	7	6	4	2	8
42	8	5	6	6	3	4	7	4	1	0	5	2	9	9	2	7	4	9	4	0
43	7	5	8	3	2	9	0	7	4	4	3	3	4	7	1	4	7	9	8	6
44	6	1	3	0	1	4	4	7	4	8	8	5	9	1	1	4	8	8	7	0
45	0	1	9	3	8	9	5	8	9	1	6	5	8	3	7	6	1	0	8	1
46	5	6	9	5	5	3	6	2	4	1	8	8	2	1	3	3	0	0	2	8
47	1	6	1	7	6	7	2	6	2	7	4	2	6	0	9	7	8	8	5	3
48	6	6	8	3	1	4	9	6	9	7	9	5	7	1	8	1	4	0	4	2
49	7	3	5	8	3	2	1	8	3	3	1	3	4	0	1	0	3	7	7	3

Random sampling numbers—5

	80	81	82	83	84	85	86	87	88	89	90	91	92	93	94	95	96	97	98	99
00	7	8	6	0	1	8	8	0	4	4	9	9	1	1	5	6	8	9	6	6
01	5	1	9	9	2	8	5	7	0	9	6	7	6	9	9	2	0	7	5	7
02	2	7	7	3	9	4	6	7	0	1	0	1	2	4	0	9	8	4	6	3
03	2	9	0	6	2	2	9	7	7	9	3	1	3	5	9	0	0	4	3	2
04	5	0	0	3	0	2	1	4	8	6	5	7	6	5	4	6	9	8	0	1
05	7	0	1	6	4	6	5	8	9	5	3	1	3	9	1	8	4	6	9	7
06	3	3	0	8	9	9	6	4	4	6	0	1	0	9	7	3	8	4	3	3
07	7	1	9	1	7	5	0	1	9	6	4	2	1	2	7	6	0	8	6	5
08	4	3	0	6	6	4	8	8	8	3	8	2	1	1	8	5	2	7	6	2
09	3	8	1	1	6	1	9	4	8	8	3	0	8	3	4	2	1	4	0	2
10	9	8	7	8	8	4	9	4	7	3	7	9	9	5	8	2	3	7	4	1
11	3	2	3	6	6	8	6	4	0	9	8	5	3	3	2	9	9	7	1	1
12	8	7	5	7	3	0	9	8	6	1	4	6	7	5	6	0	1	1	6	4
13	1	5	6	2	8	1	4	7	0	6	9	1	0	5	6	2	2	9	2	7
14	7	8	5	5	9	9	0	4	0	0	7	4	5	4	4	6	5	0	1	2
15	4	6	5	4	0	7	1	0	0	6	5	9	1	5	6	0	9	6	9	3
16	9	7	8	8	0	3	1	9	8	2	8	7	2	4	3	2	7	4	9	1
17	5	0	8	6	5	7	9	0	0	4	5	9	6	3	6	4	4	1	5	9
18	4	4	4	0	7	6	8	1	5	9	6	4	2	6	1	0	7	7	9	5
19	7	0	0	4	3	7	5	8	2	1	9	1	9	7	6	7	2	3	8	7
20	2	6	1	2	3	7	4	5	1	0	8	7	3	8	4	1	0	4	7	2
21	0	9	1	1	9	4	8	2	2	6	7	5	7	6	8	0	4	6	6	0
22	8	6	8	8	0	0	6	8	6	4	6	4	4	2	8	8	5	9	3	7
23	5	5	6	0	0	7	1	5	0	6	0	2	3	1	4	8	6	8	4	3
24	4	6	8	1	1	9	3	6	8	9	5	9	5	8	0	5	7	5	3	1
25	6	9	4	2	6	4	1	6	2	0	5	7	2	1	8	5	6	9	6	5
26	8	9	1	1	6	1	6	0	0	8	1	5	6	7	6	9	6	3	7	4
27	3	8	3	2	6	9	6	4	2	5	2	2	4	2	5	9	6	9	5	7
28	5	9	8	2	6	7	1	3	2	4	6	9	6	2	8	2	7	4	8	9
29	7	1	2	7	6	7	2	1	6	1	0	4	2	8	4	3	4	3	8	4
30	3	2	2	5	2	5	6	0	5	6	1	0	1	2	3	5	9	3	1	1
31	6	9	0	3	5	9	3	4	5	2	4	2	6	5	7	4	4	5	7	4
32	4	4	0	8	8	4	8	0	7	0	8	7	8	7	7	9	9	7	5	8
33	2	2	2	1	4	1	3	3	2	7	4	0	4	0	1	6	0	2	7	3
34	3	3	0	5	8	2	5	5	1	4	9	8	4	9	8	4	0	8	0	7
35	6	6	1	6	7	7	1	7	4	8	3	5	1	4	7	7	6	3	4	9
36	4	8	0	5	3	3	8	0	2	6	2	0	6	5	0	4	5	1	2	0
37	3	1	2	7	3	4	3	5	9	3	0	3	3	3	7	7	8	0	0	4
38	7	3	7	6	5	4	8	1	6	4	6	6	2	0	7	3	4	4	8	8
39	2	8	5	3	1	4	8	5	6	5	0	6	8	4	2	6	0	7	8	2
40	6	9	9	3	9	5	9	0	9	5	0	7	3	2	1	7	3	2	0	2
41	9	9	2	8	5	0	6	3	4	1	3	6	7	9	0	0	7	3	7	2
42	4	7	9	3	3	5	4	5	0	3	8	6	9	6	3	8	3	0	5	1
43	6	3	0	6	7	0	3	9	0	6	7	7	1	8	7	0	3	0	7	6
44	6	0	4	7	1	2	8	2	8	0	2	1	2	9	1	7	5	2	4	2
45	2	5	2	3	2	5	8	6	6	8	1	3	8	3	5	7	6	3	8	5
46	3	2	9	4	3	3	9	1	8	9	8	6	4	9	8	0	2	8	4	6
47	9	6	0	5	2	7	1	0	4	6	4	9	3	4	5	4	2	3	4	8
48	3	5	4	9	7	9	6	1	7	0	3	4	2	4	6	9	9	4	9	1
49	2	0	3	9	3	2	7	4	5	7	6	6	6	9	0	0	5	8	5	9

Appendix F

Some definitions

Common statistical terms

Frequency distribution. A table constructed from a series of records of individuals (whatever the characteristic measured) to show the frequency with which there are present individuals with some defined characteristic or characteristics.

Arithmetic mean. The sum of the values recorded in a series of observations/the number of observations. $\bar{x} = \text{Sum}(x)/n$.

Weighted mean. The average of two or more means, or rates, each mean, or rate, being weighted by the number of observations upon which it is based. Thus if the mean of 150 observations is 10.5 and of another 200 observations is 8.3, the weighted average is $[(150 \times 10.5) + (200 \times 8.3)]/(150 + 200)$.

Geometric mean. The *n*th root of the product of the observations, where *n* is the number of observations, $\sqrt[n]{(\text{product}(x))}$.

Median. The centre value of a series of observations when the observations are ranked in order from the lowest value to the highest (with an even number of observations the mean of the two central observations is usually taken).

Range. The distance between the lowest and highest values observed.

Deviation. The difference between an observation and the arithmetic mean of the observations $(x - \bar{x})$.

Variance. The sum of the squares of deviations, divided by $(n - 1)$ where *n* is the number of observations.

$$\text{Sum}(x - \bar{x})^2/(n - 1)$$

Standard deviation. The square root of the variance.

$$\sqrt{(\text{Sum}(x - \bar{x})^2/(n - 1))}$$

Coefficient of variation. The standard deviation expressed as a percentage of the mean, or $(\text{SD/mean}) \times 100$.

Covariance. The sum of products of deviations (of two variables measured on the same sample), divided by $(n-1)$ where n is the number of observations.

$$\text{Sum}(x-\bar{x})(y-\bar{y})/(n-1)$$

Parameter. This term is much misused in the medical literature nowadays (perhaps even past the 'point of no return'). Its statistical meaning is the *true value in the population* of a statistical measure, as distinct from the estimate of it that can be calculated from a sample. Thus the population mean, for example, is a parameter of which the sample mean is an estimate.

Sampling

Simple random sampling. Drawing a sample from a population by a random method, e.g. by the use of random sampling numbers, which gives every individual in the population an *equal and independent* chance of appearing in the sample.

Stratified random sampling. Drawing a sample from a population which has first been divided into subgroups or strata. From each subgroup a sample is drawn by a random method which gives every individual *in the subgroup an equal and independent chance* of appearing in the sample. The chances can deliberately be made to vary from one subgroup to another but in that event each such chance must have a *known* value.

Two (or more) stage sampling. A process of sampling a population in a series of consecutive steps, e.g. a town may be divided into a number of areas and a number of those areas drawn by a random process; within these drawn areas the schools may be listed and a number of these schools drawn by a random process. The pupils within those schools are then the sample to be examined (or further stage sampling can be applied by the random selection of a sample of the pupils).

Systematic sampling. Drawing a sample from a population by a systematic procedure, e.g. by taking every *n*th patient entering a ward.

Standard errors

The *standard error* of any statistical value is an estimate on certain assumptions, of the standard deviation that that value would show in taking repeated samples from the same population. In other words, it shows how much variation might be expected to occur merely by chance in the various characteristics of samples drawn equally randomly from one and the same population. In practice, the values are calculated as follows.

Standard error of the mean. The standard deviation of the observations in the sample/the square root of the number of observations.

Standard error of a proportion. $\sqrt{(p \times q/n)}$ where p is the proportion in the sample having the characteristic which is being discussed (e.g. the proportion dying), q is the proportion not having the characteristic (e.g. the proportion surviving), and n is the number of observations. $p+q$ must equal 1.

Standard error of the difference between two means.

$$\sqrt{\left(\frac{(n_1-1)s_1^2+(n_2-1)s_2^2}{n_1+n_2-2}\left(\frac{1}{n_1}+\frac{1}{n_2}\right)\right)}$$

where s_1 and s_2 are the standard deviations of the observations in the two samples, and n_1 and n_2 are the numbers of observations.

Confidence intervals and significance tests

Confidence interval. The interval between a pair of values so constructed that there is a given degree of confidence that a particular parameter lies within the interval. For example, if a sample of n gives mean $\bar{x}$ and standard deviation s, the confidence interval for the population mean is

$$\bar{x} \pm t\,s/\sqrt{n}$$

where t is taken from a t-distribution with $n-1$ degrees of freedom, and corresponding to a tail area probability of 1 minus the desired confidence.

Null hypothesis. A hypothesis that is assumed to be true until such time as observations indicate that it is unlikely to be. For example, the hypothesis that a particular treatment has the *same* effect as a placebo.

Significance test. A test that rejects the null hypothesis if an observed difference (or a more extreme one) would have a small probability *if* the null hypothesis were true. A non-significant result should never be interpreted as 'The treatment has no effect' but only as 'The evidence has not demonstrated that the treatment has an effect'.

Correlation and regression

Scatter diagram. A graph upon which each individual measured is entered as a point or dot, the position of each point being determined by the values observed in the individual for the two characteristics measured (e.g. height and weight, each dot representing the associated height and weight).

Regression coefficient. The amount of change that will on the average take place in one characteristic when the other characteristic changes by a unit (e.g. as age increases by one year the average increase in weight at ages 4–14 years is, say, 2 kg). If one characteristic is termed x and the other y, then the coefficient showing how much y changes, on the average, for a unit change in x is equal to

$$\frac{\text{Sum}(x-\bar{x})(y-\bar{y})}{\text{Sum}(x-\bar{x})^2}$$

Regression equation. The equation to the straight line describing the association between two characteristics and enabling the value of one characteristic to be estimated when the value of the other is known. To estimate the value of y from a known value of x the required equation is

$$(y-\bar{y}) = \text{regression coefficient} \times (x-\bar{x})$$

Where the regression coefficient is as in the previous definition. A regression line to estimate from x from y may also be calculated and is a different line.

Correlation coefficient. A measure of the degree of association between two characteristics in a series of observations, calculated as

$$\frac{\text{Sum}(x-\bar{x})(y-\bar{y})}{\sqrt{(\text{Sum}(x-\bar{x})^2 \times \text{Sum}(y-\bar{y})^2)}}$$

Rates of birth and death

Birth rate. The live births occurring (or sometimes registered) in the calendar year/the estimated total population of the area at the middle of the year (usually expressed per 1000 of population).

General fertility rate. The live births occurring (or sometimes registered)/ the female population of childbearing ages (usually taken as 15–44 years). If the age of the mother is given on the birth certificate, fertility rates at particular ages may also be calculated.

Crude death rate or mortality rate. The total deaths occurring (or sometimes registered) in the calendar year/the estimated total population of the area at the middle of the year (usually expressed per 1000 of population).

Standardised death rate or mortality rate. The death rate at all ages calculated for comparative purposes in such a way that allowance is made for the age and sex composition of the population involved.

Infant mortality rate. The deaths under 1 year of age occurring (or sometimes registered) in the calendar year/total live births occurring (or registered) in the calendar year (usually expressed per 1000). The neonatal rate relates to the deaths that take place in the first 28 days of life.

Stillbirth rate. A stillbirth in England and Wales applies to any child 'which has issued forth from its mother after the twenty-eighth week of pregnancy and which did not at any time after being completely expelled from its mother breathe or show other signs of life'. The rate is number of stillbirths occurring, or registered/total births occurring or registered (usually expressed per 1000).

Perinatal mortality rate. Stillbirths plus deaths in the first week of life per 1000 total births occurring, or registered.

Maternal mortality rate. The deaths ascribed to puerperal causes/the total live births and stillbirths (usually expressed per 1000. When a record of stillbirths is not available the denominator may consist of live births only).

Proportional mortality rate. The ratio of deaths from a given cause to the total deaths (usually expressed as a percentage, e.g. in one country deaths from tuberculosis at all ages form 7 per cent of all deaths, in another country deaths of males from cancer of the lung may form 25 per cent of all deaths of males from cancer).

Morbidity rates

(Morbidity, illness and sickness are regarded as synonymous.)

The incidence rate. The number of illnesses (number of spells of illness or number of persons sick as applicable) *beginning* within a specified period of time related to the average number of persons exposed to risk during that period (or at its mid-point).

The period prevalence rate. The number of illnesses (number of spells of illness or number of persons sick as applicable) *existing at any time* within a specified period of time related to the average number of persons exposed to risk during that period (or at its mid-point).

The point prevalence rate. The number of illnesses (number of spells of illness or number of persons sick as applicable) *existing at a specified point of time* related to the number of persons exposed to risk at that point of time.

The average duration of sickness. The total number of days of illness in a defined period of time (a) divided by the average number of persons exposed to risk during that period (or at its mid-point) to give the *average duration of sickness per person*; (b) divided by the number of persons sick during the period of time to give the *average duration of sickness per sick person*; or (c) divided by the number of spells of sickness during the period of time to give the *average duration of sickness per illness*.

Life table symbols

(In each case x can take any value within the human span of life.)

q_x is the probability of dying between any two ages x and $x+1$; it corresponds to the ratio of deaths that take place between the two ages to the population starting that year of life (e.g. if there are 150 persons who reach their 30th birthday and 5 die before reaching their 31st, then the probability of dying is 5/150).

p_x is the probability of living; it corresponds to the ratio of those who survive a year of life to those starting that year of life (the probability of living in the example above is therefore 145/150). $p_x + q_x$ must equal 1.

l_x is the number in the life table living at each age, x, e.g. if of 10 000 males at birth 9717 survive at age 5; 9717 is the value of l_5.

d_x is the number of deaths that occur in the life table between any two adjacent ages, e.g. if the number of living at age 3 is 9730 and at 4 is 9723, d_3 is 7.

$\mathring{e}_x$ is the expectation of life at age x and is the *average* length of subsequent life lived by those who have reached x, e.g. the average length of life lived after 5 by the 9717 male survivors at age 5 might be a further 65.1 years.

Clinical trials and surveys

Retrospective inquiry. Working backwards from already affected persons to discover features in their *history* that may have led to the appearance of that effect.

Prospective inquiry. Looking forward from unaffected persons whose characteristics are defined to observe the *future* incidence of effect in relation to those characteristics.

Case-control inquiry. An inquiry in which each case of the relevant effect is paired-off with one or more people who do not suffer it, usually for a retrospective analysis.

Cohort inquiry. An inquiry in which a fixed group of people have various characteristics recorded for subsequent comparison with a particular effect, usually in a prospective manner.

Between-patients trial. A comparison between the effects of treatments in two or more groups of patients, each group being treated differently.

Cross-over or within patient trial. A comparison of the effects of two or more treatments in the same patient, the treatments being applied at different points of time in the course of the illness.

Single-blind trial. A trial in which the nature of the treatment is either not known by the patient or not known by the doctor assessing its effects.

Double-blind trial. A trial in which the nature of the treatment is not known by the patient nor by the doctor assessing its effects.

Factorial trial. A trial in which two, or more, treatments are used singly and in unison so that possible interactions can be measured.

Appendix G

Ethics and Human Experimentation

(a) Responsibility in Investigations on Human Subjects

STATEMENT BY THE MEDICAL RESEARCH COUNCIL

(Printed in the Report of the Medical Research Council for 1962–63. Cmd. 2382)

During the last fifty years, medical knowledge has advanced more rapidly than at any other period in its history. New understandings, new treatments, new diagnostic procedures and new methods of prevention have been and are being, introduced at an ever-increasing rate; and if the benefits that are now becoming possible are to be gained, these developments must continue.

Undoubtedly the new era in medicine upon which we have now entered is largely due to the marriage of the methods of science with the traditional methods of medicine. Until the turn of the century, the advancement of clinical knowledge was in general confined to that which could be gained by observation, and means for the analysis in depth of the phenomena of health and disease were seldom available. Now, however, procedures that can safely, and conscientiously, be applied to both sick and healthy human beings are being devised in profusion, with the result that certainty and understanding in medicine are increasing apace.

Yet these innovations have brought their own problems to the clinical investigator. In the past, the introduction of new treatments or investigations was infrequent and only rarely did they go beyond a marginal variation on established practice. Today, far-ranging new procedures are commonplace and such are the potentialities that their employment is no negligible consideration. As a result, investigators are frequently faced with ethical and sometimes even legal problems of great difficulty. It is in the hope of giving some guidance in this difficult matter that the Medical Research Council issue this statement.

A distinction may legitimately be drawn between procedures undertaken as part of patient-care which are intended to contribute to the benefit of the individual patient, by treatment, prevention or assessment, and those procedures which are undertaken either on patients or on healthy subjects solely for the purpose of contributing to medical knowledge and are not themselves designed to benefit the particular individual on whom they are performed. The former fall within the ambit of patient-care and are governed by the ordinary rules of professional conduct in medicine; the latter fall within the ambit of investigations on volunteers.

Important considerations flow from this distinction.

Procedures contributing to the benefit of the individual

In the case of procedures directly connected with the management of the condition in the particular individual, the relationship is essentially that between doctor and patient. Implicit in this relationship is the willingness on the part of the subject to be guided by the judgement of his medical attendant. Provided, therefore, that the medical attendant is satisfied that there are reasonable grounds for believing that a particular new procedure will contribute to the benefit of that particular patient, either by treatment, prevention or increased understanding of his case, he may assume the patient's consent to the same extent as he would were the procedure entirely established practice. It is axiomatic that no two patients are alike and that the medical attendant must be at liberty to vary his procedures according to his judgement of what is in his patients' best interests. The question of novelty is only relevant to the extent that in reaching a decision to use a novel procedure the doctor, being unable to fortify his judgement by previous experience, must exercise special care. That it is both considerate and prudent to obtain the patient's agreement before using a novel procedure is no more than a requirement of good medical practice.

The second important consideration that follows from this distinction is that it is clearly within the competence of a parent or guardian of a child to give permission for procedures intended to benefit that child when he is not old or intelligent enough to be able himself to give a valid consent.

A category of investigation that has occasionally raised questions in the minds of investigators is that in which a new preventive, such as a vaccine, is tried. Necessarily, preventives are given to people who are not, at the moment, suffering from the relevant illness. But the ethical and legal considerations are the same as those that govern the introduction of a new treatment. The intention is to benefit an individual by protecting him against a future hazard; and it is a matter of professional judgement whether the procedure in question offers a better chance of doing so than previously existing measures.

In general, therefore, the propriety of procedures intended to benefit the individual—whether these are directed to treatment, to prevention or to assessment—are determined by the same considerations as govern the care of patients. At the frontiers of knowledge, however, where not only are many procedures novel but their value in the particular instance may be debatable, it is wise, if any doubt exists, to obtain the opinion of experienced colleagues on the desirability of the projected procedure.

Control subjects in investigations of treatment or prevention

Over recent years, the development of treatment and prevention has been greatly advanced by the method of the controlled clinical trial. Instead of waiting, as in the past, on the slow accumulation of general experience to determine the relative advantages and disadvantages of any particular measure, it is now often possible to put the question to the test under conditions which will not only yield a speedy and more precise answer, but also limit the risk of untoward effects remaining undetected. Such trials are, however, only feasible when it is possible to compare suitable groups of patients and only permissible when there is a genuine doubt within the profession as to which of two treatments or preventive regimes is the better. In these circumstances it is justifiable to give to a proportion of the patients the novel procedure on the understanding that the remainder receive the procedure previously accepted as the best. In the case when no effective treatment has previously been devised then the situation should be fully explained to the participants and their true consent obtained.

Such controlled trials may raise ethical points which may be of some difficulty. In general, the patients participating in them should be told frankly that two different procedures are being assessed and their co-operation invited. Occasionally, however, to do so is contra-indicated. For example, to awaken patients with a possibly fatal illness to the existence of such doubts about effective treatment may not always be in their best interest; or suspicion may have arisen as to whether a particular treatment has any effect apart from suggestion and it may be necessary to introduce a placebo into part of the trial to determine this. Because of these and similar difficulties, it is the firm opinion of the Council that controlled clinical trials should always be planned and supervised by a group of investigators and never by an individual alone. It goes without question that any doctor taking part in such a collective controlled trial is under an obligation to withdraw a patient from the trial, and to institute any treatment he considers necessary, should this, in his personal opinion, be in the better interests of his patient.

Procedures not of direct benefit to the individual

The preceding considerations cover the majority of clinical investigations. There remains, however, a large and important field of investigations on human subjects which aims to provide normal values and their variation so that abnormal values can be recognised. This involves both ill persons and 'healthy' persons, whether the latter are entirely healthy or patients suffering from a condition that has no relevance to the investigation. In regard to persons with a particular illness, such as metabolic defect, it may be necessary to know the range of abnormality compatible with the activities of normal life or the reaction of such persons to some change in circumstances such as an alteration in diet. Similarly it may be necessary to have a clear understanding of the range of a normal function and its reaction to changes in circumstances in entirely healthy persons. The common feature of this type of investigation is that it is of no direct benefit to the particular individual and that, in consequence, if he is to submit to it he must volunteer in the full sense of the word.

It should be clearly understood that the possibility or probability that a particular investigation will be of benefit to humanity or to posterity would afford no defence in the event of legal proceedings. The individual has rights that the law protects and nobody can infringe those rights for the public good. In investigations of this type it is, therefore, always necessary to ensure that the true consent of the subject is explicitly obtained.

By true consent is meant consent freely given with proper understanding of the nature and consequences of what is proposed. Assumed consent or consent obtained by undue influences is valueless and, in this latter respect, particular care is necessary when the volunteer stands in special relationship to the investigator as in the case of a patient to his doctor, or a student to his teacher.

The need for obtaining evidence of consent in this type of investigation has been generally recognised, but there are some misunderstandings as to what constitutes such evidence. In general, the investigator should obtain the consent himself in the presence of another person. Written consent unaccompanied by other evidence that an explanation has been given, understood, and accepted is of little value.

The situation in respect of minors and mentally subnormal or mentally disordered patients is of particular difficulty. In the strict view of the law parents and guardians of minors cannot give consent on their behalf to any procedures which are of no particular benefit to them and which may carry some risk of harm. Whilst English law does not fix any arbitrary age in this context, it may safely be assumed that the Courts will not regard a child of 12 years or under (or 14 years or under for boys in Scotland) as having the capacity to consent to any procedure which may involve him in an injury. Above this age the reality of any purported consent which may have been obtained is a question of fact and as with an adult the evidence would, if necessary, have to show that irrespective of age the person concerned fully understood the implications to himself of the procedures to which he was consenting.

In the case of those who are mentally subnormal or mentally disordered the reality of the consent given will fall to be judged by similar criteria to those which apply to the making of a will, contracting a marriage or otherwise taking decisions which have legal force as well as moral and social implications. When true consent in this sense cannot be obtained, procedures which are of no direct benefit and which might carry a risk of harm to the subject should not be undertaken.

Even when true consent has been given by a minor or a mentally subnormal or mentally disordered person, considerations of ethics and prudence still require that, if possible, the assent of parents or guardians or relatives, as the case may be, should be obtained.

Investigations that are of no direct benefit to the individual require, therefore, that his true consent to them shall be explicitly obtained. After adequate explanation, the consent of an adult of sound mind and understanding can be relied upon to be true consent. In the case of children and young persons the question whether purported consent was true consent would in each case depend upon facts such as the age, intelligence, situation, and character of the subject and the nature of the investigation. When the subject is below the age of 12 years, information requiring the performance of any procedure involving his body would need to be obtained incidentally to and without altering the nature of a procedure intended for his individual benefit.

Professional discipline

All who have been concerned with medical research are aware of the impossibility of formulating any detailed code of rules which will ensure that irreproachability of practice which alone will suffice where investigations on human beings are concerned. The law lays down a minimum code in matters of professional negligence and the doctrine of assault. But this is not enough. Owing to the special relationship of trust that exists between a patient and his doctor, most patients will consent to any proposal made. Further, the considerations involved in a novel procedure are nearly always so technical as to prevent their being adequately understood by one who is not himself an expert. It must, therefore, be frankly recognised that, for practical purposes, an inescapable moral responsibility rests with the doctor concerned for determining what investigations are, or are not, proposed to a particular patient or volunteer. Nevertheless, moral codes are formulated by man and if, in the everchanging circumstances of medical advance, their relevance is to be maintained, it is to the profession itself that we must look, and in particular to the heads of departments, the specialised Societies, and the editors of medical and scientific journals.

In the opinion of the Council, the head of a department where investigations on human subjects take place has an inescapable responsibility for ensuring that practice by those under his direction is irreproachable.

In the same way the Council feel that, as a matter of policy, bodies like themselves that support medical research should do everything in their power to ensure that the practice of all workers whom they support shall be unexceptionable and known to be so.

So specialised has medical knowledge now become that the profession in general can rarely deal adequately with individual problems. In regard to any particular type of investigation, only a small group of experienced men who have specialised in this branch of knowledge are likely to be competent to pass an opinion on the justification for undertaking any particular procedure. But in every branch of medicine specialised scientific societies exist. It is upon these that the profession in general must mainly rely for the creation and maintenance of that body of precedents which shall guide individual investigators in case of doubt, and for the critical discussion of the communications presented to them on which the formation of the necessary climate of opinion depends.

Finally, it is the Council's opinion that any account of investigations on human subjects should make clear that the appropriate requirements have been fulfilled and, further, that no paper should be accepted for publication if there are any doubts that such is the case.

The progress of medical knowledge has depended, and will continue to depend, in no small measure upon the confidence which the public has in those who carry out investigations on human subjects, be these healthy or sick. Only in so far as it is known that such investigations are submitted to the highest ethical scrutiny and self-discipline will this confidence be maintained. Mistaken, or misunderstood, investigations could do incalculable harm to medical progress. It is our collective duty as a profession to see that this does not happen and so to continue to deserve the confidence that we now enjoy.

(b) Declaration of Helsinki

Recommendations guiding physicians
in biomedical research involving human subjects

Adopted by the 18th World Medical Assembly,
Helsinki, Finland, June 1964,
amended by the 29th World Medical Assembly,
Tokyo, Japan, October 1975,
the 35th World Medical Assembly,
Venice, Italy, October 1983,
and
the 41st World Medical Assembly,
Hong Kong, September 1989

Introduction

It is the mission of the physician to safeguard the health of the people. His or her knowledge and conscience are dedicated to the fulfilment of this mission.

The Declaration of Geneva of the World Medical Association binds the physician with the words, 'The health of my patient will be my first consideration,' and the International Code of Medical Ethics declares that, 'A physician shall act only in the patient's interest when providing medical care which might have the effect of weakening the physical and mental condition of the patient'.

The purpose of biomedical research involving human subjects must be to improve diagnostic, therapeutic and prophylactic procedures and the understanding of the aetiology and pathogenesis of disease.

In current medical practice most diagnostic, therapeutic or prophylactic procedures involve hazards. This applies especially to biomedical research.

Medical progress is based on research which ultimately must rest in part on experimentation involving human subjects.

In the field of biomedical research a fundamental distinction must be recognised between medical research in which the aim is essentially diagnostic or therapeutic for a patient, and medical research, the essential object of which is purely scientific and without implying direct diagnostic or therapeutic value to the person subjected to the research.

Special caution must be exercised in the conduct of research which may affect the environment, and the welfare of animals used for research must be respected.

Because it is essential that the results of laboratory experiments be applied to human beings to further scientific knowledge and to help suffering humanity, the World Medical Association has prepared the following recommendations as a guide to every physician in biomedical research involving human subjects. They should be kept under review in the future. It must be stressed that the standards as drafted are only a guide to physicians all over the world. Physicians are not relieved from criminal, civil or ethical responsibilities under the laws of their own countries.

I Basic Principles

1 Biomedical research involving human subjects must conform to generally accepted scientific principles and should be based on adequately performed laboratory and animal experimentation and on a thorough knowledge of the scientific literature.

2 The design and performance of each experimental procedure involving human subjects should be clearly formulated in an experimental protocol which should be transmitted for consideration, comment and guidance to a specially appointed committee independent of the investigator and the sponsor provided that this independent committee is in conformity with the laws and regulations of the country in which the research experiment is performed.

3 Biomedical research involving human subjects should be conducted only by scientifically qualified persons and under the supervision of a clinically competent medical person. The responsibility for the human subject must always rest with a medically qualified person and never rest on the subject of the research, even though the subject has given his or her consent.

4 Biomedical research involving human subjects cannot legitimately be carried out unless the importance of the objective is in proportion to the inherent risk to the subject.

5 Every biomedical research project involving human subjects should be preceded by careful assessment of predictable risks in comparison with

foreseeable benefits to the subject or to others. Concern for the interests of the subject must always prevail over the interests of science and society.

6 The right of the research subject to safeguard his or her integrity must always be respected. Every precaution should be taken to respect the privacy of the subject and to minimise the impact of the study on the subject's physical and mental integrity and on the personality of the subject.

7 Physicians should abstain from engaging in research projects involving human subjects unless they are satisfied that the hazards involved are believed to be predictable. Physicians should cease any investigation if the hazards are found to outweigh the potential benefits.

8 In publication of the results of his or her research, the physician is obliged to preserve the accuracy of the results. Reports of experimentation not in accordance with the principles laid down by this Declaration should not be accepted for publication.

9 In any research on human beings, each potential subject must be adequately informed of the aims, methods, anticipated benefits and potential hazards of the study and the discomfort it may entail. He or she should be informed that he or she is at liberty to abstain from participation in the study and that he or she is free to withdraw his or her consent to participation at any time. The physician should then obtain the subject's freely-given informed consent, preferably in writing.

10 When obtaining informed consent for the research project the physician should be particularly cautious if the subject is in a dependent relationship to him or her or may consent under duress. In that case the informed consent should be obtained by a physician who is not engaged in the investigation and who is completely independent of this official relationship.

11 In case of legal incompetence, informed consent should be obtained from the legal guardian in accordance with national legislation. Where physical or mental incapacity makes it impossible to obtain informed consent, or when the subject is a minor, permission from the responsible relative replaces that of the subject in accordance with national legislation.

Whenever the minor child is in fact able to give a consent, the minor's consent must be obtained in addition to the consent of the minor's legal guardian.

12 The research protocol should always contain a statement of the ethical consideration involved and should indicate that the principles enunciated in the present Declaration are complied with.

II Medical research combined with professional care (Clinical research)

1. In the treatment of the sick person, the physician must be free to use a new diagnostic and therapeutic measure, if in his or her judgement it offers hope of saving life, reestablishing health or alleviating suffering.
2. The potential benefits, hazards and discomfort of a new method should be weighed against the advantages of the best current diagnostic and therapeutic methods.
3. In any medical study, every patient—including those of a control group, if any—should be assured of the best proven diagnostic and therapeutic method.
4. The refusal of the patient to participate in a study must never interfere with the physician–patient relationship.
5. If the physician considers it essential not to obtain informed consent, the specific reasons for this proposal should be stated in the experimental protocol for transmission to the independent committee (I,2).
6. The physician can combine medical research with professional care, the objective being the acquisition of new medical knowledge, only to the extent that medical research is justified by its potential diagnostic or therapeutic value for the patient.

III Non-therapeutic biomedical research involving human subjects (Non-clinical biomedical research)

1. In the purely scientific application of medical research carried out on a human being, it is the duty of the physician to remain the protector of the life and health of that person on whom biomedical research is being carried out.
2. The subjects should be volunteers—either healthy persons or patients for whom the experimental design is not related to the patient's illness.
3. The investigator or the investigating team should discontinue the research if in his/her or their judgement it may, if continued, be harmful to the individual.
4. In research on man, the interest of science and society should never take precedence over considerations related to the wellbeing of the subject.

Table I The standardised normal distribution (i.e. with mean = 0, standard deviation = 1).

	Two-tailed					
P	0.50	0.20	0.10	0.05	0.02	0.01
z	0.674	1.282	1.645	1.960	2.326	2.576
	Two-tailed extreme values					
P	0.001	0.0001	0.00001	0.000001		
z	3.291	3.891	4.417	4.892		

	One-tailed					
P	0.50	0.20	0.10	0.05	0.02	0.01
z	0.000	0.842	1.282	1.645	2.054	2.326
	One-tailed extreme values					
P	0.001	0.0001	0.00001	0.000001		
z	3.090	3.719	4.265	4.753		

Table II The *t* distribution.

Degrees of freedom	*Two-tailed P*					
	0.50	0.20	0.10	0.05	0.02	0.01
1	1.00	3.08	6.31	12.71	31.82	63.66
2	0.82	1.89	2.92	4.30	6.96	9.92
3	0.76	1.64	2.35	3.18	4.54	5.84
4	0.74	1.53	2.13	2.78	3.75	4.60
5	0.73	1.48	2.02	2.57	3.36	4.03
6	0.72	1.44	1.94	2.45	3.14	3.71
7	0.71	1.41	1.89	2.36	3.00	3.50
8	0.71	1.40	1.86	2.31	2.90	3.36
9	0.70	1.38	1.83	2.26	2.82	3.26
10	0.70	1.37	1.81	2.23	2.76	3.17
11	0.70	1.36	1.80	2.20	2.72	3.11
12	0.70	1.36	1.78	2.18	2.68	3.05
13	0.69	1.35	1.77	2.16	2.65	3.01
14	0.69	1.35	1.76	2.14	2.62	2.98
15	0.69	1.34	1.75	2.13	2.60	2.95
16	0.69	1.34	1.75	2.12	2.58	2.92
17	0.69	1.33	1.74	2.11	2.57	2.90
18	0.69	1.33	1.73	2.10	2.55	2.88
19	0.69	1.33	1.73	2.09	2.54	2.86
20	0.69	1.33	1.72	2.09	2.53	2.85
21	0.69	1.32	1.72	2.08	2.52	2.83
22	0.69	1.32	1.72	2.07	2.51	2.82
23	0.69	1.32	1.71	2.07	2.50	2.81
24	0.68	1.32	1.71	2.06	2.49	2.80
25	0.68	1.32	1.71	2.06	2.49	2.79
26	0.68	1.31	1.71	2.06	2.48	2.78
27	0.68	1.31	1.70	2.05	2.47	2.77
28	0.68	1.31	1.70	2.05	2.47	2.76
29	0.68	1.31	1.70	2.05	2.46	2.76
30	0.68	1.31	1.70	2.04	2.46	2.75
31	0.68	1.31	1.70	2.04	2.45	2.74
32	0.68	1.31	1.69	2.04	2.45	2.74
33	0.68	1.31	1.69	2.03	2.44	2.73
34	0.68	1.31	1.69	2.03	2.44	2.73
35	0.68	1.31	1.69	2.03	2.44	2.72
36	0.68	1.31	1.69	2.03	2.43	2.72
37	0.68	1.30	1.69	2.03	2.43	2.72
38	0.68	1.30	1.69	2.02	2.43	2.71
39	0.68	1.30	1.68	2.02	2.43	2.71
40–43	0.68	1.30	1.68	2.02	2.42	2.70

Table II The *t* distribution—(*cont.*)

Degrees of freedom	*Two-tailed P* 0.50	0.20	0.10	0.05	0.02	0.01
44	0.68	1.30	1.68	2.02	2.41	2.69
45–46	0.68	1.30	1.68	2.01	2.41	2.69
47–48	0.68	1.30	1.68	2.01	2.41	2.68
49–51	0.68	1.30	1.68	2.01	2.40	2.68
52–53	0.68	1.30	1.67	2.01	2.40	2.67
54–55	0.68	1.30	1.67	2.00	2.40	2.67
56	0.68	1.30	1.67	2.00	2.39	2.67
57–63	0.68	1.30	1.67	2.00	2.39	2.66
64–65	0.68	1.29	1.67	2.00	2.39	2.65
66–68	0.68	1.29	1.67	2.00	2.38	2.65
69–72	0.68	1.29	1.67	1.99	2.38	2.65
73–76	0.68	1.29	1.67	1.99	2.38	2.64
77–78	0.68	1.29	1.66	1.99	2.38	2.64
79–84	0.68	1.29	1.66	1.99	2.37	2.64
85–95	0.68	1.29	1.66	1.99	2.37	2.63
96–98	0.68	1.29	1.66	1.98	2.37	2.63
99–101	0.68	1.29	1.66	1.98	2.36	2.63
102–127	0.68	1.29	1.66	1.98	2.36	2.62
128–131	0.68	1.29	1.66	1.98	2.36	2.61
132–151	0.68	1.29	1.66	1.98	2.35	2.61
152–158	0.68	1.29	1.65	1.98	2.35	2.61
159–170	0.68	1.29	1.65	1.97	2.35	2.61
171–201	0.68	1.29	1.65	1.97	2.35	2.60
202–246	0.68	1.29	1.65	1.97	2.34	2.60
247–258	0.68	1.28	1.65	1.97	2.34	2.60
259–432	0.68	1.28	1.65	1.97	2.34	2.59
433–472	0.68	1.28	1.65	1.97	2.33	2.59
473–481	0.68	1.28	1.65	1.96	2.33	2.59
482–537	0.67	1.28	1.65	1.96	2.33	2.59
538+	0.67	1.28	1.65	1.96	2.33	2.58

Table III The χ^2 distribution.

Degrees of freedom	*P* 0.50	0.20	0.10	0.05	0.02	0.01
1	0.45	1.64	2.71	3.84	5.41	6.63
2	1.39	3.22	4.61	5.99	7.82	9.21
3	2.37	4.64	6.25	7.81	9.84	11.34
4	3.36	5.99	7.78	9.49	11.67	13.28
5	4.35	7.29	9.24	11.07	13.39	15.09
6	5.35	8.56	10.64	12.59	15.03	16.81
7	6.35	9.80	12.02	14.07	16.62	18.48
8	7.34	11.03	13.36	15.51	18.17	20.09
9	8.34	12.24	14.68	16.92	19.68	21.67
10	9.34	13.44	15.99	18.31	21.16	23.21
11	10.34	14.63	17.28	19.68	22.62	24.72
12	11.34	15.81	18.55	21.03	24.05	26.22
13	12.34	16.98	19.81	22.36	25.47	27.69
14	13.34	18.15	21.06	23.68	26.87	29.14
15	14.34	19.31	22.31	25.00	28.26	30.58
16	15.34	20.47	23.54	26.30	29.63	32.00
17	16.34	21.61	24.77	27.59	31.00	33.41
18	17.34	22.76	25.99	28.87	32.35	34.81
19	18.34	23.90	27.20	30.14	33.69	36.19
20	19.34	25.04	28.41	31.41	35.02	37.57
21	20.34	26.17	29.62	32.67	36.34	38.93
22	21.34	27.30	30.81	33.92	37.66	40.29
23	22.34	28.43	32.01	35.17	38.97	41.64
24	23.34	29.55	33.20	36.42	40.27	42.98
25	24.34	30.68	34.38	37.65	41.57	44.31
26	25.34	31.79	35.56	38.89	42.86	45.64
27	26.34	32.91	36.74	40.11	44.14	46.96
28	27.34	34.03	37.92	41.34	45.42	48.28
29	28.34	35.14	39.09	42.56	46.69	49.59
30	29.34	36.25	40.26	43.77	47.96	50.89

For degrees of freedom (*df*) greater than 30, find

$$z = \sqrt{(2\chi^2)} - \sqrt{(2df - 1)}$$

and refer to Table I (one-tailed).

For example, if $\chi^2 = 62$ on 41 degrees of freedom:

$$z = \sqrt{124} - \sqrt{81} = 11.14 - 9 = 2.14$$

Table I then gives $0.02 > P > 0.01$.

Table IV e^{-m} for Poisson distributions.

This is the probability of an observation of 0 from a Poisson distribution with mean m.

m	*Probability*	m	*Probability*	m	*Probability*
0.01	0.990050				
0.02	0.980199	0.20	0.818731	2.0	0.135335
0.03	0.970446	0.30	0.740818	3.0	0.497871
0.04	0.960789	0.40	0.670320	4.0	0.0183156
0.05	0.951229	0.50	0.606531	5.0	0.00673795
0.06	0.941765	0.60	0.548812	6.0	0.00247875
0.07	0.932394	0.70	0.496585	7.0	0.000911882
0.08	0.923116	0.80	0.449329	8.0	0.000335463
0.09	0.913931	0.90	0.406570	9.0	0.000123410
0.10	0.904837	1.0	0.367879	10.0	0.000045400
0.11	0.895834	1.1	0.332871		
0.12	0.886920	1.2	0.301194		
0.13	0.878095	1.3	0.272532		
0.14	0.869358	1.4	0.246597		
0.15	0.860708	1.5	0.223130		

For $m<0.01$, the probability is $1-m+\frac{1}{2}m^2$.

Table V The Wilcoxon signed-rank statistic, w, for paired observations. w is the *larger* sum of signed ranks.

Number of paired observations showing differences	*Two-tailed*					
	P					
n	0.50	0.20	0.10	0.05	0.02	0.01
2	3	—	—	—	—	—
3	5	—	—	—	—	—
4	8	10	—	—	—	—
5	11	13	15	—	—	—
6	15	18	19	21	—	—
7	19	23	25	26	28	—
8	24	28	31	33	35	36
9	29	35	37	40	42	44
10	35	41	45	47	50	52
11	42	49	53	56	59	61
12	49	57	61	65	69	71
13	56	65	70	74	79	82
14	65	74	80	84	90	93
15	73	84	90	95	101	105
16	82	94	101	107	113	117
17	92	105	112	119	126	130
18	102	116	124	131	139	144
19	113	128	137	144	153	158
20	124	141	150	158	167	173
21	136	154	164	173	182	189
22	149	167	178	188	198	205
23	162	182	193	203	214	222
24	175	196	209	219	231	239
25	189	212	225	236	249	257
26	203	227	241	253	267	276
27	218	244	259	271	286	295
28	234	261	276	290	305	315
29	250	278	295	309	325	335
30	267	296	314	328	345	356
31	284	315	333	349	366	378
32	302	334	353	369	388	400
33	320	354	374	391	410	423
34	338	374	395	413	433	447
35	358	395	417	435	457	471

If $n>35$, calculate $x = n(n+1)/4$

$$y = \sqrt{(n(n+1)(2n+1)/24)}$$
$$z = (w-x)/y$$

and refer z to Table I (two-tailed).

For example, if $w = 460$ and $n = 36$:

$$x = 36 \times 37/4 = 333$$
$$y = \sqrt{(36 \times 37 \times 73/24)} = 63.65$$
$$z = (460 - 333)/63.65 = 1.995$$

and Table I gives $0.02<P<0.05$.

Table VI The Wilcoxon rank sum statistic, W, for two independent samples.

W is the sum of ranks in the *smaller* of the two groups (i.e. of size m).

The samples are significantly different at the 0.05 or 0.01 level if W is less than or equal to the lower of the two values given for that P-value, or is greater than or equal to the upper of those two values.

Two-tailed

Sample sizes	*P*	
m, *n*	0.05	0.01
2, 8	3, 19	—
2, 9	3, 21	—
2, 10	3, 23	—
2, 11	3, 25	—
2, 12	4, 26	—
2, 13	4, 28	—
2, 14	4, 30	—
2, 15	4, 32	—
2, 16	4, 34	—
2, 17	5, 35	—
2, 18	5, 37	—
2, 19	5, 39	3, 41
2, 20	5, 41	3, 43
2, 21	6, 42	3, 45
2, 22	6, 44	3, 47
2, 23	6, 46	3, 49
2, 24	6, 48	3, 51
2, 25	6, 50	3, 53
2, 26	7, 51	3, 55
2, 27	7, 53	4, 56
2, 28	7, 55	4, 58
3, 5	6, 21	—
3, 6	7, 23	—
3, 7	7, 26	—
3, 8	8, 28	—
3, 9	8, 31	6, 33
3, 10	9, 33	6, 36
3, 11	9, 36	6, 39
3, 12	10, 38	7, 41
3, 13	10, 41	7, 44
3, 14	11, 43	7, 47
3, 15	11, 46	8, 49
3, 16	12, 48	8, 52
3, 17	12, 51	8, 55
3, 18	13, 53	8, 58
3, 19	13, 56	9, 60
3, 20	14, 58	9, 63
3, 21	14, 61	9, 66
3, 22	15, 63	10, 68
3, 23	15, 66	10, 71
3, 24	16, 68	10, 74
3, 25	16, 71	11, 76
3, 26	17, 73	11, 79
3, 27	17, 76	11, 82

Sample sizes	*P*	
m, *n*	0.05	0.01
4, 4	10, 26	—
4, 5	11, 29	—
4, 6	12, 32	10, 34
4, 7	13, 35	10, 38
4, 8	14, 38	11, 41
4, 9	14, 42	11, 45
4, 10	15, 45	12, 48
4, 11	16, 48	12, 52
4, 12	17, 51	13, 55
4, 13	18, 54	13, 59
4, 14	19, 57	14, 62
4, 15	20, 60	15, 65
4, 16	21, 63	15, 69
4, 17	21, 67	16, 72
4, 18	22, 70	16, 76
4, 19	23, 73	17, 79
4, 20	24, 76	18, 82
4, 21	25, 79	18, 86
4, 22	26, 82	19, 89
4, 23	27, 85	19, 93
4, 24	27, 89	20, 96
4, 25	28, 92	20, 100
4, 26	29, 95	21, 103
5, 5	17, 38	15, 40
5, 6	18, 42	16, 44
5, 7	20, 45	16, 49
5, 8	21, 49	17, 53
5, 9	22, 53	18, 57
5, 10	23, 57	19, 61
5, 11	24, 61	20, 65
5, 12	26, 64	21, 69
5, 13	27, 68	22, 73
5, 14	28, 72	22, 78
5, 15	29, 76	23, 82
5, 16	30, 80	24, 86
5, 17	32, 83	25, 90
5, 18	33, 87	26, 94
5, 19	34, 91	27, 98
5, 20	35, 95	28, 102
5, 21	37, 98	29, 106
5, 22	38, 102	29, 111
5, 23	39, 106	30, 115
5, 24	40, 110	31, 119
5, 25	42, 113	32, 123

Table VI The Wilcoxon rank sum statistic, W, for two independent samples—(*cont.*)

Two-tailed

Sample sizes m, n	P 0.05	P 0.01
6, 6	26, 52	23, 55
6, 7	27, 57	24, 60
6, 8	29, 61	25, 65
6, 9	31, 65	26, 70
6, 10	32, 70	27, 75
6, 11	34, 74	28, 80
6, 12	35, 79	30, 84
6, 13	37, 83	31, 89
6, 14	38, 88	32, 94
6, 15	40, 92	33, 99
6, 16	42, 96	34, 104
6, 17	43, 101	36, 108
6, 18	45, 105	37, 113
6, 19	46, 110	38, 118
6, 20	48, 114	39, 123
6, 21	50, 118	40, 128
6, 22	51, 123	42, 132
6, 23	53, 127	43, 137
6, 24	54, 132	44, 142
7, 7	36, 69	32, 73
7, 8	38, 74	34, 78
7, 9	40, 79	35, 84
7, 10	42, 84	37, 89
7, 11	44, 89	38, 95
7, 12	46, 94	40, 100
7, 13	48, 99	41, 106
7, 14	50, 104	43, 111
7, 15	52, 109	44, 117
7, 16	54, 114	46, 122
7, 17	56, 119	47, 128
7, 18	58, 124	49, 133
7, 19	60, 129	50, 139
7, 20	62, 134	52, 144
7, 21	64, 139	53, 150
7, 22	66, 144	55, 155
7, 23	68, 149	57, 160
8, 8	49, 87	43, 93
8, 9	51, 93	45, 99
8, 10	53, 99	47, 105
8, 11	55, 105	49, 111
8, 12	58, 110	51, 117
8, 13	60, 116	53, 123
8, 14	62, 122	54, 130
8, 15	65, 127	56, 136
8, 16	67, 133	58, 142
8, 17	70, 138	60, 148
8, 18	72, 144	62, 154
8, 19	74, 150	64, 160
8, 20	77, 155	66, 166
8, 21	79, 161	68, 172
8, 22	81, 167	70, 178
9, 9	62, 109	56, 115
9, 10	65, 115	58, 122
9, 11	68, 121	61, 128
9, 12	71, 127	63, 135
9, 13	73, 134	65, 142
9, 14	76, 140	67, 149
9, 15	79, 146	69, 156
9, 16	82, 152	72, 162
9, 17	84, 159	74, 169
9, 18	87, 165	76, 176
9, 19	90, 171	78, 183
9, 20	93, 177	81, 189
9, 21	95, 184	83, 196
10, 10	78, 132	71, 139
10, 11	81, 139	73, 147
10, 12	84, 146	76, 154
10, 13	88, 152	79, 161
10, 14	91, 159	81, 169
10, 15	94, 166	84, 176
10, 16	97, 173	86, 184
10, 17	100, 180	89, 191
10, 18	103, 187	92, 198
10, 19	107, 193	94, 206
10, 20	110, 200	97, 213

Table VI The Wilcoxon rank sum statistic, W, for two independent samples—(*cont.*)

Two-tailed

Sample sizes m, n	P 0.05	P 0.01	*Sample sizes* m, n	P 0.05	P 0.01
11, 11	96, 157	87, 166	13, 13	136, 215	125, 226
11, 12	99, 165	90, 174	13, 14	141, 223	129, 235
11, 13	103, 172	93, 182	13, 15	145, 232	133, 244
11, 14	106, 180	96, 190	13, 16	150, 240	136, 254
11, 15	110, 187	99, 198	13, 17	154, 249	140, 263
11, 16	113, 195	102, 206			
11, 17	117, 202	105, 214	14, 14	160, 246	147, 259
11, 18	121, 209	108, 222	14, 15	164, 256	151, 269
11, 19	124, 217	111, 230	14, 16	169, 265	155, 279
12, 12	115, 185	105, 195	15, 15	184, 281	171, 294
12, 13	119, 193	109, 203			
12, 14	123, 201	112, 212			
12, 15	127, 209	115, 221			
12, 16	131, 217	119, 229			
12, 17	135, 225	122, 238			
12, 18	139, 233	125, 247			

If $m+n>30$, calculate $x = m(m+n+1)/2$

$$y = \sqrt{(m\,n\,(m+n+1)/12)}$$

remembering that, if m and n are different, then m must always be the *smaller* of the two,

$$z = (W-x)/y$$

If z is negative, ignore the minus sign.
Refer z to Table I (two-tailed).

For example, if $W = 260$, $m = 12$ and $n = 20$:

$$x = 12 \times 33/2 = 198$$
$$y = \sqrt{(12 \times 20 \times 33/12)} = 25.69$$
$$z = (260 - 198)/25.69 = 2.413$$

and Table I gives $0.01<P<0.02$.

Table VII Kendall's rank correlation, r_K.
If r_K is negative, ignore its minus sign when using this table.

Number of observations *n*	Two-tailed *P* 0.50	0.20	0.10	0.05	0.02	0.01
3	0.90	—	—	—	—	—
4	0.60	0.90	—	—	—	—
5	0.30	0.65	0.70	0.90	—	—
6	0.30	0.50	0.70	0.75	0.80	0.90
7	0.30	0.50	0.60	0.70	0.80	0.90
8	0.25	0.40	0.55	0.60	0.70	0.75
9	0.20	0.35	0.45	0.55	0.65	0.70
10	0.17	0.35	0.45	0.50	0.57	0.63
11	0.17	0.33	0.40	0.47	0.53	0.57
12	0.17	0.30	0.37	0.43	0.53	0.55
13	0.17	0.29	0.35	0.43	0.50	0.55
14	0.15	0.27	0.35	0.39	0.46	0.50

If $n>14$, calculate

$$y = \sqrt{(2(2n+5)/(9n(n-1)))}$$
$$z = r_K/y$$

If z is negative, ignore the minus sign.
Refer z to Table I (two-tailed).

For example, if $r_K = 0.295$ and $n = 20$:

$$y = \sqrt{(2 \times 45/(9 \times 20 \times 19))} = 0.1622$$
$$z = 0.295/0.1622 = 1.819$$

and Table I gives $0.05<P<0.10$.

Index